Neurovegetative Transmission Mechanisms

Journal of Neural Transmission
Supplementum XI

Neurovegetative Transmission Mechanisms

Proceedings of the International Neurovegetative Symposium Tihany, June 19—24, 1972

Edited by B. Csillik and J. Ariëns Kappers

Springer-Verlag
Wien New York

With 138 Figures

Softcover reprint of the hardcover 1st edition 1974

Library of Congress Catalog Card Number 73-9232

ISBN-13: 978-3-7091-8343-4 e-ISBN-13: 978-3-7091-8341-0
DOI: 10.1007/978-3-7091-8341-0

Introduction

As a former president of the International Society for Neurovegetative Research and as a co-editor it is a pleasure to write a short introduction to the present volume which contains the proceedings of an international symposium, held at Tihany, Hungary, June 19—24, 1972, under the auspices of the Society mentioned above. The meeting, which was organized by Professor B. Csillik, Szeged, Hungary, who also acted as president, took place at the Tihany Biological Institute, most beautifully situated at lake Balaton.

Such international symposia of the International Society for Neurovegetative Research have been organized at irregular intervals in several different countries. Members of the International Society are, automatically, the members of National Societies for Neurovegetative Research, active in some European countries and in Japan, as well as individuals taking interest in the aim of the International Society. From its foundation, in 1953, on the goal of the Society has always been the advancement of neurovegetative research as performed by both, workers in the basic sciences and medical practitioners doing scientific work. Therefore, symposia of the International Society for Neurovegetative Research are international forums where pure scientific work as well as clinical investigations and clinical experience can be discussed. In this way, the Society has always been trying to bridge the gap between pure and applied science in the medical field which is so often felt to exist, especially at the present time often suffering from overspecialization.

The International Society owes much to Professor Csillik who, notwithstanding many difficulties, did succeed in organizing, in an excellent way, the Tihany symposium. He has been able to bring together, from several different countries, a number of representative workers in the broad field of neurovegetative research, and to constitute an interesting and valuable programme concerning the topic of the symposium: neurovegetative transmission mechanisms. As a chief editor he has also been responsible for the difficult task to finally select the papers to be published in these proceedings. I am sure that all participants have an excellent recollection of this congress and are most grateful to Professor and Mrs. Csillik for the way in which it was realized.

The International Society has always had a journal to its disposal in which papers on neurovegetative research were published by members as well as by non-members. First, this journal was named "Acta Neurovegetativa". Then the name changed into "Journal of Neuro-Visceral Relations" recently changing again into "Journal of Neural Transmission". During its long history, the publisher of the journal has always been the Springer-Verlag in Vienna to which we are most grateful for publishing also these proceedings as a supplementum to the journal.

Amsterdam, January 1974 **J. Ariëns Kappers**

Contents

Journal of Neural Transmission, Suppl. XI, 1—11 (1974)

Ultrastructure of Synapses in the Waking State

A Laboratory Report on Recent Advances* **

K. Akert, R. B. Livingston***, **H. Moor,** and **P. Streit**

Brain Research Institute of the University of Zürich and Laboratory of Electronmicroscopy, Department of General Botany, Federal Institute of Technology, Zürich, Switzerland

With 4 Figures

Summary

Membrane faces of synapses from pigeon optic tectum as well as from rat and cat spinal cord have been examined in a large number of freeze-etched replicas. The data have been compared with those of conventional thin section electronmicroscopy. Some progress has been made to analyse the main features of presynaptic membranes morphometrically.

The presynaptic membrane (active site) is characterized according to *Pfenninger et al.* (1972) by a curving indentation ("lifting") toward the presynaptic cytoplasm, and, furthermore, by the presence of 200 Å micropits (as seen from outside) and protuberances or craters (as seen from the inside), respectively. These membrane modulations are specific for the presynaptic area and their sites correspond closely to the holes in the presynaptic vesicular grid.

The concept of a dynamic membrane organization (*Streit et al.*, 1972) stems from the fact that the "lifting" of the membrane as well as the "wrinkling" and the relative number of open protuberances (representing vesicular attachment sites) is strongly enhanced in the waking as compared with the anaesthetized state. Thus, the "textbook appearance" of the

* Supported by grants from the Swiss National Foundation for Scientific Research Nr. 3.133.69, 3.134.69, 3.366.70 and from the Dr. Eric Slack-Gyr Stiftung in Zürich.

** The collaboration of *C. Sandri*, *K. Pfenninger*, and *E. Kawana*, and the skillful assistance by Miss *C. Berger* and *A. Fäh* is gratefully acknowledged. The drawing was prepared by Miss *R. Emch* and the manuscript by Miss *U. Fischer*.

*** On Sabbatical Leave from the Neurosciences Department, University of California at San Diego, La Jolla, CA 92037, U.S.A.

synapse with rectilinear and parallel synaptic membranes probably represents an unphysiological state, inasmuch as most of the preparations have been made with anaesthetized material. The dynamism of the presynaptic membrane as revealed by morphometric analysis of unanaesthetized preparations provides further support to the vesicle hypothesis of transmitter release.

The postsynaptic membrane is characterized by a sharply circumscribed aggregation of 80—130 Å particles (*Sandri et al.*, 1972) which are characteristic of the outer membrane leaflet. The inner leaflet contains particles of similar size (but perhaps different biochemical composition) which are more diffusely scattered over the membrane. The particle aggregations in the postsynaptic plasmalemma appear to correspond to regions of subsynaptic membrane specialization in terms of differential sensitivity to specific transmitter molecules.

Introduction

Recent communications on electronmicroscopic studies from our laboratories (*Akert* and *Sandri*, 1970; *Akert*, 1971; *Akert et al.*, 1972) have concentrated on the problem of correlating data obtained from conventionally prepared thin section material and from freeze-etched replicas with respect to synaptic membrane complexes. The present communication provides a summary of recent findings on the following two topics: (a) special features of the presynaptic membrane in the deeply nembutalized (*Pfenninger et al.*, 1971, 1972) as compared with the waking, unanaesthetized preparation (*Streit et al.*, 1972) and (b) special features of the postsynaptic membrane (*Sandri et al.*, 1972).

Materials and Methods

The studies were based on neuropil, obtained from superficial layers of the pigeon optic tectum and from rat and cat spinal cord. Tissue fixation was achieved by systemic perfusion of buffered (0.05 M phosphate, pH 7.4) 4 % paraformaldehyde followed by immersion of small blocks in 6.25 % buffered (0.1 M phosphate, pH 7.4) glutaraldehyde for 2 hours at room temperature and subsequent washing for several hours in 0.2 M phosphate buffer (pH 7.4) with 6.8 % sucrose at 4° C. Following these procedures both, freeze-etching (*Moor* and *Mühlethaler*, 1963) and thin-section electronmicroscopic techniques, were applied to equivalent materials (see *Akert et al.*, 1971).

Results

The Presynaptic Vesicular Grid

Gray (1963) was first to describe the so-called dense projections on the presynaptic membrane. A subsequent analysis was made by *Pfenninger et al.* (1969) with the aid of the bismuth iodide contrast technique. These dense projections turned out to be nodal points of a grid-like structure (Fig. 1) situated on the cytoplasmic side of the presynaptic unit membrane. The geometry of this structure resembles an "egg crate". Vacancies surrounding the pyramidal-shaped dense projections from hexagonal arrays that can accomodate six synaptic vesicles. Each dense projection shares two vesicles with each of three other dense projections which are arranged in a triangular array. A variable number of such combinations constitute a grid. The grids may form compact, oval or circular plaques. Horse-shoe shaped forms and ring-like grid arrays have also been observed. The significance of these various configurations remains to be elucidated. The number of grid loci contained within a single plaque may vary between approximately 15 and 150.

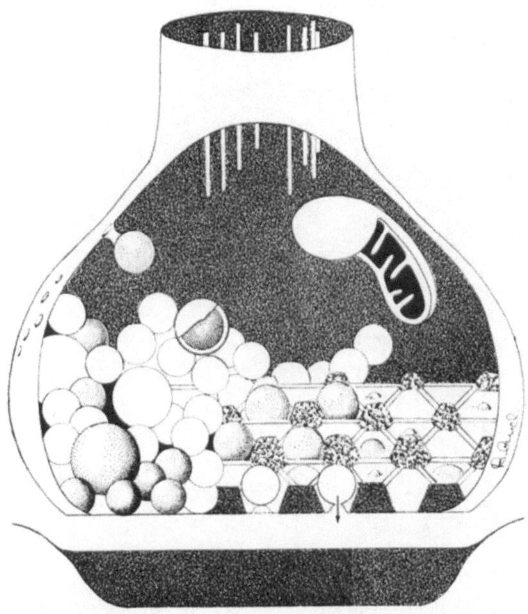

Fig. 1. Diagrammatic view of nerve terminal and synaptic contact with special emphasis on the "presynaptic vesicular grid". The grid vacancies constitute an hexagonal array supported by a filamentous meshwork in which the dense projections of Gray represent nodal points. The grid vacancies can accommodate synaptic vesicles. The initial stage of contact between vesicle and presynaptic membrane is characterized by a small inward curvature of the membrane. The phenomenon proceeds as has been observed in thin sections to full scale omega-profiles and these stages have been seen as well in freeze-etched replicas. On the left side of the terminal bulb several pits can be seen and one pinocytotic figure which is typically about twice the diameter of synaptic vesicles (from *Akert et al.*, 1972, *Structure and Function of Synapses*, Raven Press, New York)

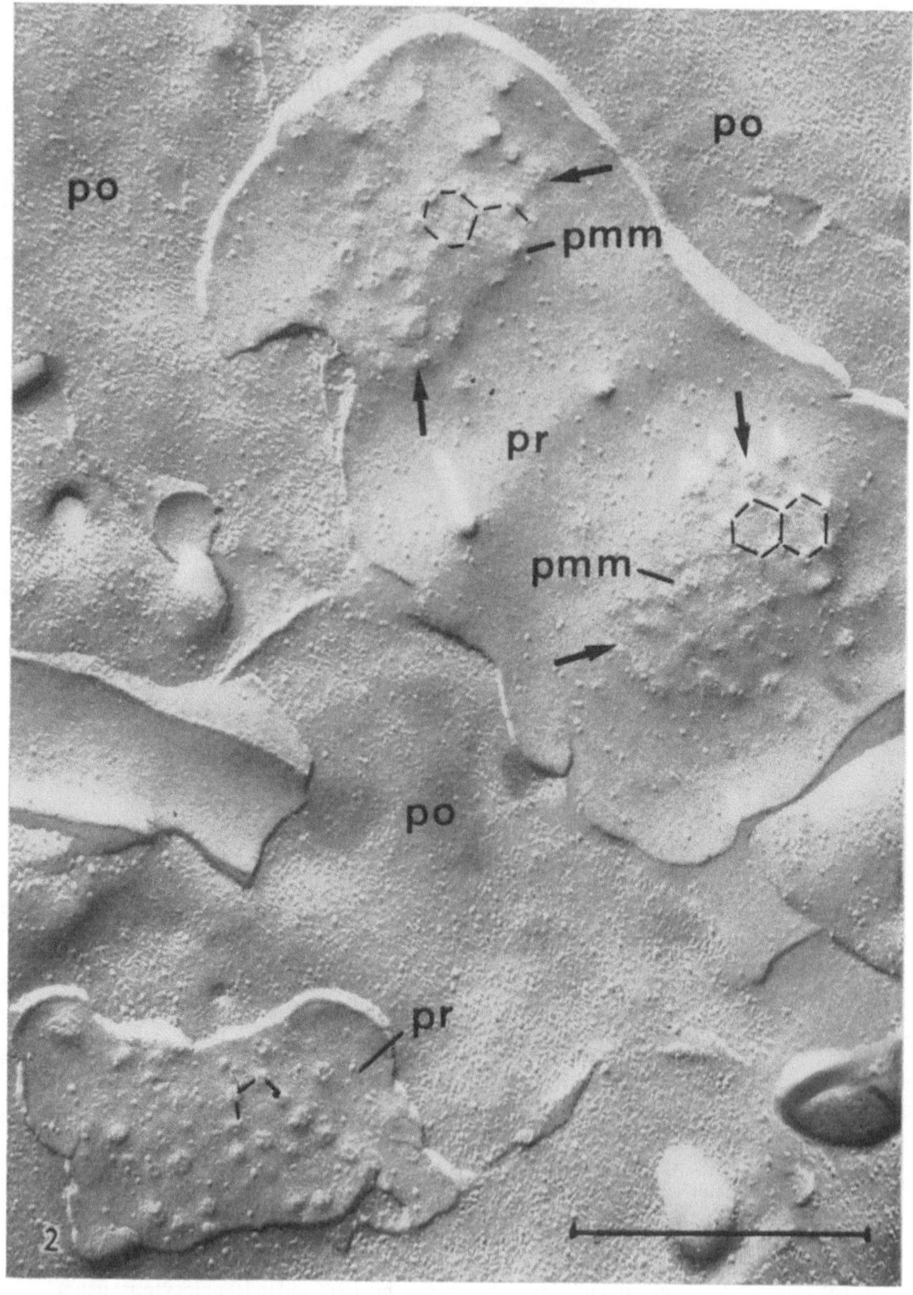
po
po
pmm
pr
pmm
po
pr
2

Vesicles which are trapped within the grid may make intimate contact and even fuse with the presynaptic membrane. Vesicles fusing with the presynaptic membrane (so-called omega-profiles) have been only rarely observed. The first convincing demonstrations were obtained in frog muscle endplates (*Couteaux* and *Pécot-Dechavassine*, 1970) and from the electroplax of torpedo (*Nickel* and *Potter*, 1970). An important reason for the raritiy of observing omegaprofiles in central nervous system synapses may be that examination of synaptic ultrastructure in the past was largely of neuropil from anaesthetized animals. Barbiturates, which are known to suppress synaptic activity, have been a commonplace choice of anaesthetic agent. *Streit et al.* (1972) have demonstrated omega-shaped fusion/fission sites more frequently (1—2 per thin section of synaptic sites) when unanaesthetized preparations were examined.

Presynaptic Membrane Faces in Freeze Fractured Neuropil

The main features of presynaptic membrane faces as seen in freeze-etched materials have been described by *Pfenninger et al.* (1971, 1972). Three structural specializations can be differentiated:

1. a slight inside (towards the cytoplasm) "lifting" or indentation of the presynaptic membrane;
2. arrays of tiny membrane modulations, about 200 Å in diameter, which are seen as pits when viewed from the extracellular space and as protuberances or craters when viewed from the cytoplasmic side. These have been termed "synaptopores" and "presynaptic membrane modulations" (pmm) in previous communications; they correspond to a "wrinkled" appearance of the presynaptic membrane as seen in thin sectioned profiles;
3. clusters of membrane particles (90—110 Å), aggregated over the lifted area and often directly attached to the pmm.

Major emphasis has been placed on the following two observations:

1. pmm often (but not always) display a hexagonal arrangement (Fig. 2) as would be expected from thin sectioned material;

Fig. 2. Freeze-etched replica of synaptic membrane faces from unanaesthetized rat spinal cord, ventral horn. A large expanse of postsynaptic membrane (po), presumably the inner leaflet of the plasmalemma of a motoneuron soma or possibly a large dendrite, with two fragments of presynaptic mebrane (pr). Three synaptic sites (arrows) are characterized by curving a "lifting" of the membrane over the entire area, and by tiny protuberances. The latter represent vesicular attachment sites which are partly "closed" and partly "open" (crater-like). Their arrangement is often but not always hexagonal. Aldehyde fixation. Primary magnification 20,000×, calibration 0.5 μ

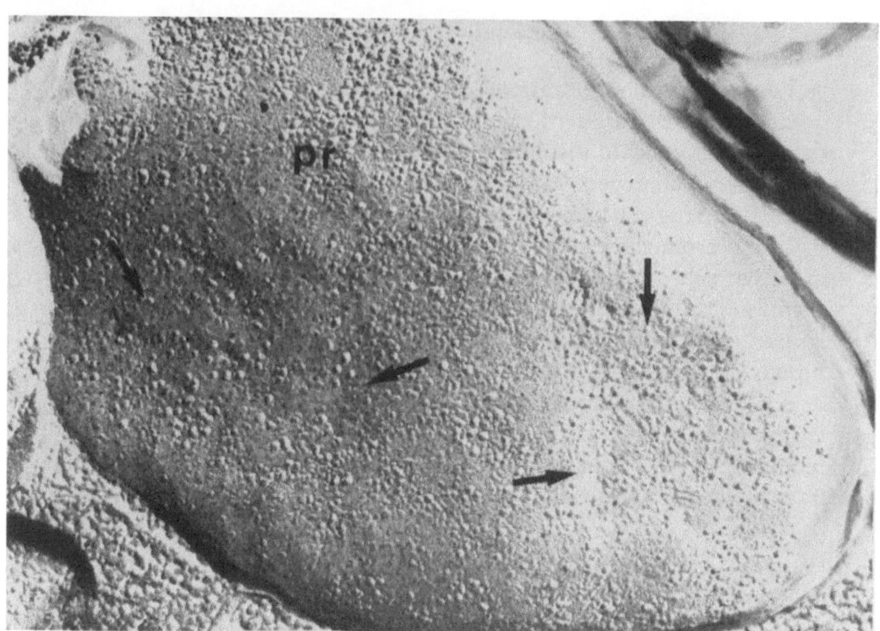

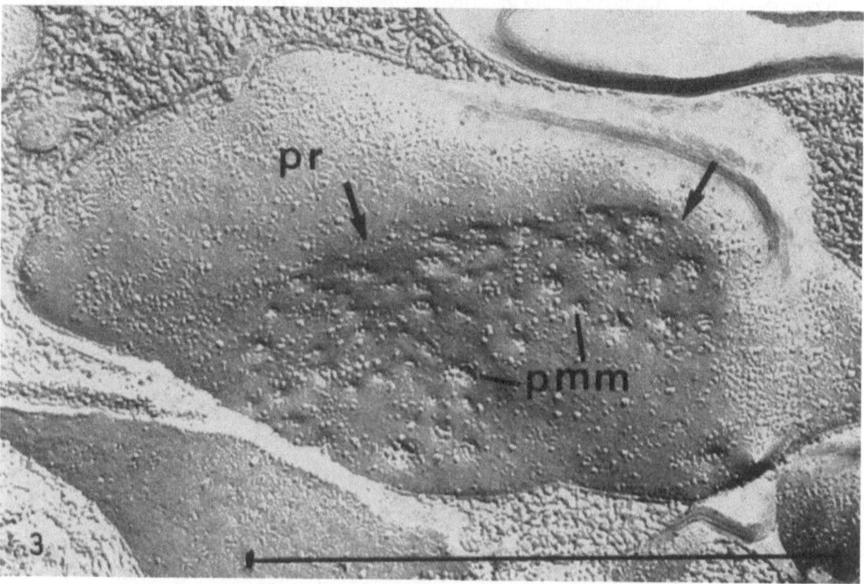

Fig. 3. Freeze-etched replicas of inner leaflets of presynaptic membranes (pr) from rat spinal cord, ventral horn. Approximately similar views of two nerve terminals are presented for comparison. Above: Deep nembutal anaesthesia (40 mg/kg, i.p.). Below: Unanaesthetized preparation. pmm, presynaptic membrane modulations representing vesicle attachment sites. Note the "lifting" of the membrane and the great prominence of pmms in the unanesthitized state. Aldehyde fixation. Primary magnification 40,000×, calibration 0.5 μ

2. the pmm are not significantly increased in frequency of occurrence but they are far more pronounced in the waking animal (Fig. 3).

Morphometric analysis of the incidence of "lifting" and "wrinkling" of the presynaptic membrane has yielded significant differences between the waking and the anaesthetized state (*Streit et al.*, 1972). Furthermore, freeze-etched craters representing vesicular attachment sites on the presynaptic membrane are more prominent and often appear open. Like the corresponding omega-profiles in thin sectioned membranes, the open craters are significantly more numerous.

It is presumed that open crater attachment sites (Fig. 2) represent vesicles which have fused with the presynaptic membrane, thus, implying direct transmitter release into the synaptic cleft. Unfortunately, such a conclusion may be premature since the morphologic evidence does not prove the directionality of the process, *i.e.* inwards (endocytosis) or outwards (exocytosis). Perhaps exocytosis may be followed by endocytosis in relation to a single fusion event. Further studies, using marker substances, are needed to clarify this point. One should also remember that plasmalemmal vesicles of approximately twice the diameter of synaptic vesicles have been observed at extrasynaptic sites in nerve terminals (*Moor et al.*, 1969; *Akert et al.*, 1971). This phenomenon seems less sensitive to nembutal anaesthesia than micro-exocytosis which takes place at the synaptic site. The two processes, which may represent not only morphologically but also biochemically different transport systems across the membrane of nerve terminals, are depicted in Fig. 1.

The small curvatures of the presynaptic membrane which are responsible for the "wrinkled" profile in thin sections, pose some interesting problems. They appear in a "closed", dome-shaped and in an "open" crater-shaped version. The former is more frequently seen in the anaesthetized state while the latter is far more numerous in the waking animal. It is conceivable that the closed form represents an initial or terminal stage of membrane to vesicle fusion or fission, whilst the open form presumably represents the actual state of membrane integration between plasmalemma and vesicle. The "necks" of vesicular attachment, as seen in the omega-shaped profiles in thin sections, are thought to be formed when vesicles have been torn off during the freeze-fracturing process. At any rate, the two forms may represent sequential stages of the same process.

Infolding or "lifting" of the presynaptic membrane has been reported to occur by *Clark et al.* (1972) in the frog neuromuscular junction after the application of black widow spider venom. This phenomenon was associated with an increased number of fusing

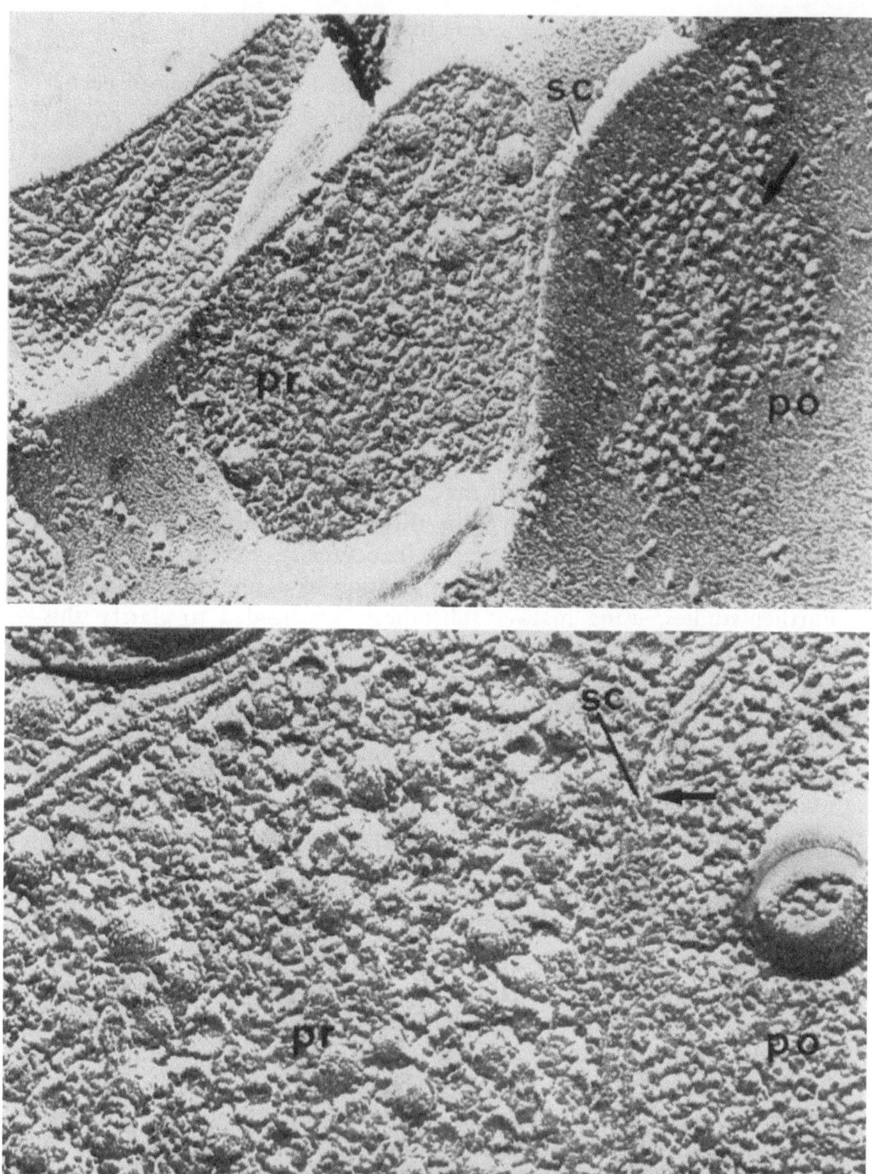

sc
pr
po
sc
pr
po
4

vesicles and with an enormous increase of miniature endplate potentials, thus suggesting a massive release of quanta of transmitter. The lifting of the membrane as shown by these authors appears more extreme than in our preparations. Yet, it is conceivable that the basic mechanism in frog neuromuscular and central synaptic junctions may be essentially similar. Obviously, these observations strongly reinforce the vesicle hypothesis of transmitter release, even though they fall short of providing complete proof.

Postsynaptic Membrane Faces in Freeze Fractured Neuropil

Both leaflets of the postsynaptic membrane, but particularly the outer one, are characterized by localized aggregations of 80—130 Å particles. Fractures that cut across profiles of both pre-and postsynaptic membranes provide evidence that such aggregations tend to be concentrated in the immediate subsynaptic region (Fig. 4). The surrounding areas of postsynaptic outer leaflet appear relatively smooth and nearly free of particles. The inner leaflet contains similar particles somewhat less densely aggregated and somewhat less strictly confined to the synaptic site. It appears to be generally accepted that the inner leaflets of cell membranes are characterized by having significantly more particles of a more variable size. They, therefore, appear "rough" whereas the outer leaflets display fewer particles and are characterized by a smoother appearance. With this generalization in mind, we can appreciate that the subsynaptic particle aggregations in the outer leaflet are especially significant. We are tempted to correlate these aggregations with the differential sensitivity of the postsynaptic membrane in this region to specific transmitter molecules. Since the particles have a diameter which likely exceeds that of the unit membrane, it may be infered that these particles protrude into the extracellular space, *i.e.* into the synaptic cleft. Unfortunately, we have been unable to verify this notion because, according to *Branton*

Fig. 4. Freeze-etched replicas of presynaptic (pr) and postsynaptic (po) membranes in synapses from pigeon optic tectum, superficial layers. Above: Oblique fracture, displaying the postsynaptic site with 80—130 Å particle aggregation (arrow) (from *Sandri et al.*, Brain Research *41:* Fig. 2, pg. 3, Elsevier Publishing Co., Amsterdam, 1972). Below: Cross fracture at right angle to synaptic cleft (sc). The extent of the active site is marked by arrows. The pre-and postsynaptic membranes are somewhat blurred by the presence in the cleft of gross particles. This picture suggests that the particles observed in the upper view, while firmly integrated into the postsynaptic membrane, may well protrude into or entirely through the synaptic cleft. Primary magnification 40,000×, calibration 0.5 μ. The animal was anaesthetized with nembutal

(1966), direct inspection of external membrane faces is not possible over extended areas with the present freeze-fracturing techniques.

References

Akert, K.: Struktur und Ultrastruktur von Nervenzellen und Synapsen. Klin. Wschr. *49*, 509—519 (1971).

Akert, K., *H. Moor*, and *K. Pfenninger*: Synaptic fine structure. In: Advances in Cytopharmacology (*Clementi, F.*, and *B. Ceccarelli*, eds.), Vol. 1, pp. 273—290. New York: Raven Press. 1971.

Akert, K., *K. Pfenninger*, *C. Sandri*, and *H. Moor*: Freeze-etching and cytochemistry of vesicles and membrane complexes in synapses of the C.N.S. In: Structure and Function of Synapses (*Pappas, G. D.*, and *D. P. Purpura*, eds.), pp. 67—86. New York: Raven Press. 1972.

Akert, K., and *C. Sandri*: Identification of the active synaptic region by means of histochemical and freeze-etching techniques. In: Excitatory synaptic mechanisms (*Andersen, P.*, and *J. K. S. Jansen*, eds.), pp. 27—41. Oslo: Universitetsforlaget. 1970.

Branton, D.: Fracture faces of frozen membranes. Proc. Nat. Acad. Sci. (Wash.) *55*, 1048—1056 (1966).

Clark, A. W., *P. Hurlbut*, and *A. Mauro*: Changes in the fine structure of the neuromuscular junction of the frog caused by black widow spider venom. J. Cell Biol. *52*, 1—14 (1972).

Couteaux, R., and *M. Pécot-Dechavassine*: Vésicules synaptiques et poches au niveau des zones actives de la jonction neuromusculaire. C.R. Acad. Sci. Ser. D. *271*, 2346—2349 (1970).

Gray, E. G.: Electronmicroscopy of presynaptic organelles of the spinal cord. J. Anat. (Lond.) *97*, 101—106 (1963).

Moor, H., and *K. Mühlethaler*: Fine structure in frozen-etched yeast cells. J. Cell Biol. *17*, 609—628 (1963).

Moor, H., *K. Pfenninger*, and *K. Akert*: Synaptic vesicles in electron micrographs of freeze-etched nerve terminals. Science *164*, 1405—1407 (1969).

Nickel, E., and *L. T. Potter*: Synaptic vesicles in freeze-etched electric tissue of Torpedo. Brain Res. *23*, 95—100 (1970).

Pfenninger, K., *K. Akert*, *H. Moor*, and *C. Sandri*: Freeze-fracturing of presynaptic membranes in the central nervous system. Phil. Trans. Roy. Soc. (London) *B 216*, 387 (1971).

Pfenninger, K., *K. Akert*, *H. Moor*, and *C. Sandri*: The fine structure of freeze-fractured presynaptic membranes. J. Neurocytol. *1*, 129—149 (1972).

Pfenninger, K., *C. Sandri*, *K. Akert*, and *C. H. Eugster*: Contribution to the problem of structural organization of the presynaptic area. Brain Res. *12*, 10—18 (1969).

Sandri, C., *K. Akert*, *R. B. Livingston*, and *H. Moor*: Particle aggregations at specialized sites in freeze-etched postsynaptic membranes. Brain Res. *41*, 1—16 (1972).

Streit, P., K. Akert, C. Sandri, R. B. Livingston, and *H. Moor*: Dynamic ultrastructure of presynaptic membranes at nerve terminals in the spinal cord of rats — anesthetized and unanesthetized preparations compared. Brain Res. *48,* 11—26 (1972).

Author's address: Prof. *K. Akert,* Institute for Brain Research, University of Zürich, August-Forel-Strasse 1, CH-8008 Zürich, Switzerland.

Discussion

Ariëns Kappers: What is your opinion on the fate of remnants of synaptic vesicles after being repeatedly used for transmitter transport?

Akert: The fate of these vesicles after usage is not clear. We like to think that they simply resolve *in situ.*

Réthely: Please could you give us some information on the occurrence of flattened synaptic vesicles in freeze-etched material?

Akert: No clearcut flattened forms of vesicles can be observed in well-preserved freeze-etched specimes. All synaptic vesicles appear to have spheric profiles under this condition, even if aldehydes were used for fixation. This observation suggests that the flattened shape, seen in conventionally prepared thin sections, is not due to the aldehydes but rather to other causes such as osmotic pressure (see *Valdivia,* J. comp. Neurol. *142,* 257, 1971).

Szentágothai: I think the importance of exocytosis of synaptic vesicles through the "pits" of the presynaptic membrane should be stressed. Since, in conventional electron microscopy, endocytosis is often seen at sites surrounding specific (synaptic) membrane attachments, one might suppose that the balance in the membrane surface is kept by loss of membrane material at the periphery of the synaptic contacts by endocytosis.

Akert: I quite agree with your suggestion regarding the homeostasis of the membrane surface at nerve terminals.

Taxi: Please could you inform us about the size of the structures interpreted as ionic channels? It looks as if the number of synaptic vesicles in freeze-etched material is restricted.

Akert: The low density of the vesicular population in our freeze-etched replicas is partly due to the great variation normally occurring in unanaesthetized neuropil, and partly to the fact that those vesicles which are split in halves are not readily distinguished against the cytoplasmic background. We do not believe that there is a basic difference between thin-section material and freeze-etch replicas with respect to this problem.

Journal of Neural Transmission, Suppl. XI, 13—42 (1974)

Synaptochemistry

Outlines and Scope of a Discipline

B. Csillik

Department of Anatomy, University Medical School, Szeged, Hungary

With 15 Figures

Summary

Synaptochemistry, a discipline based on the cytochemical entity of the neurone, aimes to reveal structural-functional correlations of impulse transmission at the molecular level. Cholinergic, monoaminergic and aminacidergic synaptochemical systems are scrutinized with respect to pre- and post-synaptic (receptor) functions and to excitation-performance coupling. Perspectives of synaptochemistry include a new interpretation of transducer function, a rational approach to the pharmacotherapy of neurovegetative and neuropsychiatric disorders and a new model of neurocellular memory.

Since impulse transmission in most, if not all, synapses of the mammalian nervous system is realized by chemical processes, identification and localization of the metabolisms of transmitter substances at the cellular, subcellular and molecular levels, became one of the principal goals of neurobiology. Due to important technical achievements and to the merging of border-lines between neuroanatomy, neurophysiology and neurochemistry, the outlines of a new discipline, that of synaptochemistry, are becoming more and more distinct. In this review it will be attempted to give a brief survey of the past history and present status of this discipline, and to trace the main features and perspectives of synaptochemical studies in the field of neurovegetative transmission mechanisms. It can also be assumed that pathological alterations of neurovegetative transmission can most readily be understood and remedied on the basis of normal synaptochemistry.

The beginnings of synaptochemistry can be traced back to the early studies by *Elliot* (1904), *Langley* (1905) and *Dale* (1955). Like synaptology itself, this discipline is based upon the Waldeyer neurone theory as represented in the classical studies of *Cajal* (1952) and *Sherrington* (1897). It is, however, the histochemical technique for localizing acetylcholinesterase (*Koelle* and *Friedenwald*, 1949) and the precise localization of this enzyme in the postsynaptic membrane of the neuromuscular junction (*Couteaux*, 1951) that can be regarded as the very origin of up-to-date synaptochemistry. These pioneering studies rendered irrefutable proof of the neuronal independence of the cellular elements coupled by the synapse, several years before ultrathin section electron microscopy finally disproved all the old suppositions and beliefs on the "reticular" construction of the nervous system (*Stöhr*, 1957). Thus, the synaptochemical identification of neuronal independence was a prerequisite for the disclosure of the presence and functional role of all of the ultrastructural specializations of the synapse, including synaptic vesicles, presynaptic expansions, membrane thickenings, intersynaptic organelles, etc.[1]. Finally, also the discovery of ephaptic junctions (*Bennett et al.*, 1963), lacking neurochemical transmission mechanisms, had to be preceded by a synaptochemical identification of neurohumoral transmission[2].

In accordance with the methodological aproaches of its parent disciplines — neuroanatomy, neurophysiology and neurochemistry — synaptochemistry utilizes:

1. light- and electron microscopic methods, supplemented by cytochemical, immunobiological and autoradiographical reactions,
2. micro-electrophoretic studies, and
3. microchemical analysis of tissue fractions, thin layers, isolated cells and cell particles.

It is only because of the humble recognition of the limited possibilities of our laboratory (and by no means an attempt to ascribe any preferential role to the structural approach of synaptochemistry),

[1] See *Akert*'s paper in this volume, p. 1—11.

[2] Interneuronal articulations as well as neuromuscular and neuroglandular junctions in mammals appear to be exclusively chemically mediated; ephaptic (electrical) transmission of impulses, frequently occurring in the nervous system of submammalian species, appear to be restricted to non-neural intercellular junctions in mammals (epithelial and muscular nexuses). The question whether ephaptic or synaptic (neurochemical) transmission represents, phylogenetically, a more ancient type of impulse transmission, is still open to discussion (see *Sakharov*'s paper in this volume, p. 43—59).

that the present paper deals mainly with the anatomical aspects of this discipline.

The main objectives of synaptochemistry are concerned with the questions of:

1. localization and metabolism of transmitters (presynaptic events),
2. molecular anatomy and structural versatility of receptors (postsynaptic events),
3. mechanisms intercalated between receptors and the function of the postsynaptic cell (excitation-performance coupling).

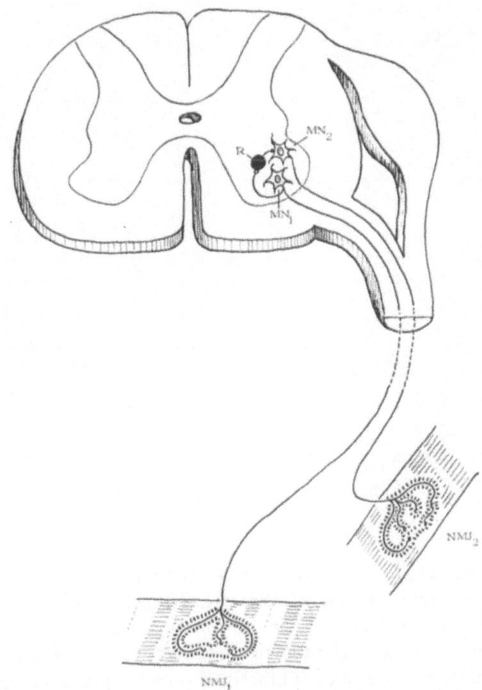

Fig. 1. Cytochemical entity of the neurone. MN_1 and MN_2 are two spinal motoneurones in the ventral horn; their axons innervate striated muscle fibres by means of neuromuscular junctions (NMJ_1 and NMJ_2). An initial axon collateral of motoneurone 1 innervates an inhibitory Renshaw cell (*R*). [In fact, scores of initial axon collaterals impinge upon Renshaw cell dendrites, constituting Renshaw elements, Figs. 5 and 10.] Axon of the (hypothetical) Renshaw cell inhibits motoneurone 2. Acetylcholinesterase activity is concentrated in perikarya of motoneurones (Fig. 7), as well as in axon terminals constituting neuromuscular junctions (Fig. 8), and in Renshaw elements (Fig. 10). Acetylcholine synthesis, as concluded from the concentration of hemicholinium, is predominant in motoneuronal perikarya, in motor endplates (Fig. 6) and in Renshaw elements (Fig. 5)

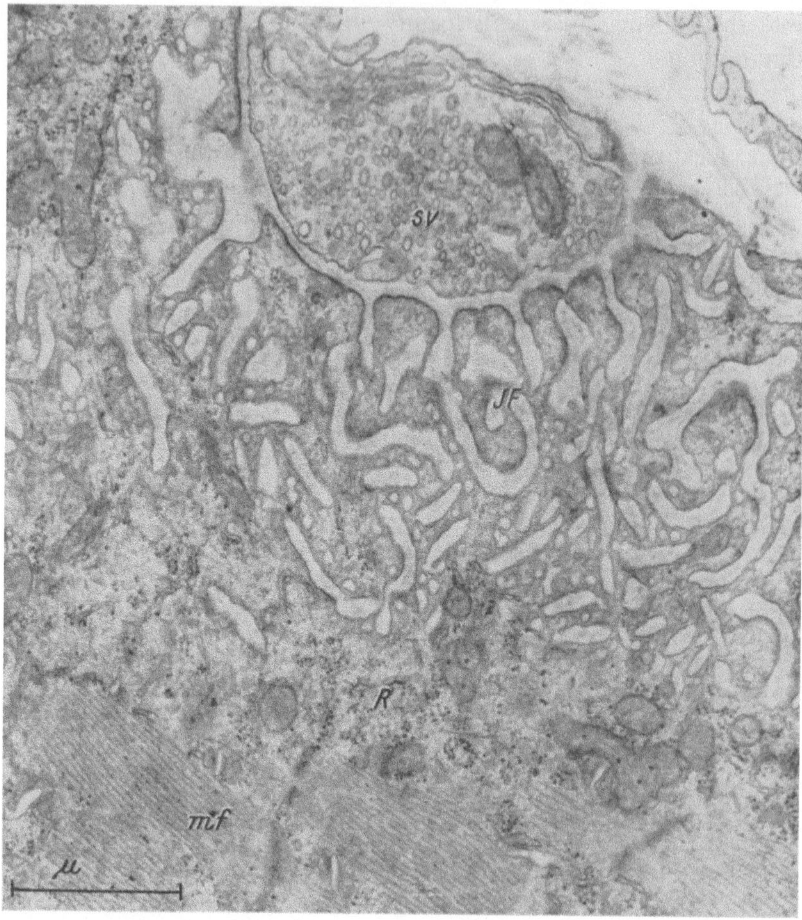

Fig. 2. Electron microscopic structure of a cholinergic synapse (motor endplate in the flexor digitorum brevis muscle of the rat). Note synaptic vesicles (*sv*) within the axon terminal, the intricate system of junctional folds (*JF*) constituting the subneural apparatus and its relation to the myofibrils (*mf*). Concentration of ribosomes (*R*) suggests the site of synthesis of postsynaptic acetylcholinesterase

Perhaps the most important contribution of synaptochemistry to the basic knowledge of neurobiology is the completion of the neurone doctrine with a new feature, *viz.*, the discovery of the "cytochemical entity" of the neurone[3]. It has already been suggested by *Dale* (1955) that the very same chemical mediator substance is released by all the axonal branches and collaterals of the neurone. Thus, *e.g.*, as pointed

[3] According to the Waldeyer theory, the neurone is an anatomical, physiological, genetical and trophical entity.

out by *Dale*, acetylcholine is released from every axon terminal of the spinal motoneurone, including not only the motor endplates, but also the initial axon collaterals staying within the spinal cord (Fig. 1). The entire problematics of the Renshaw cells goes back to this principle *(vide infra)*. In a true cytochemical entity, however, not only the mediator substance itself but also the metabolic systems responsible for the synthesis and breakdown of the transmitter have to be located within the neurone. In the following, cholinergic, monoaminergic and aminacidergic neurones will be scrutinized from this point of view.

Cholinergic Systems

Unfortunately, the transmitter of the cholinergic neurone, acetylcholine, cannot be demonstrated histochemically; until now, no effort in this respect has been successful (*Csillik*, 1969). Synaptic vesicles[4] are undoubtedly signs of synaptic activity: nevertheless, they cannot be taken as proof of the presence of acetylcholine. Synaptic vesicles

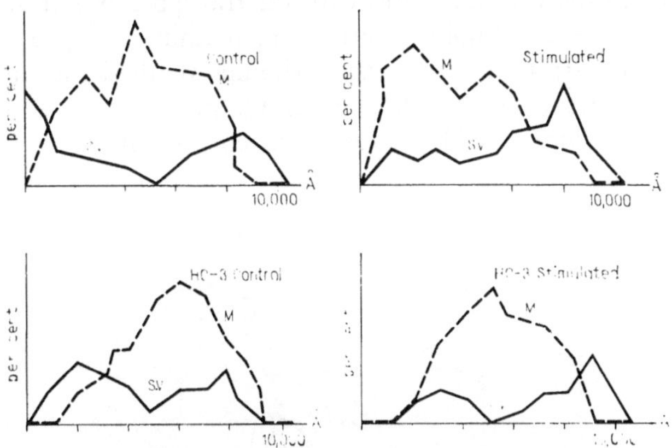

Fig. 3. Distribution of mitochondria (*M*, dashed lines) and of synaptic vesicles (*sv*, solid lines) in the motor nerve terminals in four experimental groups. Front lines (corresponding to the area near the pre-synaptic membrane) are represented at the right-hand sides of the graphs. In the control group, two peaks indicate concentrations of synaptic vesicles: one at the *front* line and one in the *back*-ground (*i.e.*, the area far from the pre-synaptic membrane). After supramaximal stimulation, vesicles are concentrated at the front line only. Hemicholinium treatment alone does not induce any considerable alteration in the distribution of synaptic vesicles, whereas in hemicholinium treated and supramaximally stimulated motor endplates the number of synaptic vesicles is considerabely decreased (*Csillik* and *Bense*, 1971)

[4] The term "synaptic vesicle" will be used to designate the spheroid, ~400 Å diameter type, lacking a dense core, and surrounded by a smooth membrane consisting of globular subunits.

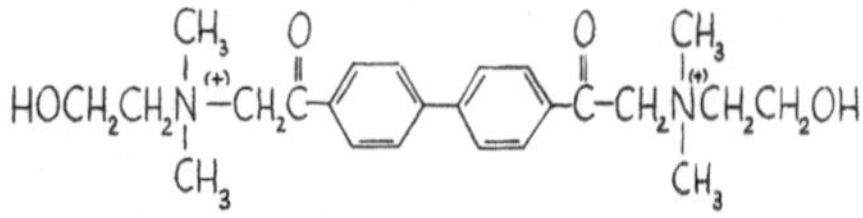

$$HOCH_2CH_2\overset{CH_3}{\underset{CH_3}{\overset{|}{\underset{|}{N^{(+)}}}}}-CH_2\overset{O}{\overset{\|}{C}}-C_6H_4-C_6H_4-\overset{O}{\overset{\|}{C}}-CH_2\overset{CH_3}{\underset{CH_3}{\overset{|}{\underset{|}{N^{(+)}}}}}CH_2CH_2OH$$

Fig. 4. Chemical structure of hemicholinium (HC-3). Note the choline moieties at both ends of the molecule, responsible for the competitive antagonistic effect of this drug in the acetylcholine synthesizing machinery

are present not only in cholinergic synapses (Fig. 2) but also in those in which other chemical substances do function, *e.g.* amino acids. On the other hand, accumulations of synaptic vesicles can be found in electrotonical (ephaptic) junctions, too. Supramaximal stimulation of cholinergic synapses results in a re-arrangement of vesicles within the terminal (*Jones* and *Kwanbunbumpen*, 1970; *Csillik* and *Bense*, 1971; *Korneliussen*, 1972) that, by careful extrapolation, can be interpreted as a sign of a function-dependent displacement of axoterminal acetylcholine. However, the number of vesicles does not decrease even after prolonged stimulation, except in animals pretreated with hemicholinium, a drug that interferes with the transport and/or activation of choline in the acetylcholine synthesizing machinery (Fig. 3).

The demonstration of the acetylcholine synthesizing enzyme is also problematic, even though considerable progress was achieved by attempting to locate -SH groups of Coenzyme A, responsible for

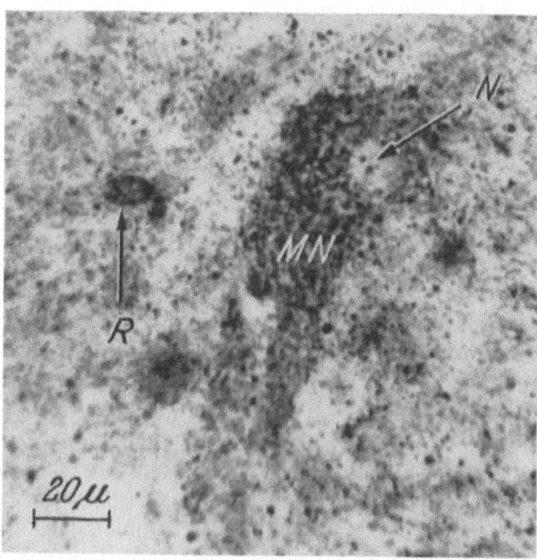

Fig. 5. Localization of ^{14}C-hemicholinium in the spinal cord (35 min after i.p. injection of 10 mg/kg). Note concentration of reduced silver grains in the cytoplasm of a spinal motoneurone (*MN*), sparing out the nucleus (*N*). The small spherical body, outlined by a heavy concentration of silver grains, is a Renshaw element (*R*)

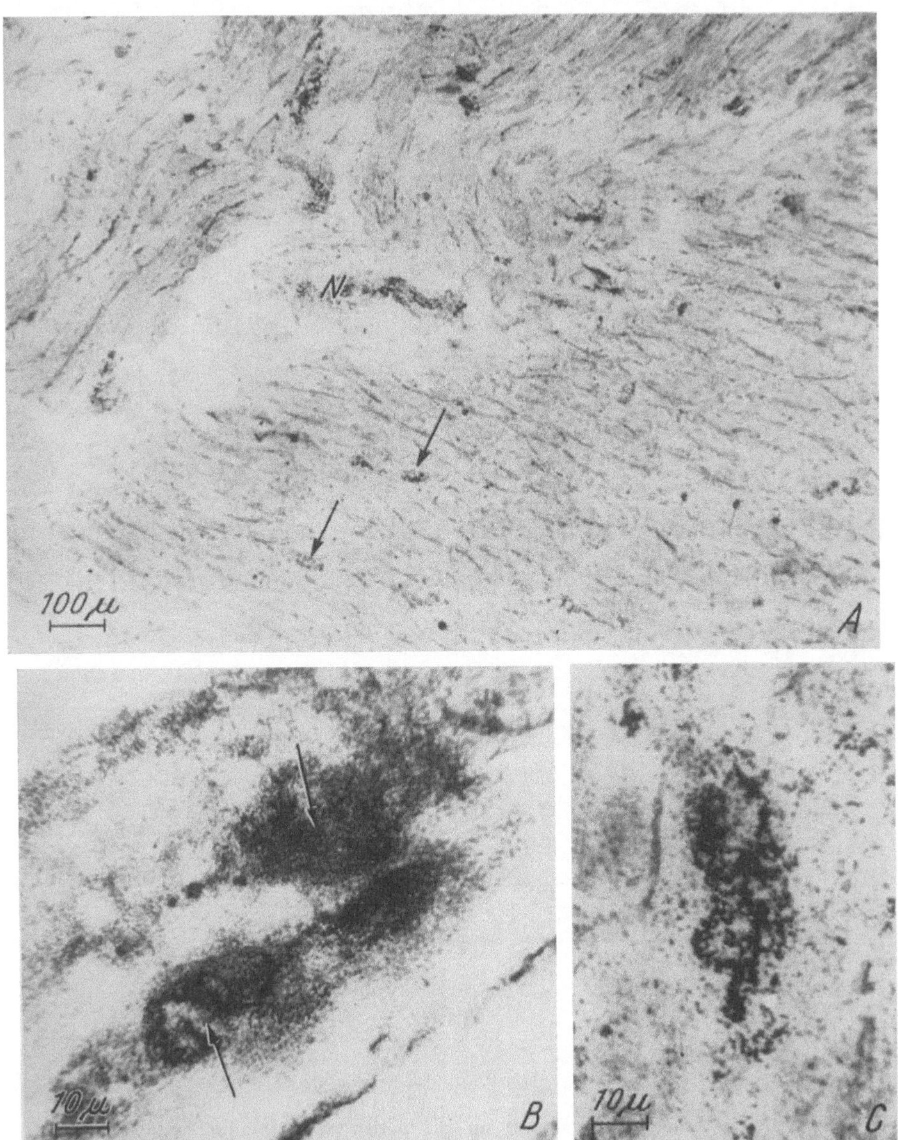

Fig. 6. Localization of ^{14}C-hemicholinium in the skeletal muscle (35 min after i.p. injection of 10 mg/kg). Non-fixed cryostat sections; exposure 10 weeks.

A: Low-power view of the flexor digitorum brevis muscle. Note concentration of reduced silver grains in an intramuscular nerve branch (*N*) and in motor endplates (arrows).

B: High-power view of motor end plates (arrows) in the sternomastoid muscle.

C: High power view of a motor endplate in the flexor digitorum brevis muscle. Reduced silver grains appear to outline pre-synaptic structures

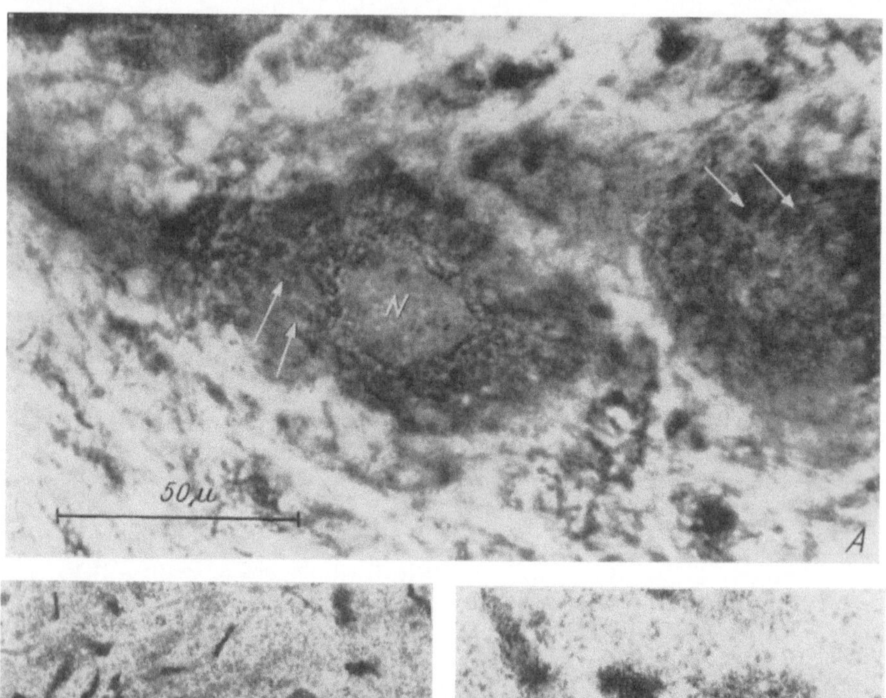

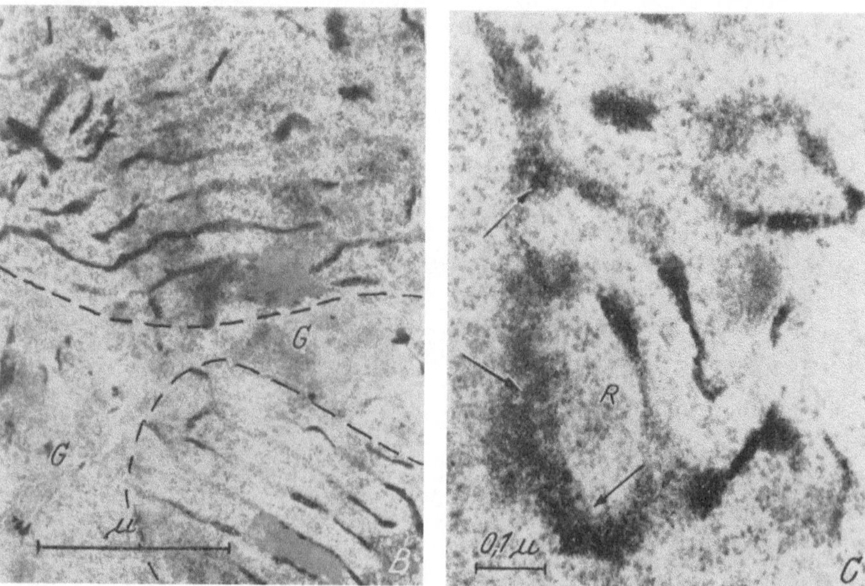

Fig. 7. Acetylcholinesterase activity in the perikarya of cholinergic nerve cells (spinal motoneurones of the rat).

A: Under the light microscope; localization of enzyme activity in cytoplasmic Nissl bodies (arrows) is apparent.

B: Electron microscopically, enzyme activity is confined to cisterns of the rough-surfaced endoplasmic reticulum. Nissl bodies are outlined by dashed lines. The Golgi apparatus between Nissl bodies does not exert any reaction.

C: Under high electron optical power: reaction product is present within the cisternae of the endoplasmic reticulum, the absence of enzyme reaction from intracisternal globular "endoplasmic units" (arrows) is apparent. These "endoplasmic units", measuring 300—350 Å in diameter, are supposed to take part in the final modelling of the tertiary molecular structure of freshly synthesized acetylcholinesterase molecules. *R*, free ribosomes

providing active acetate for acetylcholine synthesis (*Burt*, 1970; *Kása et al.*, 1970). The fact, however, that hemicholinium interferes with the acetylcholine synthesizing machinery (Fig. 4) can be used for an indirect localization of acetylcholine synthesis (*Knyihár* and *Csillik*, 1970). Autoradiographic localization of intravenously or intraventricularly injected ^{14}C-hemicholinium reveals that the most characteristic cholinergic nerve cell, the spinal motoneurone, accumulates this drug in its perikaryon, in the axon and, especially, in axon terminals. Autoradiographic grains are concentrated, however, not only in motoneuronal perikarya but also in small round bodies surrounding the motoneurones which are perhaps identical with Renshaw "elements" (Fig. 5). Preterminal branches of motor axons and, especially, axoterminal structures (motor endplates) within striated muscles show the highest accumulation of reduced silver grains (Fig. 6).

It should be kept in mind that localization of hemicholinium cannot be regarded as an equivalent of choline acetylase activity. Sites of a high choline turnover tend to concentrate this drug, irrespective of the final fate of choline. Furthermore, hemicholinium also possesses a receptor-binding capacity. Thus, after degeneration of the motor nerve, whilst no reduced silver grains can be detected at the sites of the (degenerated) endplate, a widespread membrane-bound reaction occurs at the surface of muscle fibers, probably due to the spreading of acetylcholine receptors[5].

Enzymatic breakdown of acetylcholine is, at the time being, the only metabolic event of the cholinergic cell that can be demonstrated with sufficient reliability. Both, light microscopic and electron microscopic cytochemistry prove that the site of production of acetylcholinesterase ist the endoplasmic reticulum of the cholinergic neurone (*Fukuda* and *Koelle*, 1959), *i.e.*, the reaction is confined to the Nissl bodies (Fig. 7). Acetylcholinesterase activity of motor endplates had been demonstrated at the light microscopic level at the very beginnings of synaptochemistry. Electron cytochemistry (*Csillik* and *Knyihár*, 1968) completed the picture by revealing acetylcholinesterase activity not only in the post- but also in the apposed presynaptic membrane (Fig. 8) which had been overlooked by light microscopic cytochemists[6]. The implication following from this observation is of

[5] Progress can expected by using hemicholiniums, free of receptor affinity. Such a substance, acetyl-*seco*-hemicholinium, has been synthesized recently by *V. Haarstad*.

[6] Strangely enough, in one of his early papers, *Couteaux* (1951), noted this presynaptic enzyme activity; later however he abandoned this view, perhaps because of the overwhelming evidence of postsynaptic activity.

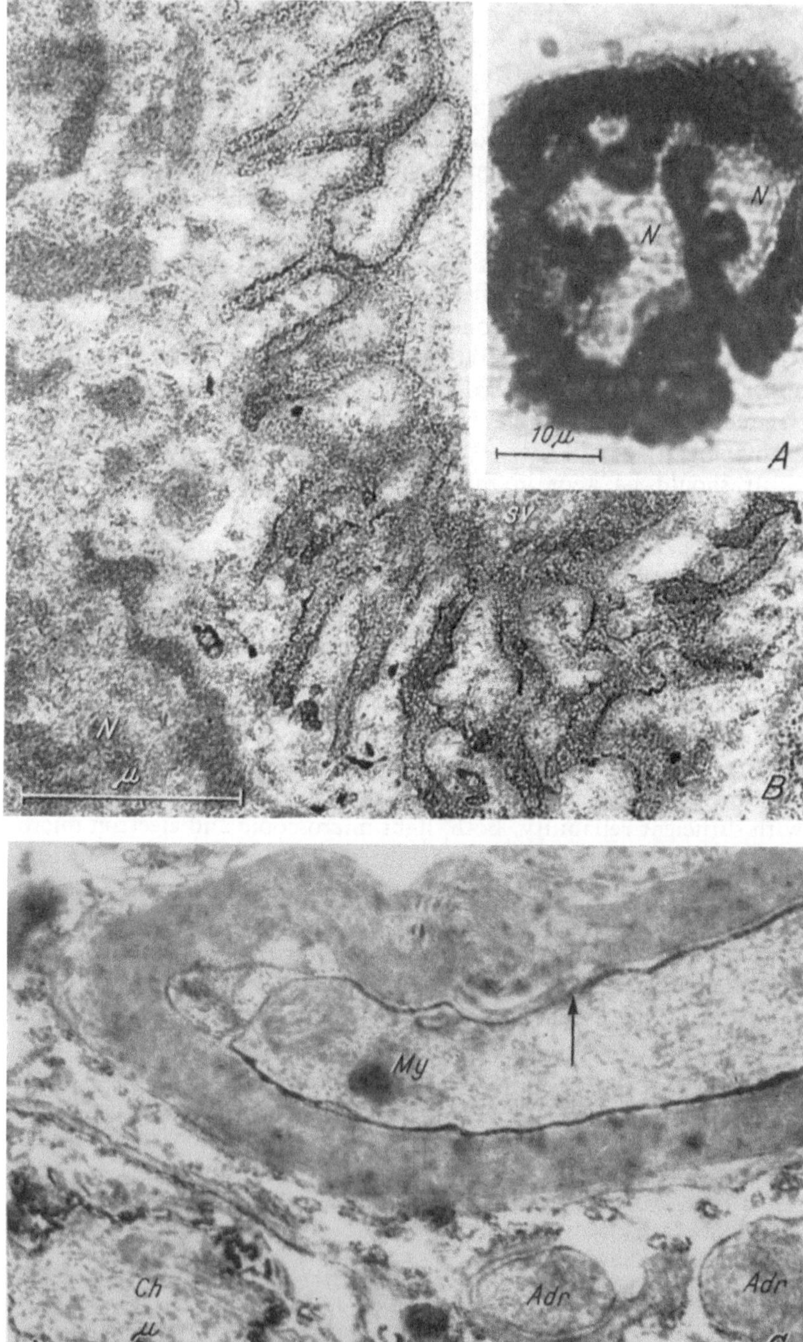
10 μ
N
N
A
SV
N
μ
B
My
Ch
μ
Adr
Adr
C

great theoretical importance. For a trivial interpretation of neurochemical transmission (*i.e.*, for a "jet-like" expulsion of the transmitter substance from the nerve terminal) a strategic localization of the hydrolyzing enzyme at the postsynaptic site would be sufficient; the fact however, that the same activity is also present presynaptically[7] suggests a more sophisticated sequence of events in chemical impulse transmission, in many ways resembling the "heretic" *Nachmansohn* (1959) ideas. On the other hand, the demonstration of presynaptic acetylcholinesterase activity in the motor endplate supplied a valuable argument for the *Dale* principle, *i.e.*, for the neurochemical entity of the neurone, since the presynaptic membrane is in anatomical continuity with the perikaryon, *via* the axolemma exerting acetylcholinesterase activity to a similar extent (Fig. 8c). It appears that the enzyme activity of the axolemmal membrane is guarenteed by the perikaryon by means of the axonal microtubules (*Kása*, 1968).

In accordance, a dual origin of neuromuscular acetylcholinesterase can be demonstrated by means of ontogenetical electron cytochemical techniques. It appears that a local synthesis of the postsynaptic enzyme by the sarcoplasmic reticulum of striated muscle cells precedes the arrival of the neuronal, presynaptic portion, transported *via* axonal microtubules and microfilaments (Fig. 9). At the same time, synthesis of pseudocholinesterase occurs within teloglial (Schwann) cells, surrounding and overlying the terminal. This enzyme, the function of which is still obscure, can be found to be concentrated in the perinuclear cisterns of the endoplasmic reticulum of these special teloglial cells. Participation of this (third) enzyme portion in the final modelling of the enzymatic and structural architecture of the neuromuscular junction cannot be excluded (Fig. 9).

[7] Predominance of postsynaptic activity is due only to the folded appearance of the postsynaptic membrane (in alpha motor endplates).

Fig. 8. A: Acetylcholinesterase activity of a motor endplate (subneural apparatus) in the gastrocnemius muscle. Note the regular arrangement of organites, constituting the subneural apparatus. The terminal motor axon is located within the groove formed by the organites. *N*, nuclei of teloglial (Schwann) cell.

B: Acetylcholinesterase activity of the motor endplate under the electron microscope. Note uniform distribution of the enzyme in pre- and postsynaptic membranes, and in the junctional folds of the latter. *sv*, synaptic vesicles within the axon terminal; *N*, nuclei of the sole plate. Uranyl-thiocholine technique.

C: Acetylcholinesterase activity of cholinergic axons (rat iris). In the myelinated nerve fibre (*My*), enzyme activity is confined to the axolemma between axoplasm and myelin lamellae. In the cholinergic non-myelinated fibre (*Ch*), enzyme activity is conspicuous in the axolemma. Two adrenergic nerve fibers (*Adr*) are completely devoid of acetylcholinesterase activity

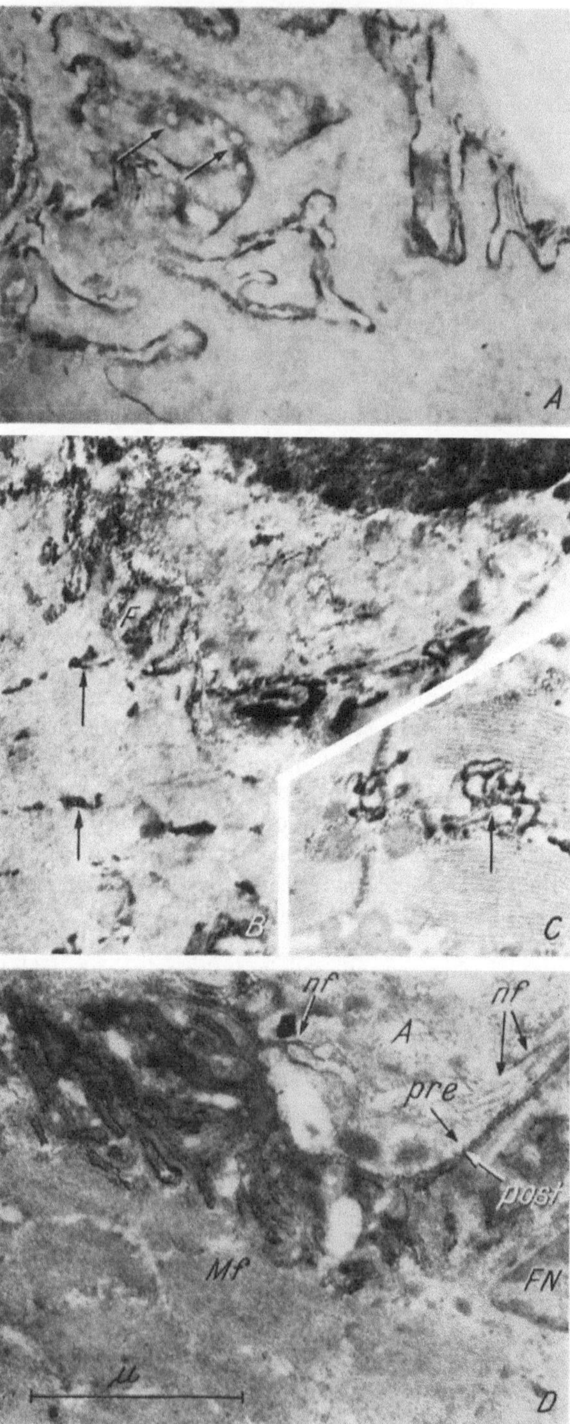
A
F
B
C
nf
nf
A
pre
post
Mf
FN
μ
D

The problem of the other axonal ending of the spinal motoneurone, *viz.*, that of the initial axon collaterals within the spinal cord is more difficult. Even though not only the existence of Renshaw cells but also the regular occurrence of initial axon collaterals have been recently questioned (*Scheibel* and *Scheibel*, 1971), both, physiological and pharmacological evidence suggest that recurrent inhibition of ventral root volleys (*Renshaw*, 1941) is exerted by cholinergic terminals impinging upon ill-defined neuronal elements in the ventral horn. Micro-iontophoretic studies performed by *Curtis* and *Ryall* (1966) revealed that, besides a nicotinic cholinergic receptor, at least two types of non-nicotinic (muscarinic) cholinergic receptors on Renshaw cells are present. Topographical mapping of Renshaw cells has been attempted by *Willis* and *Willis* (1964) and by *Thomas* and *Wilson* (1965), but the very inhibitory ("Renshaw") cells were characterized only recently (*Jankowska* and *Lindström*, 1971). Acetylcholinesterase activity of small round bodies ("Renshaw elements") in and around the motoneurone pool has been demonstrated by *Erulkar et al.* (1968), while microelectrode recording proves that stimulation of these elements results in Renshaw inhibition. However, the cytological identity of these elements was established only recently (*Csillik* and *Tóth*, 1972) using an experimental aproach. Ligature of the ventral root results in a piling-up of acetylcholinesterase within the proximal trunk, involving also initial collaterals. In such samples, acetylcholinesterase activity of Renshaw elements is markedly enhanced (Fig. 10). Careful examination of these elements reveals that enzyme activity is shown by numerous axon terminals impinging upon bulbous enlargements of peculiar dendrites, belonging to the Renshaw cells.

Accordingly, the concept of the "cytochemical entity" represented by a neurone fits well to the spinal motoneurone, being the most

---

Fig. 9. Ontogenetic development of synaptic cholinesterases in the mammalian endplate. Diaphragms of rat puppies.

A: First appearance of pseudocholinesterase activity in the perinuclear cistern of a teloglial (Schwann) cell. Note nuclear pores in tangential section (arrows). New-born rat.

B: Formation of the enzyme activity of postsynaptic folds (*F*) from the acetylcholinesterase material produced by the sarcoplasmic reticulum (arrows); first signs of a presynaptic activity transported via neurofilaments and axon tubuli are apparent.

C: Concentration of enzyme activity in the sarcoplasmic reticulum (arrow). 3-days old rat puppy.

D: Postsynaptic enzyme activity is completed; presynaptic enzyme arrives via axonal tubuli and neurofilaments (*nf*) from the parent spinal motoneurone. *A*, axon; *FN*, fundamental nucleus; *post*, post-synaptic membrane; *pre*, pre-synaptic membrane; *Mf*, myofilaments. 2 months old rat

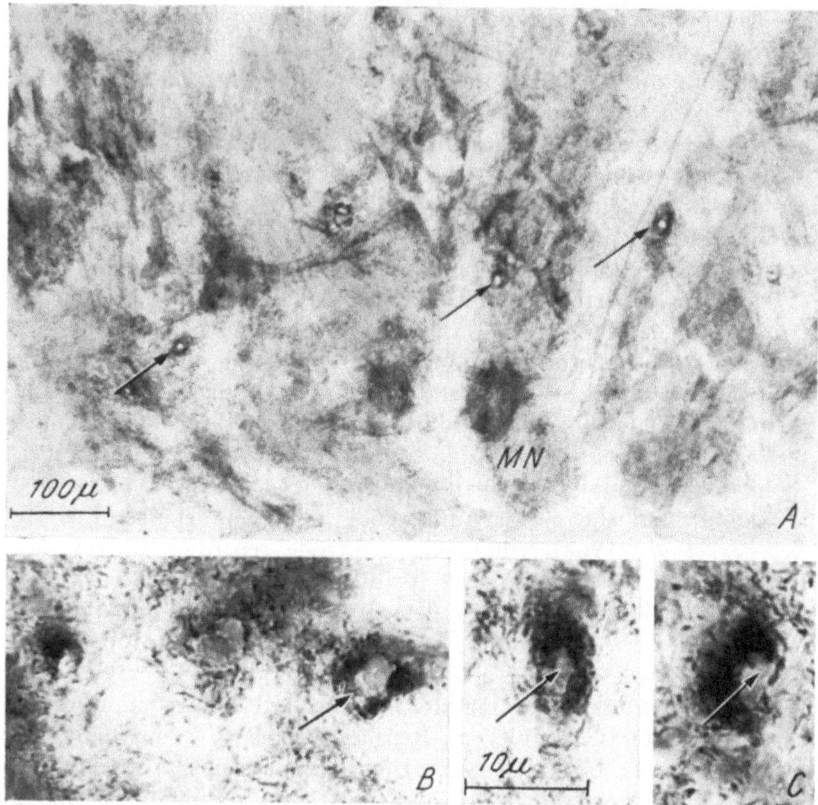

Fig. 10. Acetylcholinesterase activity of Renshaw elements. Spinal cord of the rat, 48 hours after ligation of the ventral root. *MN*, motoneurones, showing a medium intensity reaction; arrows point at Renshaw elements.

A: Low power view of the anterior horn, ×150.

B and C: High power patterns of Renshaw elements. Note the heavy concentration of acetylcholinesterase, due to axon terminals of initial axon collaterals of motoneurones, surrounding dendrite bulbs. 2200× (*Csillik* and *Tóth*, 1972)

Fig. 11. Top: Catecholamine reaction of adrenergic postganglionic axons in the rat iris. Note concentration of catecholamines in the varicosities (beads) of the nerve fibres (arrows).

Bottom: Acetylcholinesterase reaction of cholinergic nerve fibres in the rat iris. Enzyme activity is concentrated in the varicosities. The pattern is similar to that obtained with the catecholamine reaction.

Inset: Electron microscopic appearance of adrenergic (*A*) and cholinergic (*Ch*) axons in the rat iris after combined formaldehyde-glutaraldehyde fixation. Note the elongated shape of adrenergic vesicles, showing a dense core at the internal surface. Cholinergic axon contains clear synaptic vesicles

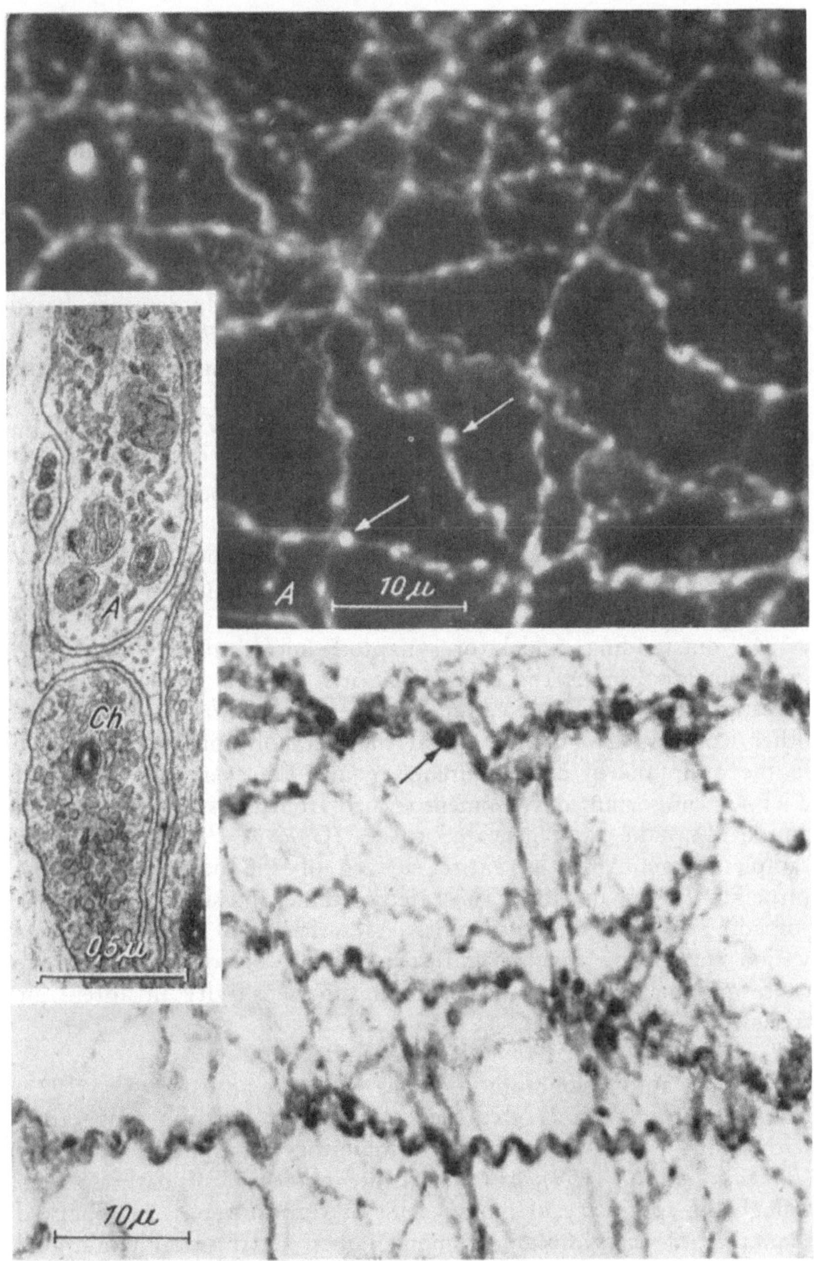
A
Ch
0,5 μ
A
10 μ
10 μ

carefully studied model of cholinergic nerve cells. The same also applies to preganglionic neurones as well as to the postganglionic neurones of the parasympathetic nervous system. Thus *e.g.*, perikarya of the ciliary ganglion as well as the axons of these nerve cells exert acetylcholinesterase activity, located at the axolemmal membrane (Fig. 8c). Varicosities of the postganglionic axons contain the highest concentrations of this enzyme (Fig. 11b).

Monoaminergic Systems

The brilliant histochemical reaction of induced fluorescence, based on the studies of *Eränkö* (1955, 1964) and of *Falck et al.* (1962), unequivocally proves the presence of norepinephrine throughout the entire extent of postganglionic neurones (Fig. 11 and 12) of the sympathetic nervous system and that of dopamine in various brain stem neurones. Apparently, the same applies to 5-hydroxy-tryptaminergic neurones in the lower brain stem. Also the small (~450 Å) dense-core vesicles in perikarya, in axons and in dendrites of monoaminergic neurones elegantly allow the identification of the transmitter, thus directly proving the cytochemical entity of monoaminergic neurons.

An outstanding model for synaptochemical studies of catecholamine metabolism is the excised rat iris (*Csillik*, 1964). Our first investigations were followed and completed by *Malmfors* (1965) and other workers, demonstrating that the metabolic processes involved in the re-uptake of catecholamines require the energizing action of ATP. An important development ensued from the study of catecholamine re-uptake by denervated tissues (*Gajó et al.*, 1970). In such samples, practically the entire surface of the denervated myoepithelium takes up extrinsic catecholamine, possibly due to the spreading of adrenergic receptors. As a matter of fact, this observation stands for the first direct structural demonstration of the cause and mechanism of denervation supersensitivity in adrenergic tissues.

Aminacidergic Systems

Recent autoradiographic studies (*Bloom* and *Iversen*, 1971; *Ehinger* and *Falck*, 1971, etc.) suggest that central neurones, operating with amino acids as excitatory or inhibitory transmitter substances (*Purpura et al.*, 1959), also constitute cytochemical entities, even though the specificity of amino acids as transmitters is less marked than that of acetylcholine or monoamines. Extrinsic glutamic acid and GABA are taken up and concentrated by nerve cells, and, according to studies performed by means of differential centrifugation (*De Robertis*, 1968) such transmitter amino acids are concentrated in

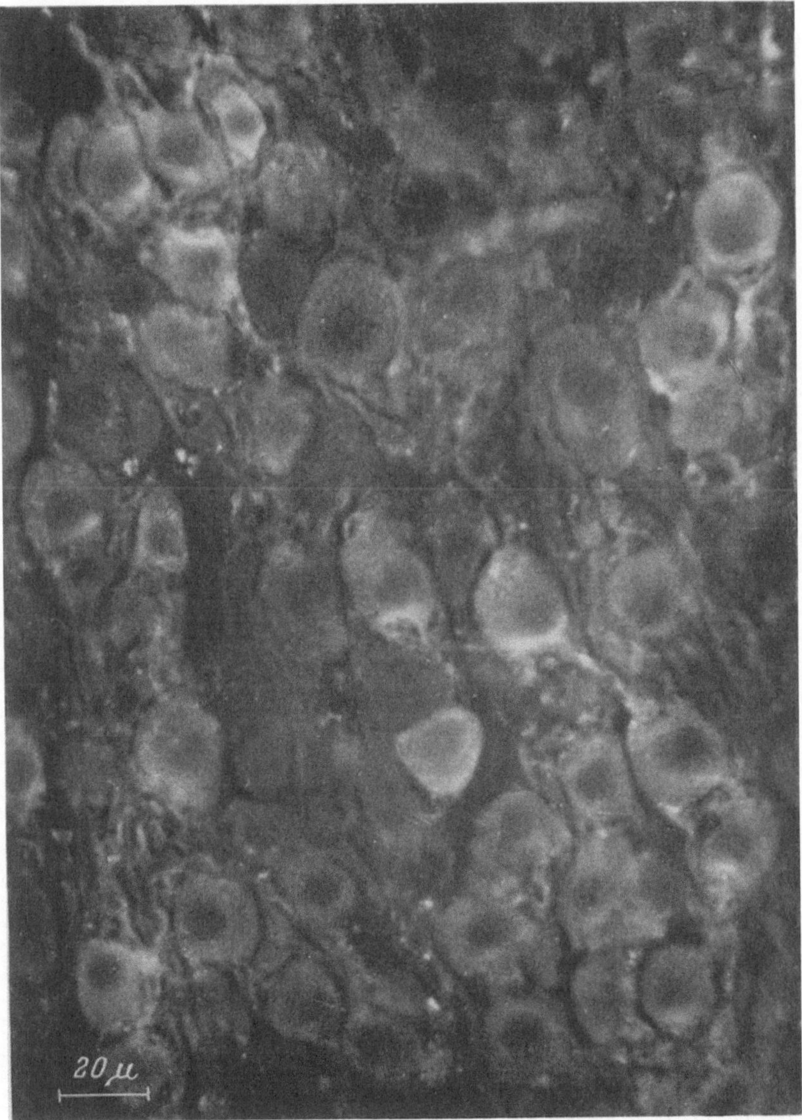

Fig. 12. Catecholamine reaction of a sympathetic ganglion (Ggl. cervicale superius of the rat; *Eränkő-Falck-Hillarp* technique in a cryostat section). Ganglion cells exhibit a medium to strong fluorescence reaction after formaldehyde treatment; intraganglionic nerve fibres (postsynaptic adrenergic axons) show up distinctly

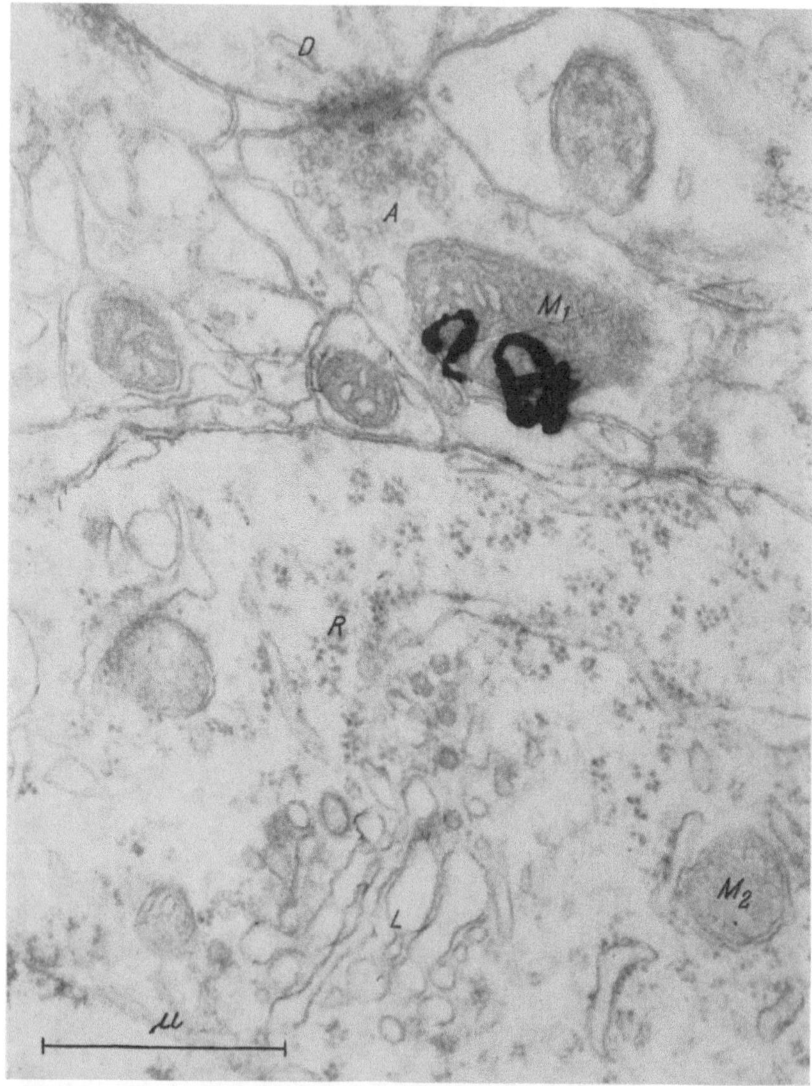

Fig. 13. Electron microscopic localization of ^{14}C-thiosemicarbazide in a hippocampal synaptosome. Reduced silver grains are seen over a mitochondrion (*M*1) within the axon terminal (*A*) synapsing with a dendrite (*D*). In the perikaryon of the nerve cell, mitochondria (*M*2), clusters of ribosomes (*R*) and dilated endoplasmic reticulum (lamellar body, *L*), are apparent

synaptosomes. Enzyme histochemical and autoradiographic studies (*Csillik et al.*, 1971) performed with the enzyme inhibitor thiosemicarbazide[8] suggest, however, that in this respect there might exist strict regional differences. Thiosemicarbazide is incorporated by nerve cell perikarya in brain stem nuclei, whereas, in the hippocampal formation, though occasional labeling of synapses occurs (Fig. 13), reduced silver grains are located mainly in glial elements.

Undoubtedly, thiosemicarbazide, like other substances when injected either intravenously or intraventricularly, has to pass glial barriers in order to reach neuronal structures. Thus, a gliocellular localization of this drug might simply represent a time-dependent intermediary state. On the other hand, consequent differences in glial versus neuronal localization in the limbic system as contrasted to brain stem nuclei, strongly speak against such an assumption. To us it seems plausible that — even though the localization of thiosemicarbazide-induced reduced silver grains cannot be taken as identical with the localization of enzymes inhibited by the drug — gliocellular localization of this inhibitor is a sign of the activity of GAD-II, a gliocellular enzyme described by *Haber et al.* (1970). If so, the possibility emerges that, in the limbic system, GABA is synthesized by glial cells and incorporated by a second step into neuronal elements.

Such an arrangement would readily explain the intriguing findings of *Sze* and *Lovell* (1970). These authors have shown that thiosemicarbazide, inhibiting GAD in tissue homogenates, does not exert any effect upon GAD in brain slices. On the other hand, *in vivo* inhibition of GAD by thiosemicarbazide results in characteristic seizures and convulsions. In view of our above experiments, it appears that, under normal conditions of life, the "driving force" of the brain fluid circulation is sufficient to establish contact between the enzyme and the inhibitor. Since this force fails to be present in brain slices, inhibition does not take place. In tissue homogenates, however, where metabolic compartments are destroyed by mechanical forces, the enzyme is mixed with the inhibitor resulting in inhibition just as occurs in the normal, living state.

The question of interaction between different transmitter substances, or rather between transmitter mechanisms, has been raised as early as 1945. Since that time, *Burn* and *Rand* (1965) and, in a more sophisticated form, *Koelle* (1962) developed general concepts on transmitter interactions (see *Vizy*'s paper in this volume, p. 61—78).

---

[8] Thiosemicarbazide inhibits enzymes operating with B_6-vitamine; biochemical studies by *Albers* (1960) and by *Roberts* (1968) prove that the most sensitive enzymes in the brain are glutamic acid decarboxylase and GABA-transaminase, *i.e.*, the two enzymes responsible for GABA production and breakdown.

Until now, synaptochemistry failed to yield any decisive proof in favour of this theory.

Receptors

The second main objective of synaptochemistry is the identification of receptor moieties in the postsynaptic membrane, and the recording of their structural-molecular alterations due to transmitter binding. The main body of investigations in this field was performed by means of tissue fractionation techniques. Generally speaking, two methodological approaches were attempted: isolated tissue fractions were brought into reaction with reversible receptor binding substances, or, in other experiments, tissue homogenates were incubated with covalent receptor-binding substances and subsequently the fraction, containing the bound substance, was isolated (*Triggle*, 1971). So far, isolation of the cholinergic receptor substance (*Changeaux et al.*, 1970; *De Robertis et al.*, 1970) appears to be the most promising.

Strikingly, very little cytochemical work has been carried out in this respect. The pioneering study of *Waser* and *Hadorn* (1961), locating tritiated curarine in motor endplates, is an example to be followed*. Polarization optical studies (*Csillik*, 1963) revealed molecular alterations of the postsynaptic membrane after binding with acetylcholine. These results were confirmed by *Zacks* (1964). More sophisticated techniques of optical analysis might reveal further details of structural re-arrangement of postsynaptic membranes after transmitter interaction. On the other hand, the possibility to locate adrenergic transmitter substances can be used in the search for extension of receptor moieties in visceral organs. An example of such studies has already been mentioned (*Gajó et al.*, 1970).

Excitation-Performance Coupling

This third aim of synaptochemical studies comprises the very postsynaptic effects within the effector cell. This question involves the problem of excitation-contraction coupling in muscle. In this connection, the outstanding role of Ca^{++}ions (*Heilbrunn* and *Wierzinski*, 1947) has been proved cytochemically in cardiac muscle (*Koshtoyantz*, 1951) and in skeletal muscle (*Sávay* and *Csillik*, 1965). Whether the situation is analogous in the excitation-secretion coupling (glandular tissue) and in excitation-excitation or excitation-inhibition coupling (in excitatory and in inhibitory interneuronal synapses, respectively) is still problematic. On the other hand, an increasing body of evidence

* *Note added to the proof.* Recent autoradiographic and fluorescent microscopic studies performed with labelled α-bungarotoxin indicate that localization of the acetylcholine receptor in the motor end plate is strikingly similar to that of acetylcholinesterase.

suggests that neurotransmitter substances, once attached to the receptor molecules of the postsynaptic membrane, may exert their action upon the performance of the target cell by means of interfering with adenyl cyclase activity of the latter. Thus, the postsynaptic "second messenger" appears to be cyclic 3'-5'-AMP, at least in monoaminergic synapses (*Sutherland* and *Robison*, 1966). Synaptochemical studies might prove or disprove ideas concerning the generality of this regulatory effect in postsynaptic events of cyclic AMP. A cytochemical technique for the visualization of adenyl cyclase activity has been proposed by *Reik et al.* (1970). Its specificity should be re-investigated in order to achieve results comparable to biochemical studies (see *Gerebtzoff*'s paper in this volume, p. 181—185).

Perhaps even more puzzling is the role of prostaglandins, the cyclic eicosa-tetraenic and eicosa-trienic acid derivates, regulating postsynaptic effects by interfering with the action of adenyl cyclase (*Horton*, 1972). The physiological studies of *Clegg* (1966), performed on isolated organs, suggest that prostaglandins increase the affinity of monoamines to adrenergic receptors resulting in an initial potentiation and in a subsequent inhibition. Since such effects were observed in a score of visceral organs, the role of prostaglandins in regulation of neurovegetative transmission to smooth muscle cells might be a fundamental, general phenomenon. Inactivation of prostaglandins by a specific dehydrogenase has been demonstrated histochemically. Combined micro-electrophoretic studies (*Siggins et al.*, 1971) suggest that catecholamine-induced inhibition of cerebellar Purkinje cells is suspended by the local action of prostaglandins.

Transducers

In many respects, also sensory end organs ("receptors" in the histological sense of the word, or "transducers" in up-to-date physiology) appear to fall into the category of synapses. As a matter of fact, pharmacological evidence (*Jancsó*, 1968) suggests that we are dealing here with a special kind of neurochemical transmission between the "transducer" (sensory, receptor) cell and the afferent axon. Thus, also the transmission mechanism operating between sensory cells and sensory axon endings appear to be accessible to the techniques of synaptochemistry. Meissner's corpuscles show cholinesterase activity (*Csillik et al.*, 1954). Here, the enzyme is concentrated within the special sensory cells constituting a *sui generis* type of junction with the sensory axon terminal. The analogy between sensory endorgans and synapses is even more striking when taking the fact into account that enzyme activity of Meissner's corpuscles survives nerve degeneration in a manner comparable to that observed in motor endplates.

Nevertheless, the question of the *efferent* innervation of sensory cells and sense organs is an intriguing one. The bestknown example of this is the cholinergic efferent ("Rasmussen's") bundle in the vestibulo-cochlear nerve (*Gacek et al.*, 1965). It is tempting to speculate on the generality of the efferent innervation of receptor organs. If actually intercalated, as a rule, between efferent and afferent axons, sensory cells would rather transmit and modulate than generate impulses in a clear-cut synaptochemical way.

Perspectives

It is felt that pharmacotherapy of nervous disorders will achieve the level of an objective science only if abnormal alterations are understood at the molecular level of structural organization. There is, however, another aspect of synaptochemistry which might be of importance in understanding the basic function of the nervous system.

Short- and long-term memory, conditioning and learning are unique features of nervous systems, based upon an interrelated function of larger cell assemblies. In addition the individual neurones in this chain should have properties rendering them capable of reiterating repetitive impulses with increased efficiency. There is now general agreement that this property is related to protein metabolism of the nerve cell, but regarding the importance of various constituents of the protein synthesizing machinery, different theories of memory take extremely differing positions (*Glassman*, 1969).

It is commonplace that neuroproteins are synthesized in the endoplasmic reticulum (Nissl granules) of the perikaryon, whereas the axon, at least in mammalian neurones, is devoid of a protein synthesizing machinery. Thus the axon, having a 100—1000 times larger volume than the perikaryon, relies on proteins synthesized by the Nissl substance and transported by axonal flow mechanisms[9].

Since axons and, especially, nerve endings, do not contain lysosomes, the question arises as to the fate of the proteins once arrived at the axon terminal. Probably one (smaller) part is returned to the perikaryon by means of retrograde axonal transport. The bulk of the proteins, however, has to be disintegrated within the terminal by means of a non-lysosomal proteolytic mechanism in order to keep constant the size and shape of the terminal[10]. In fact, the presence of

[9] Or, perhaps, of proteins synthesized by glial and/or Schwann cells, supposing there exists a leakage through these cellular elements.

[10] Experiments aimed to prove a function-dependent "growth" of synaptic endbulbs (in order to explain the increased efficiency of the synapse after repetitive impulses by an increased contact area with the postsynaptic cell) failed to give consistent results (*Eccles*, 1964).

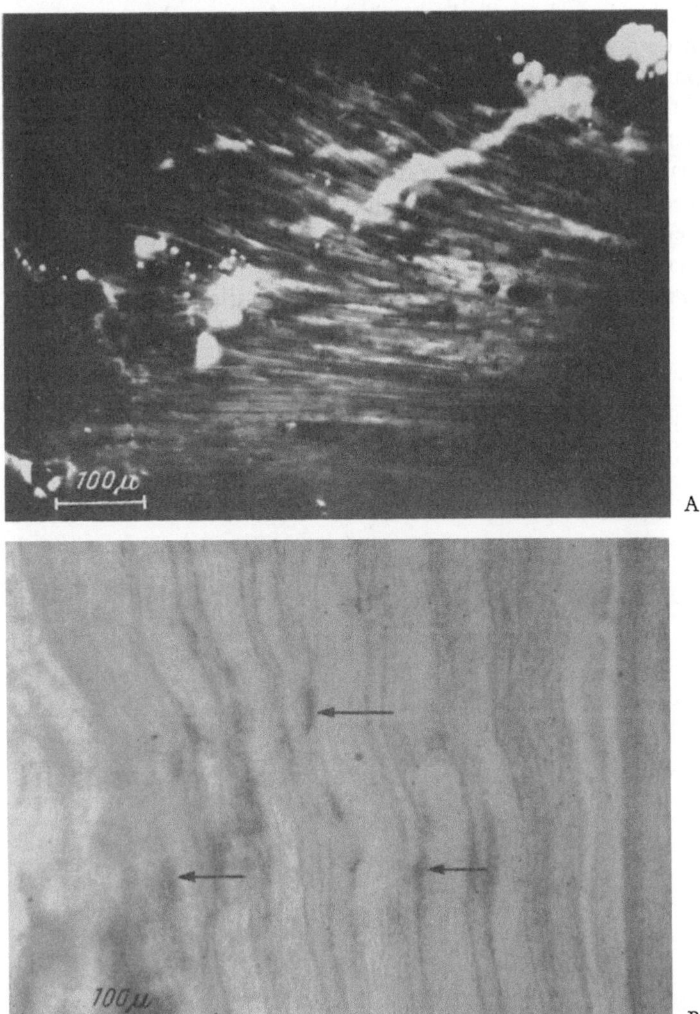

Fig. 14. Function-dependent proteolytic activity in the neuromuscular synapse.

A: Gelatine film digestion (white spots) in a row of motor endplates. Flexor digitorum muscle after supramaximal stimulation (50 cpm square pulses, 1.2 V, 50 min).

B: Increased leucyl aminopeptidase reaction in motor endplates of the diaphragm, 10 min after i.p. injection of the cholinesterase inhibitor Neostigmine, 0.5 mg/kg. Arrows point at endplates (*Poberai et al.*, 1972)

a non-lysosomal, neutral proteinase in the central nervous system has been demonstrated (*Marks* and *Lajtha*, 1965); this enzyme might be responsible for the theoretically postulated axoterminal proteolysis.

Strangely enough, this hypothetical axoterminal proteolysis cannot be demonstrated by cytochemical techniques under resting conditions, probably because of the restricted sensitivity of the methods employed. After prolonged stimulation, however, cytochemical signs of axoterminal proteolysis become apparent. Such a function-dependent axoterminal proteolysis can be readily demonstrated by means of the *Adams-Tuquan* (1961) gelatine film digestion technique (Fig. 14a),

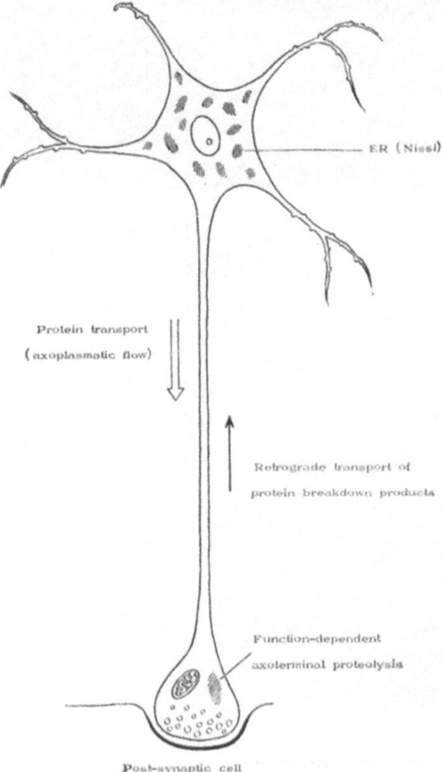

Fig. 15. Protein metabolism of the nerve cell. Proteins are synthesized by the endoplasmic reticulum within the cytoplasm and in major dendrites (*ER* = Nissl). Proteins are transported by axonal flow mechanisms to the terminal. Here, the proteolytic breakdown of neuroproteins happens by axoterminal proteolytic processes, activated by the synaptic function. Breakdown products of function-dependent axoterminal proteolysis either pass the synaptic cleft and exert post-synaptic effects, or are transported back to the perikaryon by means of a retrograde transport. A regulatory function of these breakdown products upon perikaryal protein synthesis is anticipated

and also the leucyl-β-aminopeptidase reaction (*Poberai et al.*, 1972) reveals a function-dependent activity in neuromuscular junctions (Fig. 14b).

At present, the number of observations in different areas of the nervous system is not sufficient to claim a general validity of this function-dependent increase of axoterminal proteolysis. However, two major conclusions appear to be pertinent at the present state of our studies. First of all, small molecules (oligopeptides and amino acids) resulting from the activity of proteinase, may traverse the synaptic cleft. Such low molecular weight substances might play a supplementary role in cholinergic junctions, coupling the effects of presynaptically released acetylcholine to postsynaptic events. The role of such small molecules (breakdown products of proteins) might be of essential importance in central aminacidergic synapses, taking a direct part in excitation (or inhibition) of the postsynaptic cell.

On the other hand, the possibility cannot be excluded that some of these small molecules, resulting from function-dependent axoterminal proteolysis, do not pass the synaptic membranes, but return to the perikaryon of the parent neurone by means of retrograde axoplasmatic transport. If such a function-dependent "reflux" of degraded neuroproteins exists, it can be anticipated that these small molecules exert some kind of regulatory action upon the metabolism (especially upon the protein synthesizing machinery) of the nerve cell[11]. Regulation of the perikaryal metabolism might involve a large scale of possibilities. A simple increase of the amino acid stores at disposal for protein synthesis may produce mass effects or rather activate more sophisticated allosteric mechanisms. Such a feed-back regulation might provide a structurally determined basis for the "cellular memory" of the neurone (Fig. 15).

References

Adams, C. W. M., and *N. A. Tuquan*: The histochemical demonstration of protease by a gelatine-silver substrate. J. Histochem. Cytochem. *9*, 469—472 (1961).

Albers, R. W.: The distribution of gamma-aminobutyrate and related enzyme systems. In: Inhibition in the nervous system and gamma-aminobutyric acid (*Roberts, Baxter, Van Harreveld, Wiersma, Ross Adey, Killiam*, eds.), pp. 196—201. Oxford: Pergamon Press. 1960.

Bennett, M. V. L., E. Aljure, Y. Nakajima, and *G. D. Pappas*: Electronic junctions between teleost spinal neurons: electrophysiology and ultrastructure. Science *141*, 262—264 (1963).

[11] A feed-back regulation of neuronal functions has been proposed recently by *Matthies* (1971).

Bloom, F. E., and *L. L. Iversen*: Localizing ^{3}H-GABA in nerve terminals of rat cerebral cortex by electron microscopic autoradiography. Nature (London) *229*, 628—630 (1971).

Burn, J. H., and *M. J. Rand*: Acetylcholine in adrenergic transmission. Ann. Rev. Pharmacol. 163—182 (1965).

Burt, A. M.: A histochemical procedure for the localization of choline acetyltransferase activity. J. Histochem. Cytochem. *18*, 408—415 (1970).

Cajal, R. Y.: Histologie du systeme nerveux de l'home et des vertébrés. Madrid. (Facsimile edition of the 1911 work.)

Changeaux, J. P., *M. Kasai*, and *C. Y. Lee*: Use of a snake venom toxin to characterize the cholinergic receptor proteins. Proc. Nat. Acad. Sci. U.S.A. *67*, 1241 (1970).

Clegg, P. C.: The effect of prostaglandins on the response of isolated smooth-muscle preparations to sympathomimetic substances. Mem. Soc. Endocr. *14*, 119—136 (1966).

Couteaux, R.: Remarques sur les methodes actuelles de detection histochimique des activités cholinestérasiques. Arch. int. Physiol. *59*, 526—537 (1951).

Csillik, B.: Submicroscopic organization of the post-synaptic membrane in the myonerural junction. J. Cell. Biol. *17*, 571—586 (1963).

Csillik, B.: Histochemical model experiments on the effect of various drugs on the catecholamine content of adrenergic nerve terminals. J. Neurochem. *11*, 351—355 (1964).

Csillik, B.: Acetylcholine. J. Neuro-Visceral Relat., Suppl. IX, 187—211 (1969).

Csillik, B., and *S. Bense*: Function-dependent alterations in the distribution of synaptic vesicles. Acta biol. Acad. Sci. hung. *22*, 131—139 (1971).

Csillik, B., and *E. Knyihár*: On the effect of motor nerve degeneration on the fine-structural localization of esterases in the mammalian motor end plate. J. Cell. Sci. *3*, 529—538 (1968).

Csillik, B., *A. M. Gerebtzoff*, *J. Kiss*, and *E. Knyihár*: Zur Histochemie der limbischen Hemmung. Histochemie *28*, 38—54 (1971).

Csillik, B., *G. Sávay*, *I. Nagy*, *O. Bondray*, and *M. Poberai*: Cholinesterase activity of sensory nerve endings. Acta physiol. hung. *6*, 379—382 (1954).

Csillik, B., and *L. Tóth*: Histochemical identification of Renshaw elements. J. Histochem. Cytochem. *20*, 385—387 (1972).

Curtis, D. R., *J. W. Phillips*, and *J. C. Watkins*: Acidic amino acids with strong excitatory actions on mammalian neurones. J. Physiol. (London) *166*, 1—14 (1963).

Curtis, D. R., and *R. W. Ryall*: The excitation of Renshaw cells by cholinomimetics. Exp. Brain Res. 2, 49—65 (1966).

Dale, H. H.: Junctional transmission of nervous effects by chemical agents. Proc. Mayo Clin. *30*, 5—20 (1955).

De Robertis, E.: Isolation of inhibitory nerve endings from brain. In: Structure and function of inhibitory neuronal mechanism (*Euler, Skoglund, Söderberg*, eds.), pp. 511—522. Oxford: Pergamon Press. 1968.

De Robertis, E., and *Fiszer de S. Plazas*: Acetylcholinesterase and acetylcholine proteolipid receptor: two different components of electroplax membranes. Biochim. Biophys. Acta *219*, 388—397 (1970).

Eccles, J. C.: The physiology of synapses. Berlin-Göttingen-Heidelberg-New York: Springer. 1964.

Ehinger, B., and *B. Falck*: Autoradiography of some suspected neurotransmitter substances: GABA, glycine, glutamic acid, histamine, dopamine and L-DOPA. Brain Research *33*, 157—172 (1971).

Elliot, T. R.: On the action of adrenaline. J. Physiol. (London) *31*, 20 p (1904).

Eränkő, O.: Distribution of fluorescing islets, adrenaline and nor-adrenaline in the adrenal medulla of the hamster. Acta endocr. (Kbh) *18*, 174—179 (1955).

Eränkő, O.: Histochemical demonstration of catecholamines by fluorescence induced by formaldehyde vapour. J. Histochem. Cytochem. *12*, 487—488 (1964).

Erulkar, S. D., *C. W. Nichols*, *M. B. Popp*, and *G. B. Koelle*: Renshaw elements: localization and acetylcholinesterase content. J. Histochem. Cytochem. *161*, 128—135 (1968).

Falck, B., *N. A. Hillarp*, *G. Thieme*, and *A. Torp*: Fluorescence of catecholamines and related compounds condensed with formaldehyde. J. Histochem. Cytochem. *10*, 348—354 (1962).

Fukuda, T., and *G. B. Koelle*: The cytological localization of intracellular neuronal acetylcholinesterase. J. biophys. biochem. Cytol. *5*, 433—440 (1959).

Gacek, R. R., *Y. Nomura*, and *K. Balogh*: Acetylcholinesterase activity in the efferent fibers of the stato-acoustic nerve. Acta oto-laryng. *59*, 541—553 (1965).

Gajó, M., *G. Kálmán*, and *B. Csillik*: Ein Beitrag zur cytochemischen Interpretation der Denervation-Überempfindlichkeit. Acta histochemica *38*, 293—304 (1970).

Glassman, R.: The biochemistry of learning: an evaluation of the role of RNA and protein. Annual Rev. Biochem. *38*, 605—646 (1969).

Haber, B., *K. Kuriyama*, and *E. Roberts*: L-Glutamic Acid Decarboxylase: a new type in glial cells and human brain gliomas. Science *168*, 598—599 (1970).

Heilbrunn, L. V., and *F. J. Wierzinski*: The action of various cations on muscle protoplasm. J. cell. comp. Physiol. *29*, 15—32 (1947).

Horton, E. W.: Prostaglandins, p. 145 ff. Berlin-Heidelberg-New York: Springer. 1972.

Jancsó, N.: Desensitisation with capsaicin and related acylamides as a tool for studying the function of pain receptors. Proc. 3rd Int. Pharmacol. Meet. Parmacology of pain, Vol. 9, pp. 33—55. Oxford-New York: Pergamon Press. 1968.

Jankowska, E., and *S. Lindström*: Morphological identification of Renshaw cells. Acta physiol. scand. *81*, 428—430 (1971).

Jones, S. F., and *S. Kwanbunbumpen*: The effect of nerve stimulation and hemicholinium on synaptic vesicles at the mammalian neuromuscular junction. J. Physiol. (London) *207*, 31—50 (1970).

Kása, P.: Acetylcholinesterase transport in the central and peripheral nervous tissue: the role of tubules in the enzyme transport. Nature (London) *218*, 1265—1267 (1968).

Kása, P., *S. P. Mann*, and *C. Hebb*: Localization of choline acetyltransferase. Nature (London) *226*, 812—816 (1970).

Knyihár, E., and *B. Csillik*: Localization of inhibitors of the acetylcholine- and GABA-synthesizing systems in the rat brain. Exp. Brain Res. *11*, 1—16 (1970).

Koelle, G. B.: A new general concept of the neurohumoral functions of acetylcholine and acetylcholinesterase. J. Pharm. (London) *14*, 65—90 (1962).

Koelle, G. B., and *J. S. Friedenwald*: A histochemical method for localizing cholinesterase activity. Proc. Soc. exp. Biol. N.Y. *70*, 617—622 (1949).

Korneliussen, H.: Ultrastructure of normal and stimulated motor endplates. Z. Zellforschung *130*, 28—57 (1972).

Koshtoyantz, H. S.: Proteins, metabolism and nervous regulation, pp. 61—62. Moscow. 1951. (In Russian.)

Langley, J. N.: On the reaction of cells and of nerve endings to certain poisons, chiefly as regards the reaction of striated muscle to nicotine and curare. J. Physiol. (London) *33*, 374 (1905).

Malmfors, T.: Studies on adrenergic nerves. The use of rat and mouse iris for direct observations on their physiology and pharmacology at cellular and subcellular levels. Acta physiol. Scand. *64*, Suppl. 248, 1—93 (1965).

Marks, N., and *A. Lajtha*: Protein breakdown in the brain. Biochem. J. *89*, 438—447 (1963).

Matthies, H.: The intracellular regulation of the interneuronal connectivity — the molecular foundation of learning processes. Thesis, Magdeburg (1971).

Nachmansohn, D.: Chemical and molecular basis of nerve activity. New York: Academic Press. 1959.

Poberai, M., *G. Sávay*, and *B. Csillik*: Function-dependent proteinase activity in the neuromuscular synapse. Neurobiology 2, 1—7 (1972).

Purpura, D. P., *M. Girado*, *T. G. Smith*, *D. A. Callan*, and *H. Grundfest*: Structure-activity determinants of pharmacological effects of amino acids and related compounds on central synapses. J. Neurochem. *3*, 238—268 (1959).

Reik, L., *G. L. Higgins*, *G. L. Petzold*, *P. Greengard*, and *R. J. Barrnett*: Hormone-sensitive adenyl cyclase: cytochemical localization in rat liver. Science *168*, 382—384 (1970).

Renshaw, B.: Influence of discharge of motoneurons upon excitation of neighbouring motoneurons. J. Neurophysiol. *4*, 167—183 (1941).

Roberts, E.: Some biochemical-physiological correlations in studies of γ-aminobutyric acid. In: Structure and function of inhibitory neuronal mechanisms (*Euler, Skoglund, Söderberg*, eds.), pp. 401—418. Oxford: Pergamon Press. 1968.

Sávay, G., and *B. Csillik*: Acetylcholine-induced calcium release in the postjunctional sarcoplasm. Symp. Biol. Hung. *5*, 149—157 (1965).

Scheibel, M. E., and *A. B. Scheibel*: Inhibition and the Renshaw cell. A structural critique. Brain Behav. Evol. *4*, 53—93 (1971).

Sherrington, C. S.: The central nervous system. A text-book of physiology. Cit.: In *Sherrington, C. S.*, 1947: The integrative action of the nervous system. Cambridge: University Press. 1897.

Siggins, G. R., A. P. Oliver, B. J. Hoffer, and *F. E. Bloom*: Cyclic adenosine monophosphate and norepinephrine: effects on transmembrane properties of cerebellar Purkinje cells. Science *171*, 192 (1971).

Stöhr, P.: Mikroskopische Anatomie des vegetativen Nervensystems. In: Handbuch der mikroskopischen Anatomie des Menschen (*Möllendorff*), Vol. 4, Part 5. Berlin-Göttingen-Heidelberg: Springer. 1957.

Sutherland, E. W., and *G. A. Robison*: The role of cyclic 3', 5'-AMP in responses to catecholamines and other hormones. Pharmacol. Rev. *18*, 145—161 (1966).

Sze, P. Y., and *R. A. Lovell*: A re-examination of the effect of thiosemicarbazide on brain GABA and GAD in vivo. Life Sci. *9*, 889—899 (1970).

Thomas, R. C., and *V. J. Wilson*: Precise localization of Renshaw cells with a new marking technique. Nature (London) *206*, 211—213 (1965).

Waser, P. G., and *I. Hadorn*: Relations of cholinergic receptors to acetylcholinesterase of endplates in denervated muscle. Bibl. Anat. *2*, 155—160 (1961).

Willis, W. D., and *J. C. Willis*: Location of Renshaw cells. Nature (London) *204*, 1214—1215 (1964).

Zacks, S. I.: The motor end-plate. Philadelphia: Saunders. 1964.

Author's address: Prof. Dr. *B. Csillik*, Department of Anatomy, University Medical School, Kossuth Lajos sug.-út 40, Szeged, Hungary.

Discussion

Sakharov: I want to point out that the experimental material proves that uptake mechanisms for amino acids are on a low level of specificity which does not necessarily mean that amino acid transmission mechanisms are less specific than cholinergic or adrenergic ones. On the other hand, nicotinic and muscarinic cholinergic receptors, being features of the postsynaptic cell, do not take part in the chemical specificity of the presynaptic cell. As a matter of fact, in *Aplysia* different branches of one axon are known to make contacts with postsynaptic cells of different cholinergic characters.

Wolleman: Probably, the specificity of the hemicholium experiments can be enhanced by using inhibitors of choline acetylase and of phospholipid synthesis, respectively. In my opinion, different binding and inhibition of pyridoxalphosphate to neuronal and glial glutamic acid decarboxylases might account for the affinity of thiosemicarbazide to glial cells. With regard to excitation-performance coupling, I should like to mention the inhibition by acetylcholine of adenyl cyclase in cardiac and smooth muscles, and the simultaneous increase of ATP-synthesis from cyclic AMP, which is enhanced by Ca^{++}.

Ungvári: I should like to ask whether in noradrenaline uptake experiments by denervated myoepithelium, the catecholamines were bound to degenerated nerve fibers or to postsynaptic structures (α or β receptors).

Csillik: In striking contrast to tissues with a normal innervation pattern, denervated myoepithelium binds extrinsic noradrenaline very strongly, which is possibly caused by the spreading of adrenergic receptors due to denervation. Thus, catecholamines are bound to postsynaptic structures.

Storm-Mathisen: The autoradiographic experiments represent, indeed, a very interesting approach since they allow the visualization of biochemical phenomena in the tissue. I want to stress the importance of specificity of the test (inhibitor) substances. Even though it is possible to find a hemicholinium derivative that blocks choline uptake without binding to the receptor, it cannot be assumed that it will bind specifically to cholinergic neurones, since all cells need choline uptake for their metabolism. In accordance with the preliminary results by *Kuhar, Roth* and *Aghajanian* (Fed. Proc. *31*, 516 Abs. 1972), *Fonnum* and I (unpublished) found a substantial fall in choline uptake in hippocampal synaptosomes after destruction of the supposedly cholinergic septo-hippocampal projection. This loss, however, does not appear to be total as it is nearly total for choline acetylase.

Csillik: I agree that the utmost caution is necessary in the interpretation of autoradiographic results — which was also stressed in my paper. The tacitly accepted relative specificity plagues also chromogenic reactions.

Storm-Mathisen: I want to mention the discrepancy between my results regarding the bimodal distribution of GAD in the hippocampal formation, *i.e.*, the presence of two peaks of activity in the molecular and in the pyramidal granular cell layers, and the concentration of thiosemicarbazide in the vicinity of the pyramidal and granular cell layers only. Thiosemicarbazide is bound to pyridoxal phosphate and not to the apoenzyme itself. Thus, the relative susceptibility of GAD to the action of thiosemicarbazide is probably due to a relatively loose binding of the coenzyme. In my opinion, GAD is a neuronal enzyme and the fact that thiosemicarbazide is bound to glial cells suggests that the distribution of this drug does not display GAD activity.

Csillik: The studies of *Albers* (1960) point to the extreme susceptibility of GAD and GABA-T for thiosemicarbazide inhibition which, in fact, might be due to a special binding of pyridoxal phosphate both to the apoenzyme and to the test drug. Numerous studies from Roberts' laboratory prove the fact that, *in vivo*, the thiosemicarbazide inhibits GABA formation by interfering with GAD activity. Aspecific binding of thiosemicarbazide appears to be ruled out by the marked regional differences in thiosemicarbazide localization in the limbic system as contrasted to brain stem nuclei.

Journal of Neural Transmission, Suppl. XI, 43—59 (1974)

Evolutionary Aspects of Transmitter Heterogeneity*

D. A. Sakharov

Institute of Developmental Biology, U.S.S.R. Academy of Sciences, Moscow, U.S.S.R

With 1 Figure

Summary

Different mechanisms of chemical transmission coexist in advanced nervous systems since, in the distant past, nerve cells originated many times and neurones of a particular ancestry retained a specific type of chemistry from their appearance on. Diversification of transmission mechanism within the inherited type would be an additional source of transmitter heterogeneity while a kind of selection could reduce the diversity of transmitters during evolution.

This hypothesis is explored in the article as well as the alternative view suggesting that neurones are of common phylohistogenetic origin and that they differentiated into various chemical types in the course of evolution due to functional specialization. The latter explanation is regarded as unsatisfactory.

Introduction

For a long time, chemical transmission mechanisms were believed to fall into two categories, cholinergic and adrenergic. Nowadays, it is widely recognized that the chemical transmission is a much more complex event. The plurality of transmission mechanisms is now taken into consideration as one of the fundamental features of the mammalian nervous system.

There is no need to review the literature on substances proved or supposed to be neural transmitters. This has recently been well summarized by *McLennan* in his monograph "Synaptic Transmission" (1970). The present article represents an attempt to answer the

* Abbreviations used in this paper: ACh (acetylcholine), CA (catecholamine), DA (dopamine), GABA (gamma-aminobutyric acid), Gl (glutamate), 5-HT (5-hydroxytryptamine), NA (noradrenaline).

question: Why do nerve cells work with the help of different transmitter substances?

In any case, it is quite certain that the chemical mosaic of the mammalian nervous system is one of the results of neural evolution. However, the way of achieving this heterogeneity is not self-evidently clear and different models of the evolutionary pattern can be proposed.

It is widely accepted that, during evolution, neurones differentiated into various chemical types due to functional specialization. In this article, the problem will be considered also from another point of view based mainly on the assumption that transmission mechanisms are not uniform due to multiple origins of the neurones (*Sakharov*, 1970 a).

Similarities

Heterogeneity is but one side of the problem, another being similarity. A striking fact is that *similar* transmitters (*e.g.* ACh, DA, 5-HT, GABA) operate in nervous systems evolved within divergent metazoan lineages.

In view of much accumulated evidence, it now seems certain that the plurality of transmission mechanisms is not a distinctive feature of higher vertebrates. In different invertebrate animals, the detailed structure of the nervous system has been shown to be precisely determined from individual to individual within a species, both transmission mechanisms and connections of cells being characteristically specified. Neurones of the same chemical types as the ones that occur in the mammalian brain are present in a number of unrelated invertebrate neural organs, such as nervous strands of flatworms, chained ganglia of annelids, and the circumoesophageal collar of gastropods (see review articles of *Florey*, 1967; *Cottrell* and *Laverack*, 1968; *Sakharov*, 1970 b; *Pitman*, 1971).

In our opinion, a theory that explains heterogeneity cannot be recognized as satisfactory until and unless it can explain both, differences and similarities. Two questions are to be answered at once, first, why different transmitter substances exist within one nervous system, and second, why similar transmitters are present in different, unrelated nervous systems.

Multiple Origins of Nerve Cells

Our idea is that cells that constitute nervous systems of contemporary animals are of different types since they are the descendants of primitive neurones that, in the distant past, originated separately and, therefore, were of distinct types from their first appearance on.

This explanation is based on two assumptions: (1) neurones arose not once, but many times; (2) the type of chemical specificity is a conservative character. Once a transmission mechanism became established, it was retained during further evolution.

The inherited chemical mechanism would clearly be modified in different directions in the course of evolution by means of reduction of some biochemical steps or addition of new ones, or otherwise. In terms of the proposed hypothesis, it seems useful to introduce the term "family" to refer to a group of genealogically related types of neurones. For instance, three types of CA-ergic neurones of vertebrates (adrenergic, NA-ergic and DA-ergic) are generally believed to be modifications of a single basic type. If actually so, they may be considered as a family. Peptidergic cells that produce oxytocin, vasopressin and related substances in lower vertebrates appear to be another example of a neurone family. They have a set of common characteristics, both chemical and ultrastructural, that indicate that the similarity is not superficial.

The system of classification of nerve cells based on genealogical grounds would be essentially natural while existing systems are invented and arbitrary.

Thus, the presence of several transmitter substances in the mammalian nervous system is, in our opinion, a relic of the primitive neural organization. An additional source of the heterogeneity would be the divergent evolution within families of nerve cells. Diversity of transmission mechanisms could have been established at the very appearance of diffuse nerve nets when no neural organs occurred. The nets are generally considered to be the initial point for processes of ganglionization which are not uniform in different lineages of metazoans. This is why a close relationship could occur between apparently unrelated nerve cells situated in unrelated neural organs. This implies that those neurones of contemporary animals which share a common and distinct set of specific chemical characteristics are of common ancestry, irrespective of relations of the animals concerned. In other words, similarities between neurones with respect to their transmission chemistry are, in our view, an indication that the neurones are *homologues*.

Alternative

Theoretically, the interpretation given above is not the only possible one. An alternative would be that the plurality of transmission mechanisms resulted from evolutionary diversification.

The idea that some ancestral cell type is a stemming point of all the types of neurones is not a new one. It is noteworthy that this view was shared by a prominent histologist who made a most interesting attempt to elaborate a natural, genealogical system of animal tissues; it was the aim of *Chlopin's* "Foundations of Histology" (1946) to call attention to multiple origins of specialized cell types (muscles, glands, etc.). Nevertheless, even in this book, the neurones have been regarded as cells of a common phylohistogenetic ancestry. This conclusion was based on the evidence that morphological and physiological properties of different neurones are remarkably similar.

The old idea provides a convenient answer to the question discussed. The following quatation is from *Lentz*'s book "Primitive Nervous Systems" (1968):

"I consider the nervous system of the primitive ancestral organism to have been composed of a single basic nerve cell type . . . The primitive nerve cell may have been capable of producing neurohumors and neurosecretory substance and thus could have given rise to both ordinary neurons and neurosecretory cells." On ordinary neurons, internuncial and motor, Lentz comments: "Their transmitter agents were restricted to a small number of neurohumors." As to neurosecretory cells, they have arisen "by becoming specialized primarily for the secretion of polypeptide hormonal substances" (pp. 117 and 118).

In accordance with this concept, it is claimed that "secretory products of neurones can exist in various combinations in individual cells . . . Combinations of humoral and hormonal substances may be of common occurrence and could represent the primitive situation" (p. 115).

Thus, chemical specification is supposed to follow the pattern of functional specialization. It remains to bring this view in logical correspondence with the fact that similar transmitters exist in unrelated animals. The only plausible explanation is to postulate that the source of resemblance is similarity in function. In other words, neurones that exhibit similar transmission chemistry should be designated as *analogous*.

Actually, many investigators believe that independent events may have been involved in the origin of a particular transmission mechanism. For instance, *Michelson* and *Zeimal* (1970), in their most useful monograph on ACh, suppose that cholinergic transmission has evolved independently several times in different phylogenetic lines since ACh is "a very convenient candidate for the role of transmitter". *Voskresenskaya* (1968) expected the sympathetic nervous system of insects to be adrenergic since it is an analogue of the sympathetic nervous system of vertebrates.

The problem thus is put into the usual evolutionary dilemma: Homology versus analogy.

By definition, structures are homologous when they have a common ancestral precursor irrespective of similarity; the latter is, nevertheless, an important basis for the recognition of homologous structures. Analogy (and allied phenomona such as parallelism and convergence) signifies the appearance of similar characters in two or more lineages due to similarity in function; resemblance of this kind is inevitably superficial. Detailed consideration of these phylogenetic terms as applied to problems of neurology is given by *Campbell* and *Hodos* (1970).

Differences between two explanations of transmitter heterogeneity are summarized in Table 1.

Table 1. *Working Hypotheses to Account for the Occurrence of Different Transmitters in a Single Nervous System and that of Similar Transmitters in Unrelated Nervous Systems*

	Proposed hypothesis	Alternative hypothesis
Origin of neurones	Multiple	Common
Transmission mechanism in primitive neurones	Single (in cells of a particular ancestry)	Plural (in any cell)
Trend of evolution of transmission mechanism	Differentiation within inherited basic type	Adaptive restriction
Source of similarities in transmission mechanisms	Inheritance from a common ancestry	Parallel evolution and convergence due to common function

Formally, both genealogical and functional hypotheses seem to be devoid of intrinsic controversies. It becomes clear from Table 1 what kind of evidence is needed to test the hypotheses. Firstly, it would be useful to know whether a correspondence exists between a given transmitter substance and certain physiological properties of junctions. Secondly, nervous systems of simple metazoans are to be considered to see what, in fact, is manifestation of neuronal primitivity. Finally, recent data on combinations of transmitter substances in individual nerve cells should be discussed.

Nature of Transmitter and Properties of Transmission

First, the direction of synaptic action of a transmitter substance, and underlying ionic mechanisms will be considered.

As far as ACh is concerned, at least five types of postsynaptic effects based on selective change of ionic permeability have been discovered: (1) depolarization by an increase of membrane permeability to Na^+; (2) depolarization by a decrease of permeability to K^+, a previously unknown mechanism recently found by *Krnjevič et al.* (1971) in cortical neurones; (3) hyperpolarization by an increase of permeability to K^+; (4) hyperpolarization by an increase of permeability to Cl^- (low internal chloride concentration); (5) depolarization by an increase of permeability to Cl^- (high internal chloride concentration).

Various types of ACh action may occur within a single nervous system. Gastropod molluscs such as *Aplysia* and *Helix* provide an excellent example. Four of the five mechanisms listed above are known to coexist in a gastropod ganglion (*Kerkut et al.*, 1970).

The nervous system of gastropods provides also a favourable type of preparation for studying varieties of membrane effects of transmitters other than ACh. 5-HT appears to be involved in transmission of both excitation and inhibition in this preparation; so far, three types of serotonin receptors and corresponding ionic mechanisms have been found in snail neurones (*Gerschenfeld*, 1971). DA excites some neurones and inhibits others (*Glaizner*, 1968). To summarize, the results of current electrophysiological studies on gastropod neurones suggest that the direction of the synaptic effect and underlying ionic mechanism are not defined by the nature of the transmitter substance.

One may suggest that this is not a general rule. Effects of GABA and glycine are believed to be invariably inhibitory while that of Gl exclusively excitatory (*McLennan*, 1970). In fact, however, the situation is not so simple. Comparative data on effects of glycine are lacking. Dual (hyperpolarizing and depolarizing) effects of both GABA and Gl have been recorded in snail neurones (*Salánki*, 1968; *Gerasimov* and *Kholodova*, 1971).

Let us consider next the efficacy of synaptic action. It is sufficiently well known that variable physiological characteristics of synapses (*e.g.* size, form and duration of postsynaptic potentials, facilitation, desensitization, plasticity, etc.) are controlled by a number of factors that may act pre-, post-, and extrasynaptically. The chemical nature of the transmitter substance is not among these factors. Junctions that share a common presynaptic chemistry and have the same type of postsynaptic ionic mechanism display widely varying physiological properties. The literature in this field is enormous and need not concern us here.

Thus, both ionic mechanism and efficacy of synaptic action may change widely, the transmitter substance being one and the same.

It must be noted, at last, that neurones of a given chemical type can held all possible posts in the nervous system. Cholinergic cells work as sensory neurones (in crabs — see *Florey*, 1967), interneurones (cell L10 in the abdominal ganglion of *Aplysia*), and motor neurones in both somatic and visceral muscles (vertebrates). Sensory neurones that contain a primary CA occur in a number of invertebrate phyla; interneurones of this chemical type are situated in insect and vertebrate brains; finally, such cells can function as motor neurones and control visceral and somatic muscles (visceral organs of vertebrates, pedal muscle of molluscs).

In the face of all these facts, it seems difficult to argue for the view that a specific chemical type of neuron could have been established in the course of neural evolution under the pressure of some functional need.

Primitive Nervous Systems

Among contemporary animals that possess a nervous system, the coelenterates are the most primitive. They consist of ectoderm and entoderm separated by largely noncellular mesogloeal material. The nerve net of coelenterates is generally considered to be the first nervous system evolved by multicellular organisms. Until recently, there was still uncertainty concerning the principles of organization of this simple nervous system. Nowadays the idea that it is syncytial is out of date. The presence of true chemical synapses, both polarized and symmetrical, has been demonstrated ultrastructurally even in the nerve nets of some of the simplest coelenterates. A striking fact is that the net is constituted of different neurones, each variety exhibiting a characteristic type of secretory vesicles (*Westfall*, 1970 a, b; *Westfall et al.*, 1971; and personal observations).

Histochemical data extend these findings. In ectoderm of sea anemones, the presence of a primary CA has been demonstrated in only one certain category of nerve cells (*Dahl et al.*, 1963). These authors regarded the cells as sensory neurones since their spindle-like bodies were shown to be situated in the external epithelium. In a sea anemone *Bunodactis stella*, axons of the epithelial CA-ergic neurones make contacts with muscle fibres (Fig. 1). The cells seem to be capable of both sensory and motor function. This is a primitive feature. Otherwise, however, CA-ergic neurones of sea anemones seem to be as specific as those of higher animals.

It is not out of place to quote here from *Burnstock*'s excellent review on the evolution of autonomic innervation in vertebrates (1969): "The structure of sympathetic ganglion cells throughout the vertebrates shows considerable uniformity ... Fluorescent histochemical and electronmicroscopic studies of the innervation of various tissues in lower vertebrates have shown that the complex nature of neuroeffector relationships is established early in vertebrate evolution, *i.e.*, the peripheral extensions of sympathetic nerves are varicose with high levels of catecholamines located within the varicosities, suggesting 'en passage' release of transmitter during transmission, as in mammals" (p. 304). The only correction would be that these specific features of CA-ergic neurons were established as early as at the level of the nerve net of coelenterates.

At the same time, coelenterate nerve cells do not show signs of predicted primitivity in the sense of transmitter composition. There

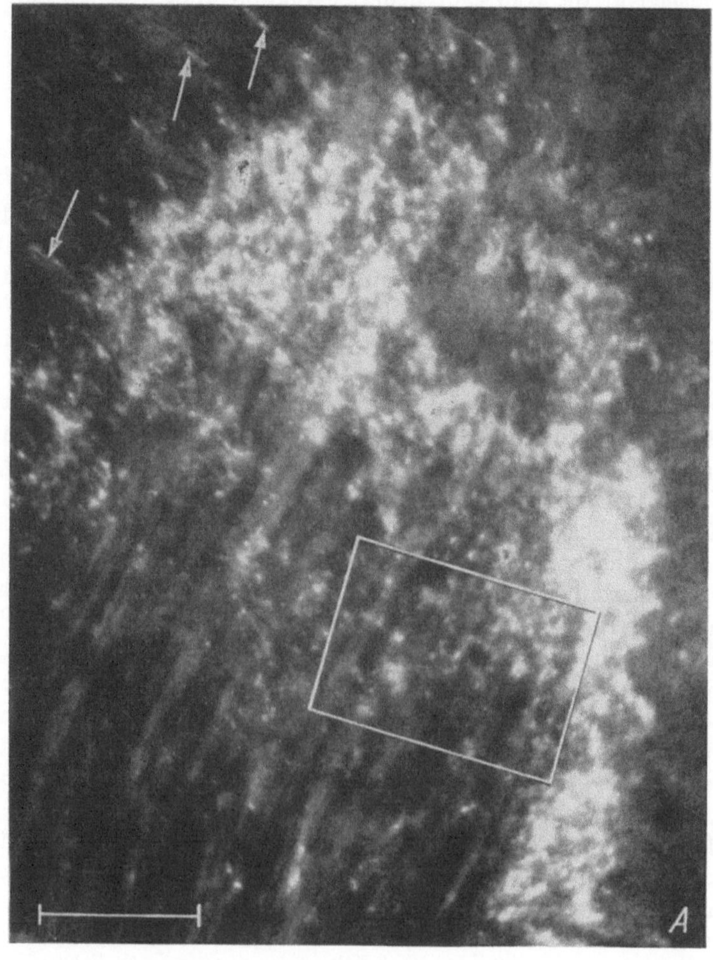

seems to be little doubt that neurones constituting the net are of different chemical types. Developmental data confirm that the ectodermal net is a compound formation. In ectodermal layer of *Hydra*, at least three varieties of nerve cells are reported to be present; two of them have been shown to differentiate independently of each other from cambial (interstitial) cells; it has been assumed that the third type of nerve cells follows a similar pattern (*Davis*, 1971).

The general impression gained from all these findings is that of a remarkable heterogeneity of the ectodermal nerve net. Even more suggestive evidence comes from the comparison between ectodermal nerve cells and those of extraectodermal origins.

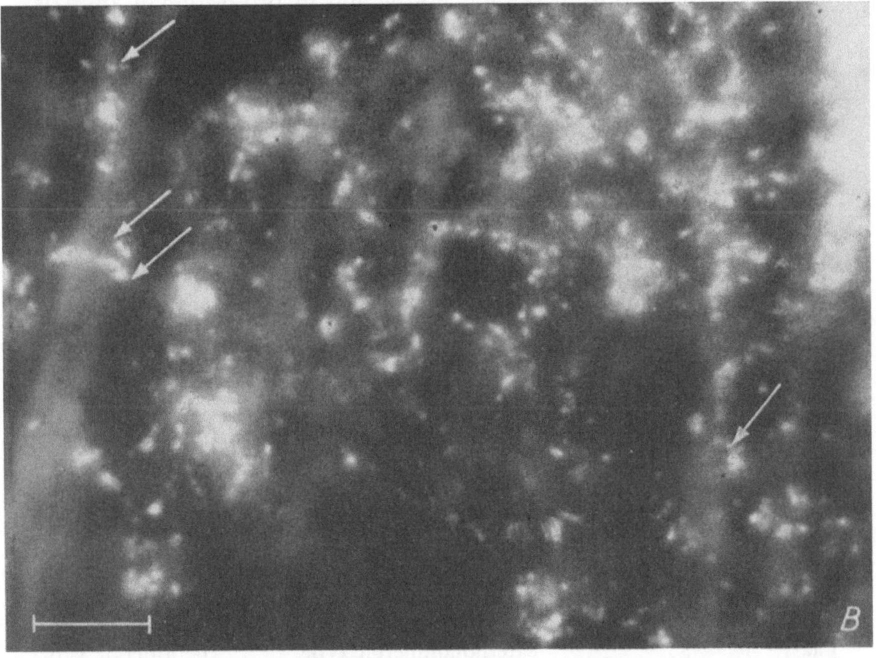

Fig. 1. Tentacle of a sea anemone, ***Bunodactis stella.*** Formaldehyde-induced fluorescence of a primary catecholamine, aqueous method (*Sakharova* and *Sakharov*, 1971).

A: In oblique section, the subepithelial plexus of varicose CA-ergic fibres is seen. Note elongated perikarya in the epithelium (arrows).

B: Part of micrograph A at higher magnification. Note contacts between varicose axons and muscle cells (arrows). Scale bar: 50 μ (A) and 10 μ (B)

By the way, the wording that "the neurones are of ectodermal origin", though transferred from one textbook to another, is far from being exact. It is true that, in vertebrates, embryonic ectoderm gives rise to supposedly all the types of nerve cells, but their evolutionary history remains obscure. The phenomenon of methorisis (*i.e.*, shift and unification of embryonic primordia of an organ that has a compound origin — *Schimkewitsch*, 1908) may interfere in neurogenesis in chordates. As a matter of fact, sources of neurones other than ectoderm exist in different invertebrate phyla. Independent systems of neurones originated in ectoderm, entoderm and, in some instances, coelome, are known to be present in Cnidaria, Ctenophora, different groups of lower Deuterostomia; primary sensory neurones have been found within the gut epithelium in protostomian animals (*Beklemischev*, 1964).

The occurrence of such a widespread, independent development points to the ease with which neurones could arise. On the other hand, this primitive neural organization is of special advantage since the neurones that constitute the ectodermal nervous system (sensory, internuncial and motor cells) have their corresponding analogues (functional counterparts) among extraectodermal nerve cells. This provides an opportunity for testing the functional hypothesis directly.

So far, no evidence is available to support the view that similar transmitters could have evolved in analogous neurones of different origins. On the contrary, there is evidence, though still fragmentary, in favour of the genealogical hypothesis. In animals that possess nerve cells of both ectodermal and extraectodermal origins, CA-ergic neurones have been found exclusively among the neurones of ectodermal origin (see *Sakharov*, 1970 b).

Table 2. *Histogenetic Sources of Two Types of Neurones in Different Classes of Echinodermata**

Class	Neurones of a catecholaminergic type	Neurones of a peptidergic type
Crinoidea	No data	Coelom
Holothurioidea	Ectoderm	No data
Asteroidea	Ectoderm	Coelom
Ophiuroidea	Ectoderm	Coelom
Echinoidea	Ectoderm	No data

* Compiled from *Cobb*, 1969; *Cottrell* and *Pentreath*, 1970; *von Hehn*, 1970; *Holland*, 1970; *Leclerce* and *Delavault*, 1971; *Pentreath* and *Cottrell*, 1971; *Sakharova* and *Sakharov*, 1971.

Echinoderms are of particular interest in this connection. Besides neurones of ectodermal origin ("ectoneural" system) and a nerve net originated from the intestinal epithelium, these organisms have so-called hyponeural and apical nervous systems that are derivatives of the coelom. The systems in question reach a quite different relative development in different classes of echinoderms. Nevertheless, the available data demonstrate remarkably similar patterns of composition of nerve cell population (Table 2).

Do Combinations of Transmitter Substances Exist?

This is an aspect of the problem about which considerable controversity still exists.

The fact that ACh was claimed to be a member of different transmitter combinations should not be surprising. ACh was the first substance identified as a neural transmitter. Due to this merit, ACh is generally held in high respect as if it were the "chief" transmitter. This peculiar attitude was quite manifest in repeated attemps to generalize cholinergic mechanisms and discover some role of ACh in non-cholinergic synapses.

Examples are widely known. For a long time, one of the main criteria for identification of chemical transmitters was "the criterion of the inactivation enzyme", that is to say, of an analogue of acetylcholinesterase. Furthermore, it is customary to interpret nerve endings containing electron lucent vesicles as cholinergic though other transmitter substances, say Gl, may have the same type of packing. Even the presence of empty vesicles in adrenergic and peptidergic endings was regarded as an indication to the presence of ACh. *Burn*'s and *Rand*'s concepts of a "cholinergic link" in non-cholinergic transmission represented an extreme degree of such an attitude.

Step by step, however, generalizations of this kind were abandoned. The presence of an inactivating enzyme turned out to be a peculiar feature of certain cholinergic junctions. The nature of agranular vesicles in adrenergic and peptidergic secretory endings has been clarified. At least, it has been demonstrated that, contrary to hypotheses of a "cholinergic link", no measurable ACh is present in adrenergic fibres (*Ehinger et al.*, 1970; *Consolo et al.*, 1972).

We referred to the history of cholinergic generalizations since these have been the main basis for the assumption that combinations of different transmitters within individual neurones could exist. These assumed combinations, in their turn, were interpreted as a primitive feature (*Lentz*, 1968; *Michelson* and *Zeimal*, 1971, and others).

Some combinations, however, do really exist. First, a substance known as a transmitter may be present in different nerve cells due to its non-synaptic functions. For instance, ATP or a related nucleotide appears to serve as synaptic transmitter in some vertebrate autonomic neurones (*Burnstock et al.*, 1970). Certainly, the presence of ATP in other nerve cells should not be surprising. Combinations of this kind are not, in fact, combinations of transmitters. Neither are compositions of another type when a substance that works as transmitter in one group of neurones serves as transmitter precursor in another. The presence of DA in NA-ergic cells is a well-known example. In terms of our hypothesis, both groups of cells are members of one family.

Evolutionary relations of neurones transmitting with the help of Gl are of particular interest in this connection. Gl is a metabolic precursor of GABA. It has been shown that, in the lobster, Gl is about equally concentrated in perikarya of both excitatory (Gl-ergic) and inhibitory (GABA-ergic) motor neurones (*Otsuka et al.*, 1967). The possibility should be borne in mind that these cells are modifications of one basic type. Studies on specific proteins may help to elucidate the problem and probably establish a homology of the neurones in question. Another possibility that should be explored is that cholinergic neurones may be members of this family too. Gl is closely related to ACh synthesis. It is intriguing that, in arthropods, muscles of the somatic type receive a Gl-ergic supply while their supposed homologue, the somatic muscle of annelids, receives a cholinergic motor innervation. At last, ultrastructural characteristics of synaptic vesicles are very similar in endings of cholinergic and Gl-ergic types.

A peculiar transmitter combination of a so far unexplained nature appears to exist in gastropod molluscs. In the visceral loop of some pulmonates, a cluster of nerve cells is situated reportedly containing in their cytoplasm both DA and 5-HT (*Kerkut et al.*, 1967). Nothing is known of the physiology of these cells which may be responsible for biphasic postsynaptic potentials recorded in the area. Though snails are regarded as "lower" animals, they are known to possess separate DA-ergic and serotonin-ergic neurones. Even in flatworms that are much more primitive, DA and 5-HT are contained in different neurones (*Welsh* and *Williams*, 1970). Thus it seems improbable that the above group of visceral neurones of snails displays a primitive level of chemical organization. Maybe this combination is of a secondary character like the situation in the mammalian pineal gland where sympathetic nerve fibres of primarily CA-ergic type reportedly contain additional 5-HT.

Pattern of Evolution of Heterogeneous Neuronal Populations

A variety of evidence considered above provides strong support for the proposed hypothesis, while the alternative explanation evidently does not work in the face of the facts.

The present hypothesis provides a basis for speculation on the pattern of evolution of chemical transmission. The conclusion that nerve cells had different transmitters from their first appearance on does not imply that transmission mechanisms did not evolve. Diversification of basic types and appearance of modified transmitter substances is but one possible aspect of this evolution. Evolutionary sophistication of biochemical mechanisms related to transmission might be another (comparative data on cholinergic transmission have been earlier analysed from this point of view, see *Sakharov* and *Turpaev*, 1968). Furtheron, a kind of selection could take place in the course of neural evolution.

Certainly, if a nerve cell population is heterogeneous, the probability of success of some types will be higher than that of others. As progressive evolution of metazoans occurred, rapidly metabolizing nerve cells were faced with a number of problems. Selection might act to eliminate certain types of neurones that were imperfectly equipped with mechanisms for manufacturing transmitter and allied substances, for transporting metabolites over long distances, etc. The diversity of transmission mechanisms could thus be reduced.

In fact, one cannot conclude from comparative data that well-developed and voluminous nervous systems are composed of nerve cell populations that are more heterogeneous than those in simple animals. Just in contrast, the author's personal impression is that, in snails, ultrastructurally distinct types of motor nerve terminals are more numerous than in mammals. Among these, junctions of peptidergic types exist (for instance, in the muscles of the perineurium). In general, a great variety of peptidergic cell types can be found in the snail nervous system. Such types are mainly eliminated from nerve cell populations of higher animals. In our opinion, this is due to the fact that high molecular transmitter substances cannot provide a progressive nervous system with prompt and sustained transmission. In higher vertebrates, a small and heterogeneous population of peptidergic neurones could survive as neurosecretory cells that have lost their synaptic functions.

The progressive types of neurones are those whose populations are expanding and whose dominance within the nervous system is increased in comparison with other types. Cholinergic and CA-ergic cells seem to be of this kind. In this connection, evolutionary trends in the vertebrate autonomic nervous system are of great interest (expansion of cholinergic and CA-ergic terminal networks; independent, in different zoological lines, displacement of CA-ergic cells into visceral organs; etc., see *Burnstock*, 1969). Cholinergic neurones of the molluscan central nervous system provide another striking example. In gastropod ganglia, but single cells of this type occur (*McCaman* and *Dewhurst*, 1970; *Giller* and *Schwartz*, 1971) while, in the advanced brain of cephalopods, cholinergic mechanisms are known to be abudantly presented. There is little doubt that, in relatively new formations of the vertebrate brain, those types of neurones flourish which are provided with the most progressive chemical mechanisms for synaptic transmission.

In spite of the continuous accumulation of data on chemical synapses, there seems to be little progress towards the conceptual understanding of transmitter diversity. This paper may help to clarify the problem on the basis of the idea that the diversity in question resulted from polygenesis (multiple origins) of neurones and, additionally, from evolutionary diversification within different nerve cell lineages. This view is in good accordance with the comparative material at our disposal. While fully realizing how far are sea anemones and land snails from neurovegetative mechanisms of the man, the author has consoled himself with the thought that any serious consideration of neurones that constitute the human nervous system inevitably leads to speculation regarding their natural history.

References

Beklemischev, W. N.: Foundations of comparative anatomy of invertebrates, Vol. 2. Moscow: Nauka. 1964. (In Russian.)

Burnstock, G.: Evolution of the autonomic innervation of visceral and cardiovascular systems in vertebrates. Pharmacol. Revs. *21*, 247—324 (1969).

Burnstock, G., G. Campbell, D. Satchell, and *A. Smythe*: Evidence that adenosine triphosphate or a related nucleotide is the transmitter substance released by non-adrenergic inhibitory nerves in the gut. Brit. J. Pharmacol. *40*, 668—688 (1970).

Campbell, C. B. G., and *W. Hodos*: The concept of homology and the evolution of the nervous system. Brain Behav. Evol. *3*, 353—367 (1970).

Chlopin, N. G.: Foundations of histology. Leningrad: Publ. House of U.S.S.R. Acad. Sci. 1946. (In Russian.)

Cobb, J. L. S.: The distribution of mono-amines in the nervous system of echinoderms. Comp. Biochem. Physiol. *28*, 967—971 (1969).

Consolo, S., S. Garattini, H. Ladinski, and *H. Theonen*: Effect of chemical sympathectomy on the content of acetylcholine, choline and choline acetyltransferase activity in the cat spleen and iris. J. Physiol. *220*, 639—646 (1972).

Cottrell, G. A., and *M. S. Laverack*: Invertebrate pharmacology. Ann. Rev. Pharmacol. *8*, 273—298 (1968).

Cottrell, G. A., and *V. W. Pentreath*: Localization of catecholamines in the nervous system of a starfish, *Asterias rubens*, and of a brittlestar, *Ophiothrix fragilis*. Comp. Gen. Pharmacol. *1*, 73—81 (1970).

Dahl, E., B. Falck, C. von Mecklenburg, and *H. Myhrberg*: An adrenergic nervous system in sea anemones. Quart. J. micr. Sci. *104*, 531—534 (1963).

Davis, L. E.: Differentiation of ganglionic cells in *Hydra*. J. exp. Zool. *176*, 107—128 (1971).

Ehinger, B., B. Falck, H. Persson, A.-M. Rosengren, and *B. Sporrong*: Acetylcholine in adrenergic terminals of the cat iris. J. Physiol. *209*, 557—565 (1970).

Florey, E.: Neurotransmitters and modulators in the animal kingdom. Fed. Proc. *26*, 1164—1178 (1967).

Gerasimov, V. D., and *Yu. D. Kholodova*: The effect of amino acids on the electrical activity in the neurons of the snail *Helix pomatia*. Zh. evol. Biokhim. Fiziol. *7*, 156—161 (1971).

Gerschenfeld, H. M.: Serotonin: Two different inhibitory actions on snail neurons. Science *171*, 1252—1254 (1971).

Giller, E., Jr., and *J. H. Schwartz*: Choline acetyltransferase in identified neurons of abdominal ganglion of *Aplysia californica*. J. Neurophysiol. *34*, 93—115 (1971).

Glaizner, B.: Pharmacological mapping of cells in the subocsophageal ganglia of *Helix aspersa*. In: Neurobiology of Invertebrates (*Salánki, J.*, ed.), pp. 267—283. Budapest: Akadémiai Kiadó. 1968.

Hehn, G. von: Über den Feinbau des hyponeuronalen Nervensystems des Seesternes (*Asterias rubens* L.). Z. Zellforsch. *105*, 137—154 (1970).

Holland, N. D.: The fine structure of the axial organ of the feather star *Nemaster rubiginosa* (Echinodermata: Crinoidea). Tissue & Cell 2, 625—636 (1970).

Kerkut, G. A., K. Ralph, R. J. Walker, G. Woodruff, and *R. Woods*: Excitation in the molluscan central nervous system. In: Excitatory synaptic mechanisms, pp. 105—117. University of Oslo Press. 1970.

Kerkut, G. A., C. B. Sedden, and *R. J. Walker*: Uptake of DOPA and 5-hydroxytryptophan by monoamine-forming neurones in the brain of *Helix aspersa*. Comp. Biochem. Physiol. *23*, 159—162 (1967).

Krnjević, K., R. Pumain, and *L. Renaud*: The mechanism of excitation by acetylcholine in the cerebral cortex. J. Physiol. *215*, 247—268 (1971).

Leclerc, M., and *R. Delavault*: Présence de fibres nerveuses dans la paroi coelomique chez *Asterina gibbosa* Pennant (Echinoderme, Astéride). C. R. Acad. Sc. Paris *272*, 3311—3313 (1971).

Lentz, T. L.: Primitive nervous systems. New Haven: Yale University Press. 1968.

McCaman, R. E,. and *S. A. Dewhurst*: Choline acetyltransferase in individual neurons of *Aplysia californica.* J. Neurochem. *17*, 1421—1426 (1970).

McLennan, H.: Synaptic transmission. Philadelphia: Saunders. 1970.

Michelson, M. J., and *E. V. Zeimal*: Acetylcholine. Leningrad: Nauka. 1970. (In Russian.)

Otsuka, M., E. A. Kravitz, and *D. D. Potter*: Physiological and chemical architecture of a lobster ganglion with particular reference to gamma-aminobutyrate and glutamate. J. Neurophysiol. *30*, 725—752 (1967).

Pentreath, V. W., and *G. A. Cottrell*: 'Giant' neurons and neurosecretion in the hyponeural tissue of *Ophiothrix fragilis* Abildgaard. J. exp. mar. Biol. Ecol. *6*, 249—264 (1971).

Pitman, R. M.: Transmitter substances in insects: A review. Comp. gen. Pharmacol. *2*, 347—371 (1971).

Sakharov, D. A.: Principal approaches to the systematization of nerve cells. Zh. obshch. Biol. *31*, 449—457 (1970a).

Sakharov, D. A.: Cellular aspects of invertebrate neuropharmacology. Ann. Rev. Pharmacol. *10*, 335—352 (1970b).

Sakharova, A. V., and *D. A. Sakharov*: Visualization of intraneuronal monoamines by treatment with formalin solutions. In: Histochemistry of nervous transmission (*Eränkö*, O., ed.), pp. 11—25. Amsterdam: Elsevier. 1971.

Sakharov, D. A., and *T. M. Turpaev*: Evolution of cholinergic transmission. In: Neurobiology of invertebrates (*Salánki, J.*, ed.), pp. 305—314. Budapest: Akadémiai Kiadó. 1968.

Salánki, J.: Studies on the effect of iontophoretically applied L-glutamate on the giant nerve cells of Gastropoda (*Helix* and *Lymnaea*). Annal. Biol. Tihany *35*, 75—81 (1968).

Schimkewitsch, W. M.: Die Methorisis als embryologisches Prinzip. Zool. Anz. *33* (1908). Cit. Knorre, A. G.: Embryonic histogenesis. Leningrad: Medicina. 1971. (In Russian.)

Voskresenskaya, A. K.: The regulating function of the invertebrate nervous system. In: Neurobiology of invertebrates (*Salánki, J.*, ed.), pp. 367 to 380. Budapest: Akadémiai Kiadó. 1968.

Welsh, J. H., and *L. D. Williams*: Monoamine-containing neurons in planaria. J. Comp. Neurol. *138*, 103—116 (1970).

Westfall, J. A.: Ultrastructure of synapses in a primitive coelenterate. J. Ultrastruct. Res. *32*, 237—246 (1970a).

Westfall, J. A.: Synapses in a sea anemone, *Metridium* (Anthozoa). Septième Congress international de microscopie électronique, Grenoble, *3*, 717 (1970b).

Westfall, J. A., S. Yamataka, and *P. D. Enos*: Ultrastructural evidence of polarized synapses in the nerve net of *Hydra.* J. Cell Biol. *51*, 318—323 (1971).

Author's address: Dr. *D. A. Sakharov*, Institute of Developmental Biology, U.S.S.R. Academy of Sciences, Moscow, U.S.S.R.

Discussion

Szentágothai: I want to point out the importance of the phylogenetic approach in understanding basic phenomena of neural activity. I may remind of my early studies, based at that time mainly on structural criteria of specificity, especially branching patterns of dendrites and axons. It would be rewarding to extend these studies to other physiological properties like the capacity of mechano-electric transduction, a property characterizing primary sensory neurones but lacking in other nerve cells.

Sakharov: I agree that transmitter chemistry is not the only manifestation of neuron specificity; in my original paper (*Sakharov,* 1970a) also other possible criteria such as biophysical properties of the cell membrane, specific receptors, etc., were taken into consideration.

Journal of Neural Transmission, Suppl. XI, 61—78 (1974)

Interaction between Adrenergic and Cholinergic Systems: Presynaptic Inhibitory Effect of Noradrenaline on Acetylcholine Release

E. S. Vizi

Department of Pharmacology, Semmelweis University of Medicine, Budapest, Hungary

With 5 Figures

Summary

Noradrenaline and adrenaline inhibit acetylcholine release due to nerve activity, through α-adrenoceptors. The inhibition of acetylcholine release by the adrenergic transmitters is, in fact, a rather economical way for the adrenergic (orthosympathetic) nervous system to counteract the parasympathetic nervous system. It is shown that there is a permanent adrenergic control of acetylcholine output. The increase in acetylcholine output after catecholamine depletion or α-blockade of nervous tissue only represents partial removal of an adrenergic restraint. The modulatory role of catecholamines on acetylcholine release mechanism seems to be of physiological importance. Furthermore, it is also shown that acetylcholine may regulate its own release by a negative feedback mechanism. Although this mechanism operates in brain cortex, no such phenomenon, however, was found in the Auerbach plexus.

In this report a special form of functional interaction between the cholinergic and noradrenergic nervous system, *viz.*, the presynaptic inhibitory action of noradrenaline on acetylcholine release from cholinergic nerve terminals (*Vizi*, 1968; *Paton* and *Vizi*, 1969) will be discussed. This phenomenon is likely to be of physiological importance.

Gastrointestinal Tract

The two types of autonomic nerves are classically considered to be reciprocal in function: parasympathetic nerves are mainly excitatory and sympathetic nerves are inhibitory. It has been shown by *Paton* and *Vizi* (1969) and *Vizi* (1968) that noradrenaline and adrenaline (10^{-7}—10^{-6} g/ml) are capable of inhibiting the ACh output due to stimulation of the nerve terminals and that the inhibitory action is mediated via α-adrenoceptors on presynaptic nerve terminals (Fig. 1). When sustained stimulation was used, the inhibitory effect of NA varied inversely with the frequency. The fewer the shocks delivered and the lower the frequency applied, the higher was the volley output in the control period and the greater was the reduction by noradrenaline.

Furthermore, *Knoll* and *Vizi* (1970, 1971) have shown using high frequency stimulation that for brief trains of shocks (3—10) adrenaline and NA both reduce release as measured directly,

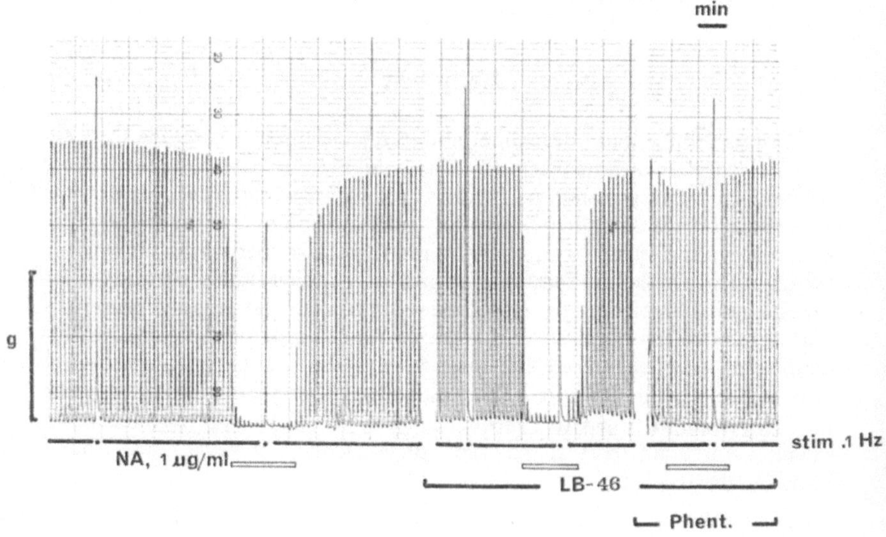

Fig. 1. Inhibitory effect of noradrenaline on contractions of a longitudinal muscle strip of guinea-pig ileum to electrical stimulation at 0.1 and 10 Hz. At 10 Hz, ten shocks were delivered as indicated by dots. Krebs solution. Field stimulation. 95% O_2 + 5% CO_2. As β-receptor blocking agent LB/46 (22 ng/ml) was used. Phentolamine (Phent.) 1 μg/ml. Note that phentolamine suspended the inhibitory action of noradrenaline and LB/46 failed to influence its effect. This fact indicates the role of the α-receptor in the presynaptic inhibitory effect of noradrenaline on acetylcholine release

Table 1. *Reduction by Sympathetic Stimulation of Acetylcholine Release from the Nerve Terminals of Rabbit Jejunum*

Condition	n	Collection period in min	ACh release* ng/g.min	P
Resting	4	5	8.6 ± 1.1 (32.0 pmol/g.min)	< 0.01
Resting + sympathetic stim. 10 Hz for 5 min	4	5	5.1 ± 0.4	

* The release of ACh is expressed as acetylcholine iodide, M. wt.: 273.1.

and depress the twitch response recorded isometrically. From the depressant action of the catecholamines on ACh release, it can also be concluded that the *in vivo* circulating catecholamines control neurogenic intestinal activity. But the increased ACh output, at rest as well as in response to stimulation, which occurred after catecholamine depletion by reserpine and guanethidine (*Vizi*, 1968; *Paton* and *Vizi*, 1969; *Beani et al.*, 1969; *Kazic*, 1971; *Vizi* and *Knoll*, 1971) and after 6-hydroxy-dopamine (*Knoll* and *Vizi*, 1970) suggest a local permanent control by noradrenergic nerves of cholinergic function as well. It was confirmed (*Beani et al.*, 1969; *Tacca et al.*, 1970; *Kosterlitz et al.*, 1970) that the inhibitory action of NA prevails when the cholinergic fibres were stimulated at low rates. However, our findings (*Knoll* and *Vizi*, 1970, 1971; *Vizi* and *Knoll*, 1971) that NA is also effective at a high rate of firing furnished further evidence for the physiological role of NA in controlling the ACh release from autonomic nervous system. In addition, it was found in intestine that NA, released from the sympathetic nerves by indirectly acting sympathemimetics (*Knoll* and *Vizi*, 1970, 1971), or by sympathetic nerve stimulation (*Beani et al.*, 1969; *Vizi*, 1971; *Vizi* and *Knoll*, 1971; *Kazic*, 1971), can reduce the ACh output. Table 1 shows the reduction by sympathetic nerve stimulation of acetylcholine release from the nerve terminals of rabbit jejunum. The pendular movement was inhibited by sympathetic stimulation and this effect was prevented by phentolamine, an α-adrenoceptor blocking agent, but not by β-adrenoceptor block (Fig. 2). This fact indicates the possibility that NA can control presynaptically the ACh output.

In addition, the findings that in the vagus-oesophagus (*Szabolcsi et al.*, in press) and vagus-stomach (*Vizi, Hallasmoller* and *Knoll*, unpublished) preparations (Fig. 3) the contractions resulting from lower frequencies of stimulation (0.1—5 Hz) were inhibited by low concentrations of NA (10^{-8} to 5×10^{-7} g/ml) without influencing the effect of acetylcholine added to the bath, indicate that in the whole gastro-intestinal tract including the oesophagus, the sympathetic transmitter, noradrenaline, controls the parasympathetic nerve-effector transmission by reducing the acetylcholine release presynaptically.

When the release of acetylcholine from intestine was enhanced by gastrin or gastrin-like hormones (*Vizi et al.*, 1972; *Vizi et al.*, 1973; *Vizi*, 1973) noradrenaline or sympathetic nerve stimulation was still able to inhibit the output (*Vizi et al.*, 1972; *Vizi*, 1973).

The direct inhibitory function (via β-adrenoceptors) of neuronal NA on the muscle cells of the intestine should be reconsidered. Also, a direct effect of circulating catecholamines both on pre-effector terminals and on the muscle cannot be excluded (*Weisbrodt et al.*, 1969).

The transmission in the autonomic nervous system depends on the presynaptic control of the cholinergic system. A "brake" system, *i.e.*, a presynaptic type of inhibition of transmission furnishes a rather economical form of antagonism. This is situated on the presynaptic (pre-effector) nerve terminals of the autonomic nervous systems. In terms of receptor terminology it is mediated via α-receptors (*Vizi*,

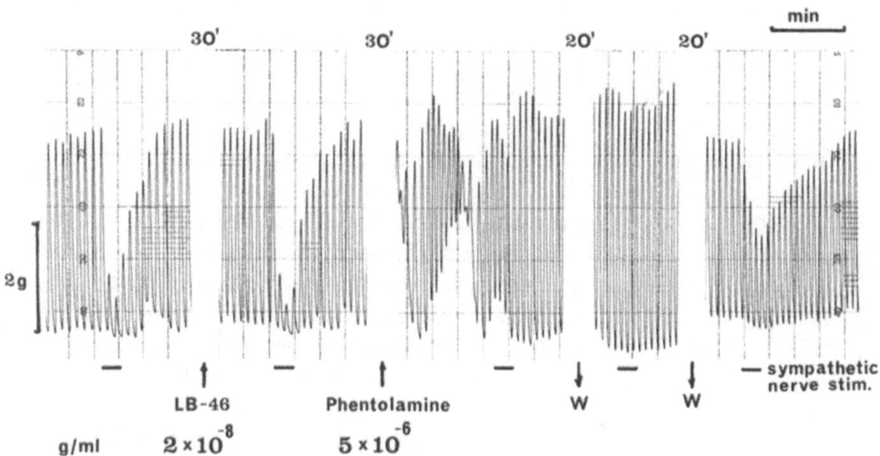

Fig. 2. Inhibitory effect of endogenous noradrenaline, released by sympathetic nerve stimulation, on pendular movement of rabbit jejunum. Finkleman preparation. Krebs solution. Isometric recording system. LB/46, dl-4-(2-hydroxy-3-isopropylaminopropoxy)-indole. Phentolamine methanesulphonate. Periarterial nerve stimulation: 10 Hz for 20 sec, 0.2 msec, 6 V

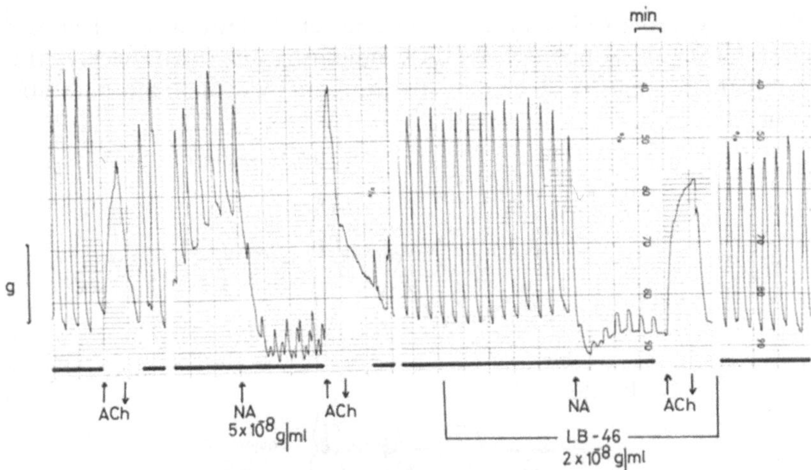

g. 3. Inhibitory effect of noradrenaline on the responses of longitudinal muscle of t stomach produced by electrical stimulation of the vagus nerve. Vagus-stomach eparation (*Paton* and *Vane*). Stimulation: 5 V, 0.2 msec, 4 Hz, 10 shocks every sec. Krebs solution. Isometric recording. ACh, acetylcholine iodide, 10^{-6} g/ml. ıe interval between the first and second part of the trace was 10 min. Note that e inhibitory effect of noradrenaline on contractions can not be blocked by β-receptor blocking agent, LB/46 (Pindolol)

)68; *Paton* and *Vizi*, 1969; *Knoll* and *Vizi*, 1970, 1971; *Kosterlitz* *al.*, 1970). The cholinergic transmission of the autonomic nervous stem can be modulated in the following ways (see Fig. 4):

a) Partly opposed by NA released from the adrenergic nerve rminals. In addition, circulating catecholamines might also act on it. ctually this happens under normal conditions.

b) Unopposed by NA. The cholinergic transmission is increased hen the "brake" is off. The partial removal of the adrenergic straint (*e.g.* depletion of NA by reserpine, by 6-OH-dopamine or by eventing the inhibitory effect of NA by α-blocking agents on the receptors of nerve terminals) leads to an enhanced ACh release oducing typical parasympathetic symptoms: diarrhoea, salivation, adycardia, increase of gastric secretion, etc.

c) Opposed by increased adrenergic activity. When the transission in the cholinergic autonomic nervous system is reduced, a lack parasympathetic tone appears resulting in sympathetic symptoms. addition, the direct effect of catecholamines on the effector cells .g., in heart) also plays an important role in producing sympathetic ne.

It seems, therefore, that the parasympathetic transmission in an ınopposed" state results in a typical parasympathetic tone (diarrhoea, hyperperistalsis of the intestine, etc.). However, when the transmission is "opposed" by NA inhibiting the ACh release, the sympathetic tone is dominant resulting in constipation, paralytic ileus, etc.

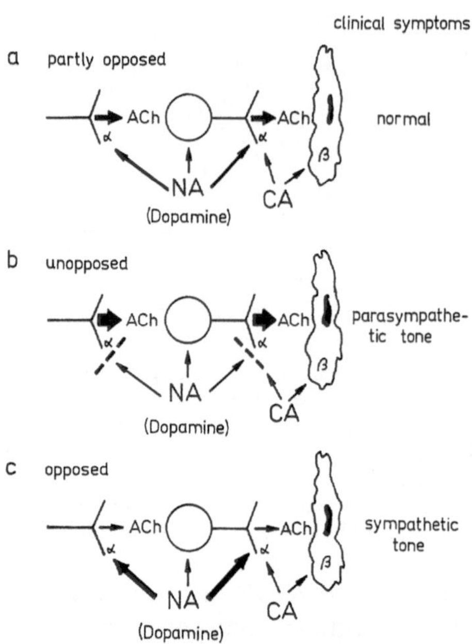

Fig. 4. Diagram of the interaction between the parasympathetic and sympathetic nervous system. Note that under normal condition the cholinergic outflow is partly reduced by noradrenaline. When the noradrenergic outflow is reduced or the effect of noradrenaline on presynaptic α-receptors is prevented (by α-receptor blocking agent etc.), there is an increased cholinergic outflow and parasympathetic tone. Sympathetic tone: increased noradrenergic outflow which, in turn, results in a reduction of the cholinergic outflow. The whole system operates as a brake system; noradrenaline modulates the acetylcholine output

Heart

Interaction between cholinergic and noradrenergic divisions in cardiac tissue has been widely discussed by different authors (cf. *Levy*, 1971).

The heart is an example of a double innervated organ, para- as well as orthosympathetically. Presumably, the transmitters released from one system can readily diffuse to the nerve terminals of the other as well as to the cardiac cells.

Noradrenaline (1 μg/ml) has been shown (*Hadházy* and *Vizi*, in press) to reduce acetylcholine release from isolated guinea-pig atria evoked by nicotine. It is difficult to ascertain whether the phenomenon observed is of physiological or pharmacological importance. Obviously, much more work and evidence are required for a full understanding of this problem. Measurement of acetylcholine output in response to vagal stimulation would be the most reasonable and simple way. However, in the presence of physostigmine sulphate (2 μg/ml) it was not possible to find (*Hadházy* and *Vizi,* in press) any increase of acetylcholine output in response to preganglionic vagal stimulation at 10 Hz. This finding might be attributed to the failure of neurochemical transmission in the parasympathetic ganglia produced by the accumulation of acetylcholine. ChE inhibition might lead to a depolarization block of ganglia. *Paton* and *Dawes* (unpublished) and *Dawes* and *Vizi* (unpublished) found a total blockage of transmission in the isolated superior cervical ganglion of the rabbit in the presence of eserine. The inhibition appeared promptly after the start of stimulation: within 10 to 30 stimuli (10 Hz), while the release of acetylcholine was maintained. In addition, *Starke* (1972) showed that the stimulation of α-adrenoceptors inhibited the myocardial responses to vagal stimulation, presumably by decreasing the release of acetylcholine.

Recently, *Hadházy* (1971, 1972) furnished evidence that the anti-vagal effect of isoprenaline is due to the stimulation of β-receptors on heart cells. A similar effect was also shown for noradrenaline added to the organ bath (*Hadházy* and *Vizi,* in press). These data, in fact, are in favour of the pharmacological importance of a direct action of noradrenaline in producing anti-vagal action. However, physiological evidence in cardiac tissue, *i.e.* anti-vagal effect of endogenous noradrenaline, is still lacking.

Ganglion

Christ and *Nishi* (1971) presented electrophysiological evidence for sympathetic ganglia and *De Groat* and *Saum* (1972) for vesical parasympathetic ganglia that adrenaline blocks ganglionic transmission by acting on an α-adrenoceptive site in the presynaptic nerve terminals.

Dawes and *Vizi* (1973) found that both noradrenaline and adrenaline significantly reduced the acetylcholine output evoked by preganglionic stimulation of isolated rabbit cervical ganglia. Noradrenaline (1 μg/ml) reduced the volley output from 3.4 ± 0.3 ng/g volley to 1.0 ± 0.2 ng/g volley when the preganglionic fibres were stimulated at a frequency of 0.3 Hz for 100 shocks (Table 2). Adrenaline had a similar effect. Alpha-receptor blockade by phentol-

Table 2. *Reduction by α-Receptor Stimulation of Acetylcholine Release from Isolated Cervical Ganglion of the Rabbit. Eserine Sulphate, 2 μg/ml. Acetylcholine Chloride*

	ACh output at 100 shocks ng/g. volley mean ± S.E.	Reduction in ACh output (%)	P	ACh output at 300 shocks ng/g. volley mean ± S.E.	Reduction in ACh output (%)	P
control	3.4 ± 0.3 (18.8 pmol/g. volley)			2.2 ± 0.3		
(—) noradrenaline bitartrate, 1 μg/ml	1.0 ± 0.2	69.4	< 0.001	1.0 ± 0.3	56.4	< 0.01
(—) noradrenaline bitartrate, 1 μg/ml + phentolamine methanesulphonate, 1 μg/ml	3.8 ± 0.25		> 0.5	2.3 ± 0.2		> 0.5
control	3.0					
+ (±)-adrenaline, 0.5 μg/ml	0.9	70.0				
control	3.5			2.0		
+ (±)-adrenaline, 0.5 μg/ml	1.4	60		1.2	40	

The effect of adrenaline and noradrenaline on the output (mean ± S.E.) of acetylcholine from the isolated superior cervical ganglion in response to preganglionic nerve stimulation at 0.3 Hz for 100 and 300 shocks (*Dawes* and *Vizi*, Br. J. Pharmacol. 1973).

amine (1 μg/ml) prevented the inhibitory effect of noradrenaline on acetylcholine release induced by preganglionic stimulation (Table 2). Therefore, it can be concluded that the inhibitory effect of adrenaline/noradrenaline on sympathetic ganglionic transmission is mainly due to their ability to reduce the release of transmitter from presynaptic nerve terminals and that this action is mediated via α-adrenoceptors. Catecholamines may be physiologically involved in ganglionic transmission, since *Lissák* (1933 a, b) showed that an adrenaline-like substance is released after preganglionic stimulation. In addition, *Schümann et al.* (1966) and *Chubb et al.* (in press, cit. *De Potter et al.*, 1972) showed that sympathetic ganglia contain a NA storage particle with similar properties (*i.e.* density and dopamine-hydroxylase content) to those found in splenic nerve trunks (*Roth et al.*, 1968; *Hörtnagel et al.*, 1969).

Cerebral Cortex

In recent years proof of the existence of cholinergic and adrenergic synapses in the cerebral cortex has accumulated. On isolated cortical slices of the rat noradrenaline inhibits the acetylcholine release induced by the inhibition of Na^+-K^+-activated ATP-ase (*Vizi*, 1972). Table 3 shows evidence that α-adrenoceptor stimulation plays an essential role in inhibiting acetylcholine release from the nerve terminals. The inhibitory effect of noradrenaline or BAY-1470 was prevented by an α-receptor blocking agent, phentolamine. Thus it seems likely that, in the cortex, a mechanism operates between adrenergic and cholinergic systems similar to that in the peripheral nervous system, noradrenaline inhibiting the release of acetylcholine from the nerve terminals.

These data, in fact, provide evidence explaining the findings of *Malcolm et al.* (1967) who showed that noradrenaline, applied to the primary sensory area of the rat cortex, inhibited the electrical responses evoked by stimulation of the contralateral forepaw. Furthermore, it was also shown that this inhibition was prevented by phenoxybenzamine, an α-receptor blocking agent.

Muscarinic Mechanism Inhibiting the Release of Acetylcholine

In the longitudinal muscle strip preparation of guinea-pig ileum which contains enteric ganglion cells, nicotine causes an increase in acetylcholine release (*Vizi* and *Knoll*, 1972; *Vizi*, 1973). This effect of nicotine is completely prevented by hexamethonium, indicating that the release of acetylcholine is due to the stimulation of

Table 3. *Reduction by α-Adrenoceptor Stimulation of Acetylcholine Release from Corticol Slices of the Rat*

Condition	No. of experiments	Collection period (min)	Acetylcholine release ng/g.min	pmol/g.min		P
1. Resting	6	20	9.4 ± 2.7	34.0		
2. Resting + Ouabain, 2×10^{-5} M	6	10	59.6 ± 15.9	218.3	2 : 1	< 0.01
3. Resting + BAY-1470, 2 μg/ml	5	20	9.2 ± 1.8	33.7	3 : 1	n.s.
4. Resting + BAY-1470, 2 μg/ml Ouabain, 2×10^{-5} M	5	10	17.2 ± 5.8	63.0	4 : 2	< 0.01
5. Resting + Phentolamine, 2 μg/ml BAY-1470, 2 μg/ml Ouabain, 2×10^{-5} M	3	10	68.1 ± 17.7	249.4	5 : 4	< 0.01

Krebs' solution. Physostigmine sulphate, 2 μg/ml. BAY-1470 is a pure α-receptor stimulant (*Kroneberg et al.*, 1967). Note that α-adrenoceptor stimulation reduced the release of acetylcholine provoked by blocking the (NA^{+}-K^{+})-activated ATP-ase by ouabain. A similar result with noradrenaline was obtained by *Vizi* (1972).

ganglion cells which in turn leads to the firing of axons resulting in release of acetylcholine from the nerve terminals. These data also indicate that nicotine has no direct effect on nerve terminals; it does not release acetylcholine. In addition, oxotremorine, a strong muscarinic stimulant, failed to increase the release of acetylcholine from the nerve terminals of Auerbach's plexus even reducing it (*Vizi* and *Knoll*, 1972). The resting release of acetylcholine was reduced from 21.8 ± 3.8 to 14.1 ± 4.1 ng/g.min when 10^{-4} M oxotremorine was added to the bath. The reduction is statistically significant, $p < 0.05$. Another interesting observation is that the rate of acetylcholine release depends on the collection period. It was observed that there is an inverse correlation between the rate of release and the duration of the collection period. During rest, the total output of acetylcholine from Auerbach's plexus, *i.e.* the concentration of acetylcholine, amounted to 105.8, 378.0, 589.0 and 1476.0 ng/g of tissue, for 1, 5, 10 and 30 min periods, respectively. Although the total amount of acetylcholine released increased with time, the rate of release decreased with the accumulation of released acetylcholine in the bathing fluid, the values being 105.8, 75.6, 58.9 and 49.2 ng/g.min for the respective collection periods. The crucial role of the extraneuronal concentration of acetylcholine in the rate of output is shown by the experiment (Table 4) in which acetylcholine was added to the bath resulting in a reduced output from the nerve terminals. A similar observation was made on isolated cortical slices (Table 4). However, the output of acetylcholine from cortical slices was much more dependent on the extraneuronal concentration of acetylcholine (Table 4). Since the total amount of acetylcholine in the organ bath increased with time, it can be established that the longer the collection period the higher was the concentration of acetylcholine in the organ bath; *i.e.*, the higher the concentration of acetylcholine in the bathing fluid, the lower the rate of release of acetylcholine. These data indicate that in the presence of cholinesterase inhibition (2 μg/ml physostigmine sulphate was present throughout the experiments) acetylcholine can reduce the release of acetylcholine. Its action is concentration-dependent, or depends on the duration of the collection period. This mechanism significantly operates in cerebral cortex (Table 4). In addition, *Szerb* and *Somogyi* (1973) recently showed, that eserine greatly depressed the evoked release of acetylcholine from cortical slices of the rat, but that this was restored to control levels when atropine was also present. It is of interest to note that, in cerebral cortex at rest, an increase in acetylcholine release has been found (*Mitchell*, 1963; *Szerb*, 1964; *Polak*, 1965; *Dudar* and *Szerb*, 1969; *Szerb et al.*, 1970; *Bourdois et al.*, 1971) after the

Table 4. *Inhibitory Effect of Extraneuronal Acetylcholine on the Release of Acetylcholine from the Nerve Terminals of Auerbach's Plexus and Cerebral Cortex of the Rat. Longitudinal Muscle Strip of Guinea-Pig Ileum (83 mg in Weight). Cortical Slices of the Rat (400 mg). Eserinised Krebs' solution, 2 μg/ml Physostigmine Sulphate. Acetylcholine Iodide*

			Longitudinal muscle strip		Cortical slice	
	Collection period in minutes	Added ACh*	Final conc. of ACh in the organ bath ng/ml	ACh release ng/g.min	Final conc. of ACh in the organ bath ng/ml	ACh released ng/g.min
Resting	2	—	4.66	98.2 (358.9 pmol/g.min)	4.08	10.2 (37.4 pmol/g.min)
Resting	10	—	15.9	67.0	12.8	6.4
Resting	30	—	35.6	50.1	15.0	2.5
Resting	2	30 ng/ml	34.1	86.4	31.9	4.75

* Recovery, 98.5%, which was measured in a parallel experiment. Note how the concentration of acetylcholine in the organ bath influences the output of acetylcholine from nerve terminals of cerebral cortex. Organ bath for longitudinal muscle strip, 3.5 ml; for cortical slices, 2 ml.

administration of atropine. In all of these experiments a cholinesterase inhibitor was used. In addition, *Tacca et al.* (1968) showed that the amount of acetylcholine released from guinea pig colon is unchanged after atropine administration. These data are in agreement with our findings that on the nerve terminals of guinea pig ileum the muscarinic inhibitory mechanism is far less significant than in the case of the cerebral cortex.

The question now arises how the reduction of acetylcholine release by acetylcholine and its increase by atropine are correlated. Do the muscarinic receptors play any role? What is the real function of the presynaptic location of acetylcholinesterase? It seems likely that this enzyme protects the presynaptic membrane and the axon from the effect of acetylcholine. In the presence of cholinesterase inhibitor, acetylcholine once released may regulate its own release by a negative feed-back mechanism which seems especially valid for cerebral cortex. ChE, situated presynaptically, protects the nerve terminal and the axon against the effect of ACh released on muscarinic receptors resulting in reduction of transmitter release. Therefore, it is suggested that the effect of atropine in enhancing ACh release might be attributed to the removal of the self-inhibitory effect of ACh on muscarinic receptors in the presence of cholinesterase inhibition.

A similar interaction was proposed by *Muscholl* and his collaborators (*Lindmar et al.*, 1968) who demonstrated a muscarinic mechanism inhibiting the release of noradrenaline from peripheral adrenergic nerve fibres.

Conclusion

Evidence, so far available, indicates a special form of interaction between the cholinergic and adrenergic nervous system: noradrenaline inhibits acetylcholine release presynaptically via α-adrenoceptor stimulation. This economical form of interaction between the cholinergic and adrenergic systems seems to operate in the gastro-intestinal tract (Fig. 5/1), and in the sympathetic ganglia (Fig. 5/2). There is also some evidence that in the brain cortex (*Vizi*, 1972; Fig. 5/3) and in the heart a similar interaction might exist. However, there is no such information on the spinal cord. In the neuromuscular junction (Fig. 5/4) no such a mechanism operates. Here, the activation of α-receptors of motor nerve endings even facilitates transmission by increasing the release of transmitter by nerve impulses (*Bowman* and *Raper*, 1967; *Kuba*, 1970).

It was also shown that acetylcholine inhibits its own release via muscarinic receptors. However, this is only of pharmacological importance since this action only occurs in the presence of cholinesterase inhibition. It can also be concluded, that the role of presynaptically situated cholinesterase is to protect the presynaptic membrane against the acetylcholine released.

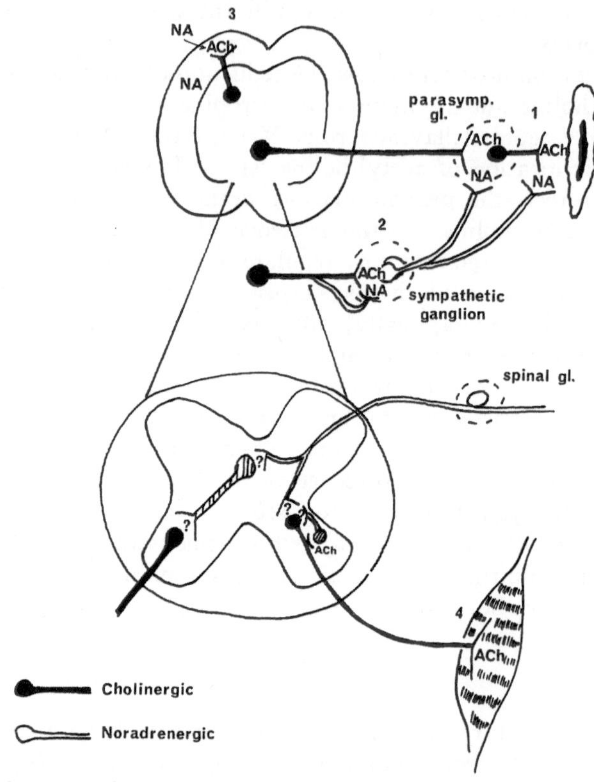

Fig. 5. Typical sites of cholinergic transmission. Otherwise see the text for further explanation

References

Beani, L., C. Bianchi, and *A. Crema*: The effect of catecholamines and sympathetic stimulation on the release of acetylcholine from the guinea-pig colon. Br. J. Pharmacol. *36*, 1—17 (1968).

Bourdois, P. S., J. F. Mitchell, and *J. C. Szerb*: Effect of atropine on acetylcholine release from cerebral cortical slices stimulated at different frequencies. Br. J. Pharmacol. *42*, 640—641 (1971).

Bowman, W. C., and *C. Raper*: Adrenotropic receptors in skeletal muscle. Ann. N.Y. Acad. Sci. *139*, 741—753 (1967).

Christ, D. D., and *S. Nishi*: Site of adrenaline blockade in the superior cervical ganglion of the rabbit. J. Physiol. (London) *213*, 107—117 (1971).

Chubb, I. W., *W. P. De Porter*, and *A. F. De Schaepdryver*: Storage of noradrenaline in sympathetic ganglia. Life Sci. (in press) cit. from *De Porter et al.* (1972).

Dawes, P. M., and E. S. Vizi: Acetylcholine release from the rabbit isolated superior cervical ganglion preparation. Br. J. Pharmacol. *48*, 225—232 (1973).

Dudar, J. D., and *J. C. Szerb*: The effect of topically applied atropine on resting and evoked cortical acetylcholine release. J. Physiol. (London) *203*, 741—762 (1969).

de Groat, W. C., and *W. R. Saum*: Sympathetic inhibition of the urinary bladder and of pelvic ganglionic transmission in the cat. J. Physiol. (London) *220*, 297—314 (1972).

Hadházy, P.: Effects of isoprenaline and aminophylline on the chronotropic responses of the isolated guinea-pig heart to vagal stimulation and acetylcholine. Br. J. Pharmacol. *42*, 364—370 (1971).

Hadházy, P.: "Anticholinergic" action of aminophylline in isolated guinea-pig atria. Eur. J. Pharmacol. *20*, 284—286 (1972).

Hadházy, P., and *E. S. Vizi*: The inhibitory action of noradrenaline on the chronotropic responses to vagal stimulation. (In press.)

Hörtnagel, A., *H. Hörtnagel*, and *H. Winkler*: Bovine splenic nerve: characterization of noradrenaline-containing vesicles and other cell organelles by density gradient centrfugation. J. Physiol. (London) *205*, 103—114 (1969).

Kazic, T.: Effect of adrenergic factors on peristalsis and acetylcholine release. Eur. J. Pharmacol. *16*, 367—373 (1971).

Knoll, J., and *E. S. Vizi*: Presynaptic inhibition of acetylcholine release by endogenous and exogenous noradrenaline at high rate of stimulation. Br. J. Pharmacol. *40*, 554—555 (1970).

Knoll, J., and *E. S. Vizi*: Effect of frequency of stimulation on the inhibition by noradrenaline of the acetylcholine output from parasympathetic nerve terminals. Br. J. Pharmacol. *42*, 263—272 (1971).

Kosterlitz, H. W., *R. J. Lydon*, and *A. J. Watt*: The effects of adrenaline, noradrenaline and isoprenaline on inhibitory α- and β-receptors in the longitudinal muscle of the guinea-pig ileum. Br. J. Pharmacol. *39*, 398—413 (1970).

Kroneberg, G., *A. Oberdorf*, and *F. Hoffmeister*: Zur Pharmakologie von 2-(2,6-Dimethylphenylamino)-4H-5,-6-Dihydro-1,3-thiazin (Bayer 1470), eines Hemmstoffes adrenergischer und cholinergischer Neurone. Naunyn-Schmiedebergs Arch. Pharmac. exp. Path. *256*, 257—280 (1967).

Kuba, K.: Effects of catecholamines on the neuromuscular junction in the rat diaphragm. J. Physiol. *211*, 551—570 (1970).

Levy, M. N.: Sympathetic-parasympathetic interactions in the heart. Circulation Res. *29*, 437—445 (1971).

Lindmar, R., K. Löffelholz, and *E. Muscholl*: Muscarinic mechanism inhibiting the release of noradrenaline from peripheral adrenergic nerve fibres by nicotinic agents. Br. J. Pharmacol. *32*, 280—294 (1968).

Lissák, K.: Effects of extracts of adrenergic fibers on frog heart. Amer. J. Physiol. *125*, 778—785 (1939 a).

Lissák, K.: Liberation of acetylcholine and adrenaline by stimulating isolated nerves. Amer. J. Physiol. *127*, 263—271 (1939 b).

Malcolm, J. L., P. Saraiva, and *J. Spear Phyllis*: Cholinergic and adrenergic inhibition in the rat cerebral cortex. Int. J. Neuropharm. *6*, 509—527 (1967).

Mitchell, J. F.: The spontaneous and evoked release of acetylcholine from the cerebral cortex. J. Physiol. *165*, 98—116 (1963).

Paton, W. D. M., and *E. S. Vizi*: The inhibitory action of noradrenaline and adrenaline on acetylcholine output by guinea-pig ileum longitudinal muscle strip. Br. J. Pharmacol. *35*, 10—28 (1969).

Polak, R. L.: Effect of hyoscine on the output of acetylcholine into perfused cerebral ventricle of cats. J. Physiol. *181*, 317—323 (1965).

Porter, W. P., I. W. De Chubb, and *A. F. De Schaepdryver*: Pharmacological aspects of peripheral noradrenergic transmission. Arch. Pharmacodyn. *196*, suppl., 258—287 (1972).

Roth, R. H., L. Stjärne, F. E. Bloom, and *N. J. Giarman*: Light and heavy norepinephrine storage particles in the rat heart and in bovine splenic nerve. J. Pharmacol. *162*, 203—212 (1968).

Schümann, H. J., K. Schmidt, and *A. Philippu*: Storage of norpinephrine in sympathetic ganglia. Life Sci. *51*, 1809—1815 (1966).

Starke, K.: Alpha sympathetic inhibition of adrenergic and cholinergic transmission in the rabbit heart. Arch. Pharmacol. *274*, 18—45 (1972).

Szabolcsi, T., E. S. Vizi, and *J. Knoll*: Inhibition by different drugs of neurochemical transmission in the guinea-pig oesophagus; morphological and functional study. (In press.)

Szerb, J. C.: The effect of tertiary and quaternary atropine on cortical acetylcholine output and on the electroencephalogram in cats. Can. J. Physiol. Pharmacol. *42*, 303—314 (1964).

Szerb, J. C., H. Malik, and *E. G. Hunter*: Relationship between acetylcholine content and release in the cats cerebral cortex. Can. J. Physiol. Pharmacol. *48*, 780—790 (1970).

Szerb, J. C., and *G. T. Somogyi*: Depression of acetylcholine release from cerebral cortical slices by cholinesterase inhibition and by oxotremorine. Nature (Biol.) *241*, 121—122 (1973).

Tacca, M. Del, S. Lecchini, C. M. Frigo, A. Crema, and *C. Benzi*: Antagonism of atropine towards endogenous and exogenous acetylcholine before and after sympathetic system blockade in isolated distat guinea-pig colon. Eur. J. Pharmacol. *4*, 188—197 (1968).

Tacca, M. Del, G. Soldani, M. Selli, and *A. Crema*: Action of catecholamines on release of acetylcholine from human taenia cóli. Eur. J. Pharmacol. *9*, 80—84 (1970).

Vizi, E. S.: The inhibitory action of noradrenaline on release of acetylcholine from guinea-pig ileum longitudinal strips. Arch. Pharmacol. *259*, 199—200 (1968).

Vizi, E. S.: Inhibition of acetylcholine release by noradrenaline released from sympathetic nerve terminals. In: Abstracts (The 34th Annual Conference of the Hungarian Physiological Society. Debrecen 1968), p. 55. Budapest: Akadémiai Kiadó. 1970.

Vizi, E. S.: Stimulation, by inhibition of (Na^+-K^+-Mg^{2+})-activated ATP-ase, of acetylcholine release in cortical slices from rat brain. J. Physiol. *226*, 95—117 (1972).

Vizi, E. S.: Acetylcholine release from guinea-pig ileum by parasympathetic ganglion stimulants and gastrin-like polypeptides. Br. J. Pharmacol. *47*, 765—777 (1973).

Vizi, E. S., G. Bertaccini, M. Impicciatore, and *J. Knoll*: Acetylcholine-releasing effect of gastrin and related polypeptides. Eur. J. Pharmacol. *17*, 175—178 (1972a).

Vizi, E. S., G. Bertaccini, M. Impicciatore, and *J. Knoll*: Evidence that acetylcholine released by gastrin and related polypeptides contributes to their effect on gastrointestinal motility. Gastroenterology *64*, 268 to 277 (1973).

Vizi, E. S., and *J. Knoll*: The effects of sympathetic nerve stimulation and guanethidine on parasympathetic neuroeffector transmission; the inhibition of acetylcholine release. J. Pharm. Pharmacol. *23*, 918—925 (1971).

Vizi, E. S., and *J. Knoll*: The mechanism of acetylcholine release by nicotine. In: Proc. of symposium on "Experimental tremor" Sarajevo, 1971, pp. 131—150. Sarajevo: Academia of Bosnia and Herzegovina. 1972.

Weisbrodt, N. W., C. C. Hug, and *P. Bass*: Separation of the effects of alpha and beta adrenergic receptor stimulation on taenia coli. J. of Pharmacol. *170*, 272—280 (1969).

Author's address: Dr. *E. S. Vizi*, Department of Pharmacology, Semmelweis University of Medicine, H-1085 Budapest, Hungary.

Discussion

Joó: I should like to mention my experiments performed in collaboration with Lever's group in Cardiff, using electron microscopic autoradiography combined with electron histochemistry. While studying the intraganglionic distribution of ^{3}H-noradrenaline in relation to the acetylcholinesterase-positive intraganglionic axonal endings, we found that ganglion cells did not take up exogenous noradrenaline, whereas there was a significant accumulation of reduced silver grains over acetylcholinesterase positive preganglionic axon terminals. These results are in favour of *Vizi*'s concept, providing morphological evidence for a noradrenaline uptake-system in the cholinergic endings of preganglionic nerves.

Ariëns Kappers: What is your opinion about the validity of *Burn*'s and *Rand*'s theory in transmitter interaction?

Sakharov: Can you please inform us about the mechanism of blocking acetylcholine synthesis in a sodium-free solution?

Akert: I want to ask whether acetylcholine has a similar presynaptic inhibitory effect upon adrenergic transmitter release. According to *Stach*, there are no morphological or histochemical proofs to suppose a presynaptic (neuromuscular) inhibition of acetylcholine release in the small intestine. Since one part of the interneuronal synapses in the plexus myentericus and the plexus submucosus is adrenergic, it can be hypothesized that an interference between the two transmitter systems takes place at this site.

Vizi: Burn and *Rand* hypothesized (Nature [London] *194*, 163—165, 1959; Ann. Rev. Pharmacol. *5*, 163—182, 1965) that when the nerve impulse reaches the noradrenergic nerve terminals there is a release of acetylcholine followed by stimulation of nicotinic receptors which, in turn, enhances the influx of calcium resulting in a release of noradrenaline. This theory is now questioned (cf. *Ferry*, Phys. Rev. *46*, 420—456 (1966). It was even shown by *Muscholl* (In "New aspects of storage and release mechanisms of Catecholamines", pp. 168—186, Springer-Verlag, 1970) that acetylcholine in low concentrations inhibits the release of noradrenaline from nerve terminals. Taking into account our results on the presynaptic inhibitory action of noradrenaline on acetylcholine release (*Vizi*, Arch. exp. Path. Pharmak. 1968; *Paton* and *Vizi*, Br. J. Pharmac. *35*, 10—28, 1969; *Knoll* and *Vizi*, Br. J. Pharmac. *42*, 263—272, 1971; *Vizi* and *Knoll*, J. Pharm. Pharmacol. *23*, 918—925, 1971) and the data of *Muscholl* it seems that there is a functional interaction between adrenergic and cholinergic nerve terminals. Both noradrenaline and acetylcholine are capable of inhibiting presynaptically the release of the other transmitter. The presynaptic type of inhibition seems to be a more economical form of interaction than the counteraction on the post synaptic membrane, on the effector cells.

The fact that noradrenaline released from nerve terminals is able to inhibit acetylcholine release supports the concept that noradrenaline physiologically controls the release of acetylcholine.

As far as the question raised by *Stach* is concerned there is morphological and histological evidence (*Jacobowitz*, J. Pharmacol. *149*, 358—364, 1965; *Norberg, K. A.*, Int. J. Neuropharmac. *3*, 379—382, 1964; *Silva et al.*, Am. J. Phys. *220*, 347—352, 1971; *Joó*, see this symposium) that the adrenergic nerve terminals terminate not far from the cholinergic ones. This enables the released noradrenaline to reach the cholinergic nerve endings by diffusion and to act there via α-adrenoceptor (*Paton* and *Vizi*, Br. J. Pharm. *35*, 10—28, 1969; *Kosterlitz et al.*, Br. J. Pharmacol. *39*, 398—413, 1970) resulting in a reduction of acetylcholine release. The intracellular sodium plays an essential role in maintaining ACh synthesis (*Birks*, Can. J. Bioch. Physiol. *41*, 2573—2597, 1963; *Paton, Vizi* and *Zar*, J. Physiol. *215*, 819—848, 1971). It is interesting that lithium, which otherwise substitutes sodium in generating action potentials, failed to replace intracellular sodium in maintaining the synthesis of acetylcholine (*Vizi et al.*, Neuropharmacology *11*, 521—530, 1972).

Journal of Neural Transmission, Suppl. XI, 79—101 (1974)

Morphologie und Histochemie von Synapsen im Bereich des Magen-Darm-Kanals

W. Stach

Anatomisches Institut der Universität Rostock, Deutsche Demokratische Republik
(Direktor: Prof. Dr. sc. med. Dr. med. dent. *G. H. Schumacher*)

Mit 10 Abbildungen

Zusammenfassung — Summary

An Hand eigener licht- und elektronenmikroskopischer sowie histochemischer Befunde wurden die folgenden Synapsenstrukturen vorgestellt und diskutiert:

1. Interneuronale Synapsen an Typ I- und II-Zellen.
2. Neuro-muskuläre Synapsen an glatten Muskelzellen des Muskelmantels, der Schleimhaut sowie der Blut- und Lymphgefäße.
3. Neuro-epitheliale Synapsen am Zylinderepithel und an den spezifischen Inselepithelzellen (A-, B-, D-Zellen) des Pankreas.
4. Außerdem wird auf die engen Beziehungen von Axonen zu Bindegewebszellen und zu den interstitiellen Zellen *(Cajal)* eingegangen.

Morphology and Histochemistry of Synapses in the Range of the Intestinal Tract

On the basis of light- and electronmicroscopic and histochemical findings there are demonstrated and discussed the following synaptic structures:

1. Interneuronal synapses at type I- and II-cells.
2. Neuro-muscular synapses at smooth muscle cells of the muscular coat, mucosa and of the blood- and lymphatic vessels.
3. Neuro-epithelial synapses at the cylindric epithel and at the specific insular epithel cells (A-, B- and D-cells) of the pancreas.
4. Moreover there are discussed the close relations of axons to connective tissue cells and interstitial cells (Cajal).

Einleitung

Es sollen in diesem Rahmen Ergebnisse vorgetragen und Befunde demonstriert werden, die vorwiegend prinzipielle Aspekte der Morphologie und Histochemie von interneuralen und peripheren Synapsen berücksichtigen.

Die Auswahl ist andererseits so getroffen worden, daß weder die quantitative noch die qualitative Seite der Innervation des Magen-Darm-Kanals vernachlässigt wird.

Material und Methoden

Die Untersuchungen wurden an Laboratoriumstieren durchgeführt (Goldhamster, Ratte, Meerschweinchen, Kaninchen, Katze, Hund, Schwein). Folgende Methoden kamen zur Anwendung: Silberimprägnation nach *Cauna* (1959), Osmiumsäure-Zinkjodid-Technik nach *Maillet* (1959), Acetylcholinesterasereaktion nach *Gomori* (1952), Paraformaldehydbedampfung nach *Falck* (1962). Weiterhin wurden elektronenmikroskopische Untersuchungen nach Glutaraldehydperfusion und nach Kaliumpermanganatfixierung (*Hökfelt* und *Jonsson*, 1968) durchgeführt.

Ergebnisse und Diskussion

1. Interneuronale Beziehungen

Bei diesen Untersuchungen müssen die morphologische Spezifität der Darmwandneurone (*Dogiel*, 1896; *Lawrentjew*, 1929, 1931) und ihre vermutlichen Funktionen (s. *Jabonero*, 1953; *Temesrèkasi*, 1955;

Symbole für Abbildung 1—10:

A, A-Zelle
a, Axon, Axonbündel
aa, Adrenerges Axon
B, B-Zelle
bm, Basalmembran
ca, Cholinerges Axon
d, Dendrit
ds, Drüsenschläuche
e, Endothel
el, Eosinophiler Leukozyt
g, Ganglion
gz, Ganglienzelle
L, Lymphgefäß
m, Glatte Muskelzelle
mm, Muscularis mucosae
mz, Mastzelle
n, „mexus"
s, Schwannsche Zelle
z, Zylinderepithelzelle
I, Typ-I-Zelle (nach *Dogiel*)
II, Typ-II-Zelle (nach *Dogiel*)
—, 1 μm in elektronenmikroskopischen Abbildungen
———, 20 μm in lichtmikroskopischen Abbildungen
- - →, Synapsenstrukturen

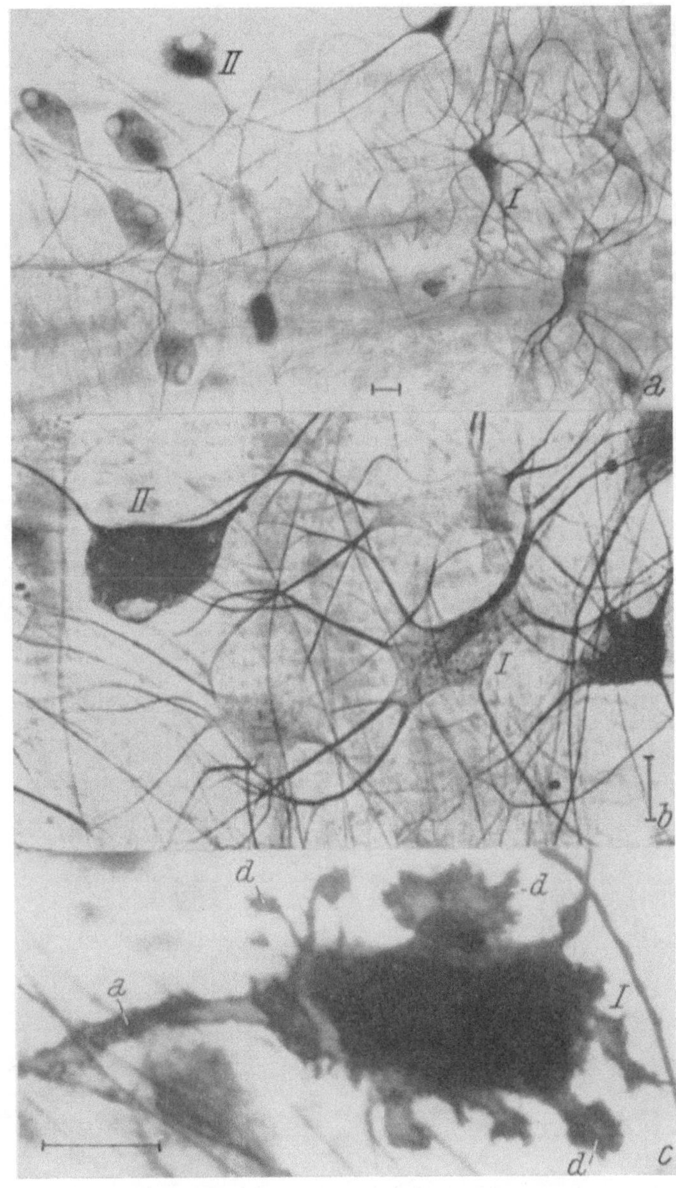

Abb. 1. Typ-I- und Typ-II-Zellen aus dem Plexus Auerbach des Duodenums vom Schwein (a, b) und aus dem Colon des Hundes (c). Silberimprägnation nach *Cauna*

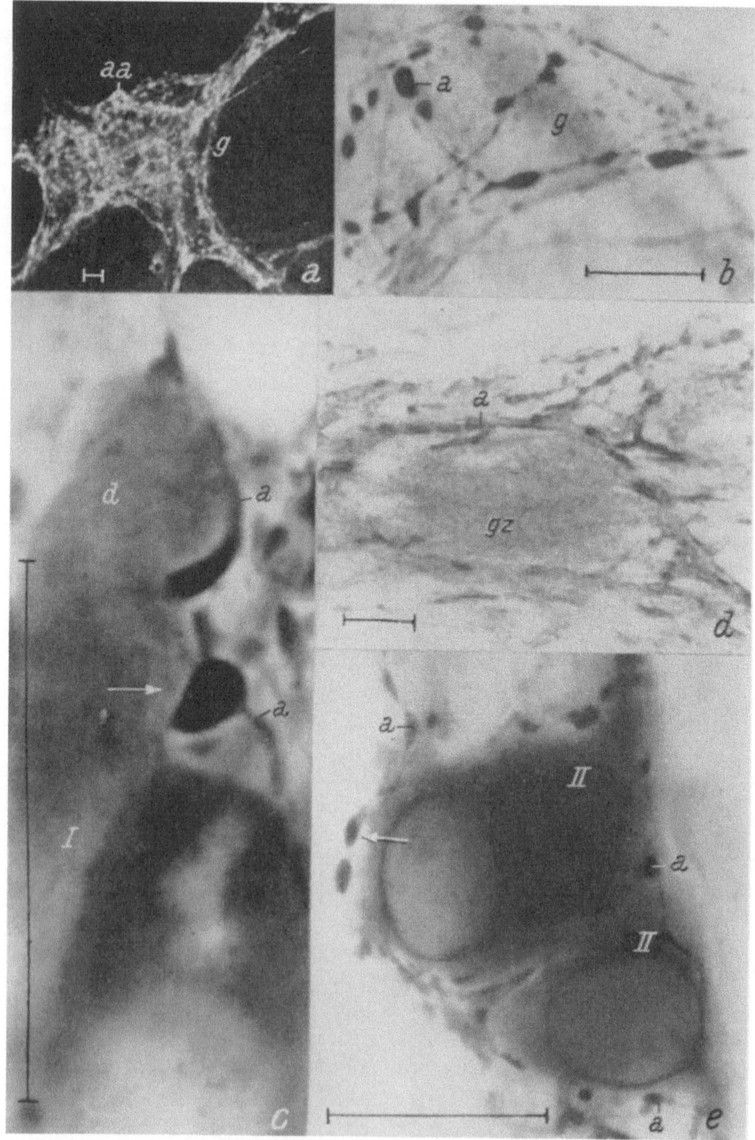

Abb. 2. Präganglionäre Fasern und Einzelaxone aus dem Plexus Auerbach von Meerschweinchen (a, Duodenum), Ratte (b, Colon) und Katze (d, Duodenum) und aus den submukösen Ganglien der Ratte (b, e, Colon). Paraformaldehydbedampfung nach *Falck* (a), Osmiumsäure-Zinkjodid nach *Maillet* (b, c, e), Acetylcholinesterasereaktion nach *Gomori* (d)

Davenport, 1966; *Milochin*, 1967; *Stach*, 1969) ebenso berücksichtigt werden wie die Beziehungen der Darmwandneurone zu vorgeschalteten Zentren und untereinander.

Der Plexus myentericus und die submukösen Nervengeflechte enthalten sowohl im Dünndarm (*Gunn*, 1968) als auch im Dickdarm (*Stach*, 1969) eine große Zahl sehr unterschiedlich dimensionierter und geformter Ganglienzellen (Abb. 1). Sie lassen sich prinzipiell in die zwei bzw. drei Dogiel-Typen differenzieren. Diese Zelltypen können weiterhin in wenigstens drei Größenklassen unterteilt werden, wobei auch die kleinsten Elemente (5—7 μm) alle Merkmale von ausdifferenzierten Neuronen aufweisen.

Die lichtmikroskopische Differenzierung ist bei optimaler Anfärbung oder Imprägnation einfach. Bei elektronenmikroskopischen Untersuchungen ist die Klassifizierung bisher weit schlechter möglich. Es sind allerdings alle Zelltypen nebeneinander vollständig dargestellt, aber vorwiegend in sehr kleinen uncharakteristischen Ausschnitten. Es ist deshalb schwierig, präganglionäre synapsenbildende Axone und Zelltypen einander zuzuordnen.

Die externen präganglionären Axone stammen aus dem Vagus (*Lawrentjew*, 1929; *Bullon*, 1945; *Castro*, 1950; *Morrison* und *Habel*, 1964; *Jabonero*, 1966); dem Sakralparasympathikus (*Lawrentjew*, 1931; *Iljina* und *Lawrentjew*, 1932; *Bullon*, 1947; *Stach*, 1969, 1971) und aus dem Sympathikus (*Botar* und Mitarbeiter, 1942; *Jabonero*, 1953; *Schofield*, 1962; *Marks* und Mitarbeiter, 1962; *Norberg*, 1964, 1967; *Jacobowitz*, 1965; *Gabella* und *Costa*, 1967; *Taxi* und *Droz*, 1967; *Furness*, 1970; *Costa* und Mitarbeiter, 1971).

Der weitaus größte Teil der nachweisbaren Synapsen dient aber offensichtlich den Beziehungen der Darmwandneurone untereinander, da deren Anzahl im Duodenum des Hundes 10 Tage nach Vagotomie bei elektronenmikroskopischer Auswertung subjektiv kaum reduziert war. Dazu kommt, daß noradrenerge Synapsen elektronenmikroskopisch nicht so häufig nachzuweisen sind wie cholinerge (s. u.).

Wir haben an allen Größenklassen der Dogiel-Typen licht- und elektronenmikroskopisch Synapsenstrukturen gefunden. Es handelt sich vorwiegend um axo-somatische und axo-dendritische aber auch um axo-axonale Synapsen (Abb. 2, 3, 4). Histochemisch findet man um die Darmwandganglienzellen einerseits cholinesterasepositive Zonen bzw. Strukturen (Abb. 2 d) und andererseits viele grün fluoreszierende (noradrenerge) Axonauftreibungen (Abb. 2 a). Elektronenmikroskopisch (Abb. 3 a) relativ gut zu differenzieren sind die Verhältnisse an großen Typ-I-Zellen. Die Synapsenstrukturen befinden sich mehr oder weniger tief zwischen den Dendriten eingefaltet am Zellkörper und häufig an verbreiterten (Dendritenlamellen nach *Lawrentjew*, 1929,

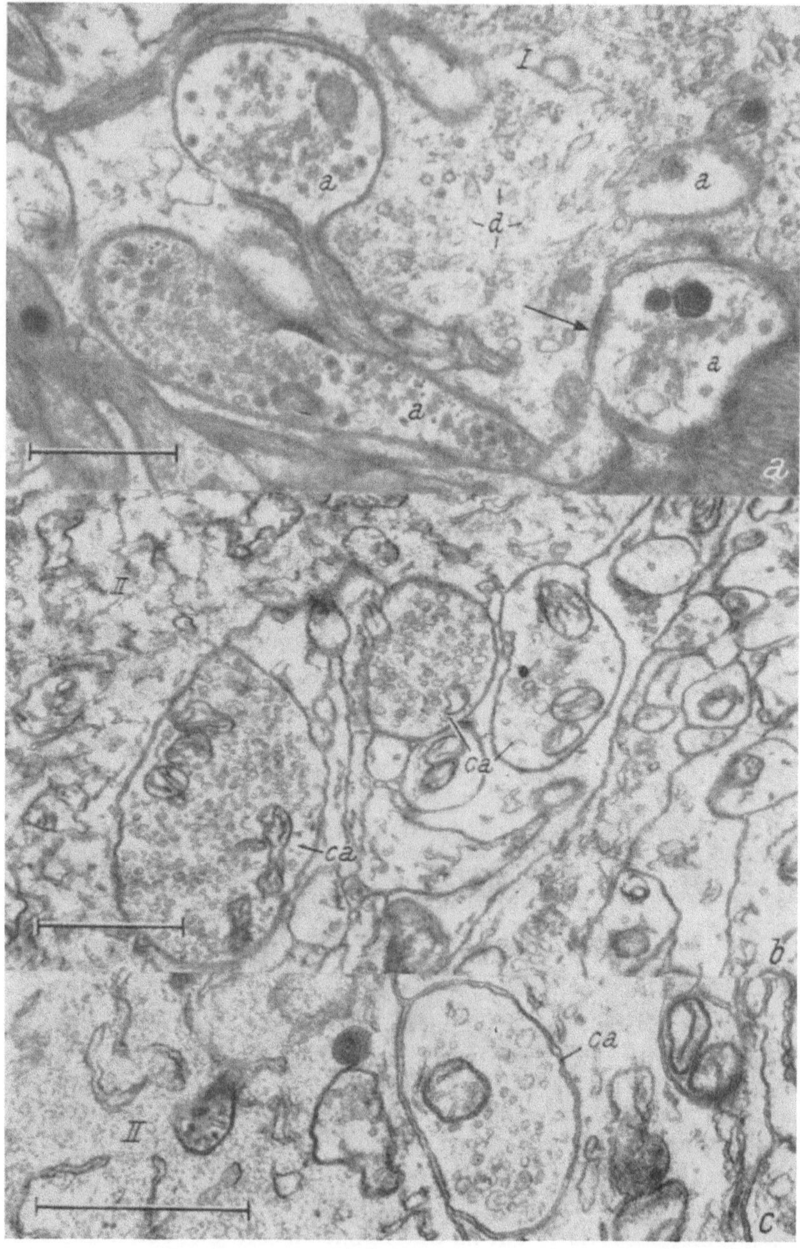

Abb. 3. Synapsenstrukturen an Typ-I- und Typ-II-Zellen aus dem Plexus Auerbach (a, b) und submucosus (c) des Rattenduodenums. Glutaraldehyd-Osmiumsäure (a), Kaliumpermanganat nach *Hökfelt* und *Jonsson* (b, c)

Abb. 1 c) oder zugespitzten (Dendritensporne, Abb. 1 c) Dendritenabschnitten (Abb. 2 c, 3 a, 4 c, d).

Axo-axonale Synapsen wurden an Typ-I-Zellen häufiger an der Neuritenwurzel als im weiteren Verlauf beobachtet. Neben axosomatischen Synapsen (Abb. 3 b, c, 4 b) wurden an Typ-II-Zellen Synapsenstrukturen auch an den Fortsatzursprüngen und im zellnahen Abschnitt der Fortsätze dargestellt (Abb. 1 b). Als Ausnahme meinen wir, auch axo-axonale Synapsen im Bereich axo-somatischer und axo-axonaler Kontaktzonen beobachtet zu haben.

Die oft gefundenen Synapsen an isolierten Fortsatzanschnitten (Abb. 4 a, c) können nur in einigen Fällen einem Zelltyp zugeordnet werden, da spezifische Kriterien fehlen.

Nach Glutaraldehydfixierung wurden neben typischen Synapsenstrukturen auch Kontaktzonen beobachtet (s. a. *Honjin* und Mitarbeiter, 1965 a; *Ono*, 1967), denen entweder die Membrandifferenzierungen oder zusätzlich noch die präsynaptischen Vesikel fehlten. Die ersteren (Abb. 3 a) halten wir für typische aber schlecht kontrastierte oder ungünstig angeschnittene Synapsen. Im zweiten Fall vermögen wir keine Funktion abzuleiten, zumal gerade im Bereich der Darmwandganglien Kontaktzonen (somato-somatisch, somato-dendritisch) zwischen Neuronen häufig vorkommen, worauf bereits *Grillo* (1966) hingewiesen hat.

Während bei Kaliumpermanganatfixierung die Membrandifferenzierungen in der Regel (*Richardson*, 1958) nahezu vollkommen fehlen (Abb. 3, 4), werden die Vesikel sehr differenziert dargestellt. In Auswertung der Modelluntersuchungen von *Hökfelt* und *Jonsson* (1968), die unter anderem der standardisierten Kaliumpermanganatfixierung in Beziehung zur Darstellung aminerger Transmitter weitgehende Spezifität zuzusprechen, handelt es sich bei Varikositäten mit kleinen granulierten Vesikeln um sympathische (noradrenerge) Axone (Abb. 4 a, 7 d, 9 d, 10 d). Das Vorhandensein kleiner klarer (agranulärer) Vesikel charakterisiert dagegen die cholinergen Achsenzylinder (Abb. 3, 4). Obgleich wir die Speicherung von Noradrenalin weder *in vivo* noch *in vitro* gesteigert haben, stellten sich bei Kaliumpermanganatfixierung zwei bzw. drei Vesikelpopulationen und damit verbunden drei Arten von präganglionären Axonen dar. Bisher konnten wir feststellen, daß der weitaus größte Teil der Synapsen von cholinergen Axonen gebildet wird.

Im Gegensatz zu den eindrucksvollen Bildern unter anderem von *Norberg* (1964, 1967), *Jacobowitz* (1965), *Gabella* und *Costa* (1967), *Furness* (1970) sowie *Costa* und Mitarbeiter (1971), die nach Paraformaldehydbedampfung (Abb. 2 d) entstehen, fanden sich bisher nur wenige noradrenerge Axonauftreibungen in Kontakt zu den Darm-

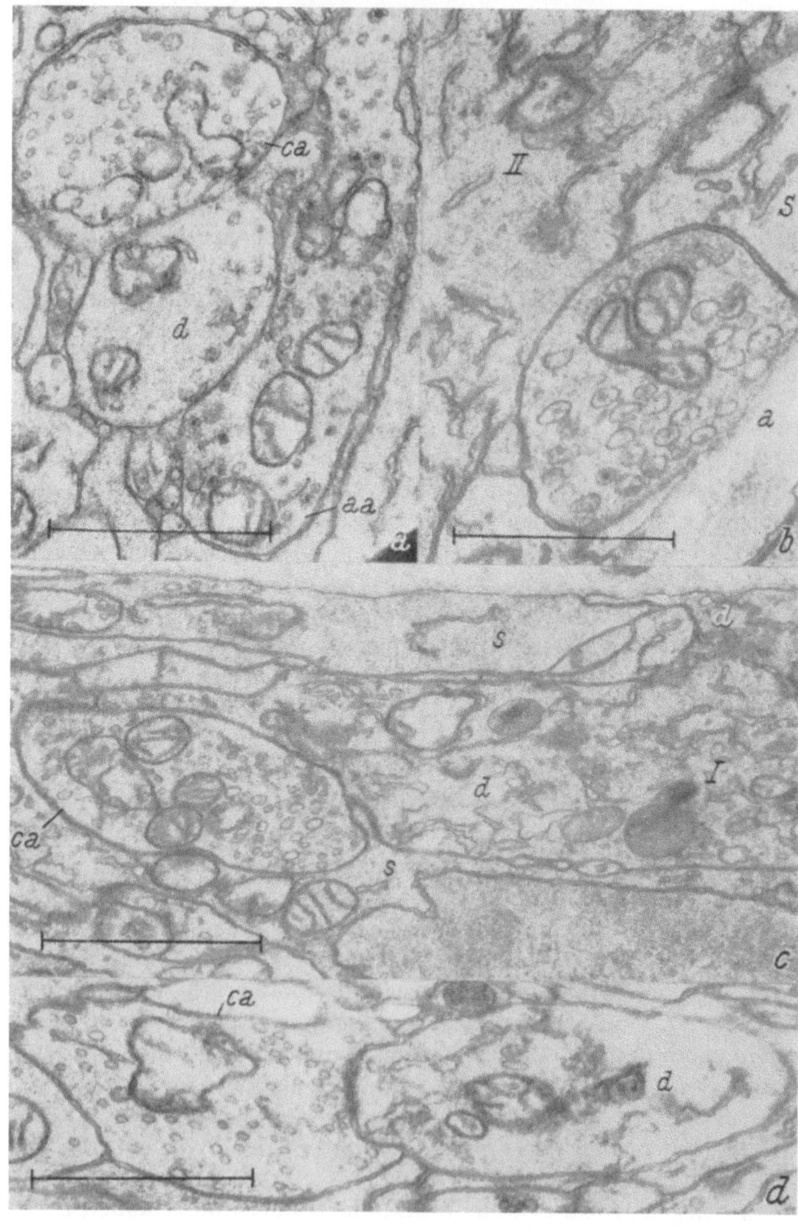

Abb. 4. Axo-somatische und axo-dendritische Synapsenstrukturen aus dem Plexus Auerbach des Duodenums von Ratte (a, b) und Goldhamster (c, d). Kaliumpermanganat nach *Hökfelt* und *Jonsson* (a—d)

wandneuronen (Abb. 4 a). Sie liegen vielmehr innerhalb der Faserstraßen zwischen den Nervenzellen und am Rande der Ganglien.

Prinzipiell muß die Auffassung (s. o.), daß auch sympathische (noradrenerge) Achsenzylinder die Ganglienzellen der Darmwand beeinflussen, auf Grund unserer licht- und elektronenmikroskopischen Befunde unterstützt werden. Eine ausschließliche oder bevorzugte Zuordnung von adrenergen oder cholinergen Synapsen zu den verschiedenen Zelltypen vermochten wir bisher nicht festzustellen. Im Gegensatz zu den Anschauungen von *Lawrentjew* (1931) sowie *Botar* und Mitarbeiter (1942) haben wir speziell an Typ-II-Zellen viele cholinerge Synapsen gefunden (Abb. 3 b, c), so daß eine Zuordnung der Typ-II-Zellen zum Sympathikus doch sehr bezweifelt werden muß.

Neben Axonen, die kleine granuläre oder agranuläre und zusätzlich einige große granulierte Vesikel enthalten, gibt es auch Axonauftreibungen, die nur mit großen granulierten Vesikeln gefüllt sind. Sie werden in den Plexus und auch in Kontakt zu Neuronen (Abb. 4 b) und Muskelzellen angetroffen. *Baumgarten* und Mitarbeiter (1970) haben aus derartigen Befunden die Existenz eines weiteren Ganglienzellentyps bzw. Transmitters abgeleitet (peptiderge Neurone).

2. *Neuro-muskuläre Beziehungen*

Jede der drei Muskelschichten ist von einem Nervengeflecht durchsetzt (Plexus muscularis superficialis, profundus und muscularis mucosae). Die Ringmuskulatur verfügt in der Regel über den dichtesten Plexus (Abb. 5 a). Es wurde allerdings auch festgestellt (*Stach*, 1969), daß z. B. bei der Katze die Längsmuskulatur des Dickdarmes über ein ebenso dichtes Geflecht verfügt. Sowohl licht- als auch elektronenmikroskopisch wurde beobachtet (*Honjin* und Mitarbeiter, 1965 b; *Taxi*, 1965; *Thaemert*, 1966; *Rogers* und *Burnstock*, 1966; *Pick*, 1967; *Stach*, 1973), daß die Geflechte vorwiegend aus prä- und terminalen Axonbündeln bestehen (Abb. 5, 6). Wir haben jedoch selbst in der Längsmuskulatur elektronenmikroskopisch vollkommen nackte Einzelaxone beobachtet.

Besonders häufig werden nackte Axone in enger Nachbarschaft zu den Stromamuskelzellen der Magen-Darm-Schleimhaut gefunden (Abb. 6 b, 9 b, c). Die synaptischen Spalten haben in Übereinstimmung mit den zitierten Autoren in der Regel eine Weite von 500—1000 Å, doch kommen auch wesentlich engere Beziehungen (um 200 Å) vor (Abb. 6 a, c). Multiaxonale Kontaktzonen (*Rogers* und *Burnstock*, 1966) kommen oft zur Darstellung (Abb. 5 d, 6 a). Viele der innervierten glatten Muskelzellen stehen in engstem Kontakt („nexus“ nach

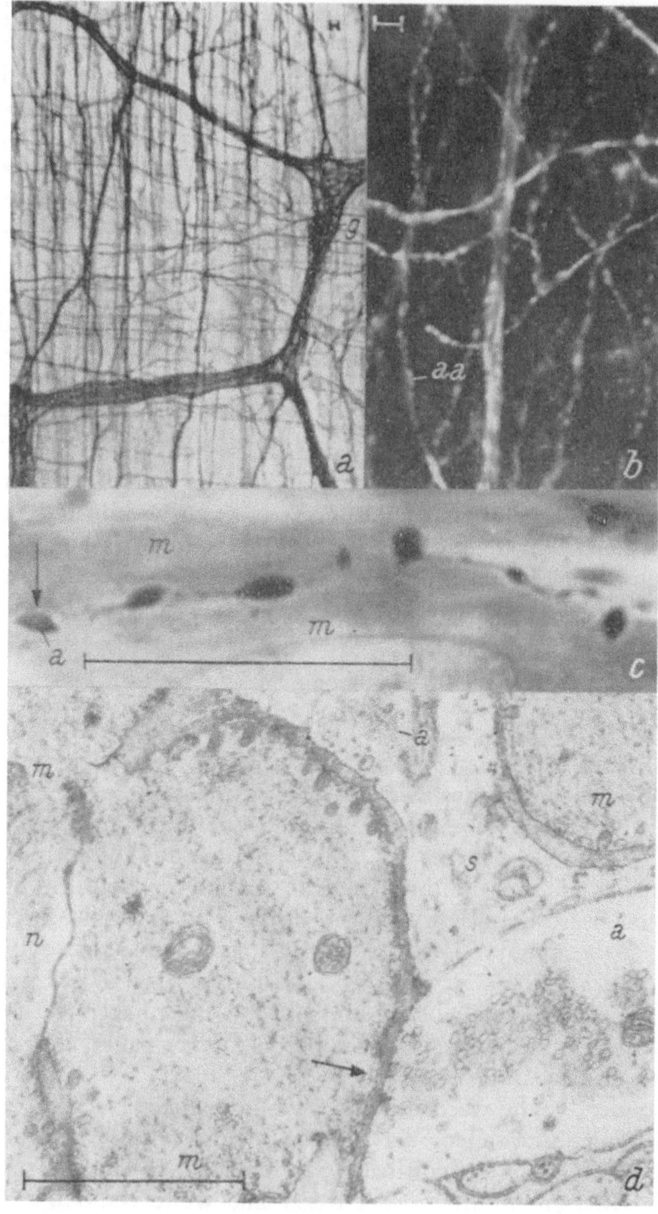

Abb. 5. Intramuskuläre Plexus und neuro-muskuläre Synapsenstrukturen. a, Kolon-Meerschweinchen; b, Duodenum-Meerschweinchen; c, Colon-Kaninchen; d, Duodenum-Ratte. Osmiumsäure-Zinkjodid nach *Maillet* (a, c), Paraformaldehydbedampfung nach *Falck* (b), Glutaraldehyd-Osmiumsäure (d)

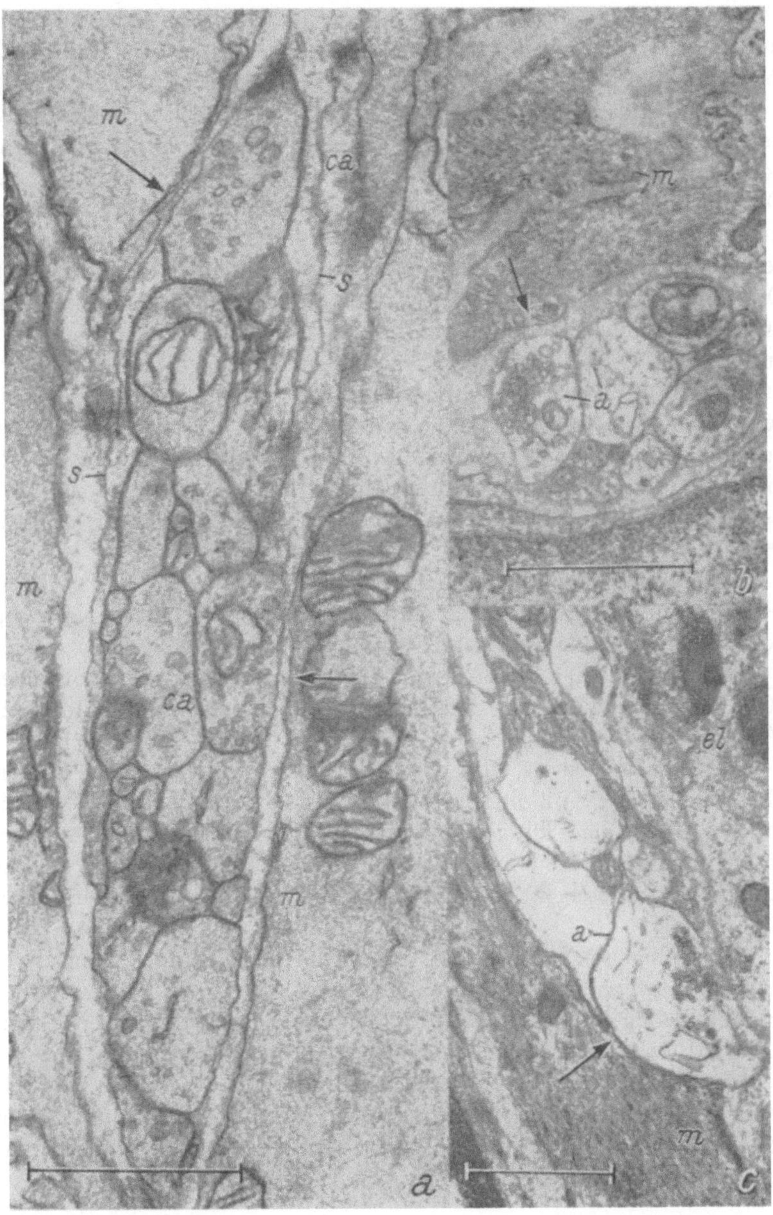

Abb. 6. Neuro-muskuläre Synapsenstrukturen. a, Ringmuskulatur-Duodenum-Ratte; b, c, Zottenstroma-Duodenum-Ratte. Kaliumpermanganat nach *Hökfelt* und *Jonsson* (a), Glutaraldehyd-Osmiumsäure (b, c)

Dewey und *Barr,* 1962) zu benachbarten Elementen (Abb. 5 d). Die funktionelle Deutung derartiger Beziehungen ist aber nicht einheitlich (*Sperelakis* und *Tarr,* 1965; *Bennett* und *Cobb,* 1969; *Bülbring,* 1970).

Histochemisch findet man (*Stach,* 1964; *Gunn,* 1971) in den Muskelschichten cholinesterasepositive Plexus und im Dickdarm mehr adrenerge Axone (*Gabella* und *Costa,* 1967; *Furness,* 1970) als im Dünndarm (Abb. 5 b). Neben den in Schichten und im Schleimhautstroma konzentrierten Muskelzellen, werden auch die Gefäßmuskelzellen, vor allem die der arteriellen Strombahn innerviert (Abb. 7). Die Innervation nimmt, bezogen auf die einzelne glatte Muskelzelle, mit abnehmendem Gefäßdurchmesser zu (*Stach,* 1969; *Stach* und Mitarbeiter, 1970). An Arteriolen, terminalen Arteriolen und Sphinkterkapillaren im Bereich des Verdauungstraktes kann jede glatte Muskelzelle als innerviert betrachtet werden (Abb. 7 b). Es handelt sich bei den Nerven, die die Gefäßmuskelzellen des Verdauungskanals innervieren, mit großer Wahrscheinlichkeit (s. *Gabella* und *Costa,* 1967; *Lever* und Mitarbeiter, 1968; *Furness,* 1970) ausschließlich um adrenerge Achsenzylinder.

Unsere, auf licht- und elektronenmikroskopischer Basis durchgeführten histochemischen Untersuchungen weisen ebenfalls eindeutig noradrenerge Axone nach (Abb. 7 a, c).

Die synaptischen Spalten mit einer Weite von 500 Å bis über 1000 Å werden von den Basalmembranen ausgefüllt bzw. begrenzt (Abb. 7 c, d).

An den mehrmals um das Gefäßlumen gewundenen glatten Muskelzellen der Sphinkterkapillaren sind jeweils mehrere Kontaktzonen ausgebildet. Die Sphinkterkapillaren finden sich besonders häufig in den inneren Schichten der Submucosa, aus denen sie zur Schleimhaut aufsteigen. Sie gehören zu den am stärksten innervierten Gefäßstrecken.

Wir haben im Bereich der Magen-Darm-Wand auch innervierte Kapillaren gefunden. Die Beziehungen von Axonen zu Perizyten und Endothelzellen sind in vielen Fällen eindeutiger (Spaltweiten um 500 Å) als zu den Muskelzellen der arteriellen Gefäße.

Letztlich soll erwähnt werden, daß nicht nur die glatten Muskelzellen der Blutgefäße, sondern auch die den kleinen Lymphgefäßen der Duodenalschleimhaut eindeutig zugeordneten Muskelzellen, innerviert sind (*Stach,* 1973).

3. *Neuro-epitheliale Beziehungen*

Hier sollen neben Beziehungen von Axonen zum Drüsen- und Zottenepithel auch die Verhältnisse in den Langerhansschen Inseln be-

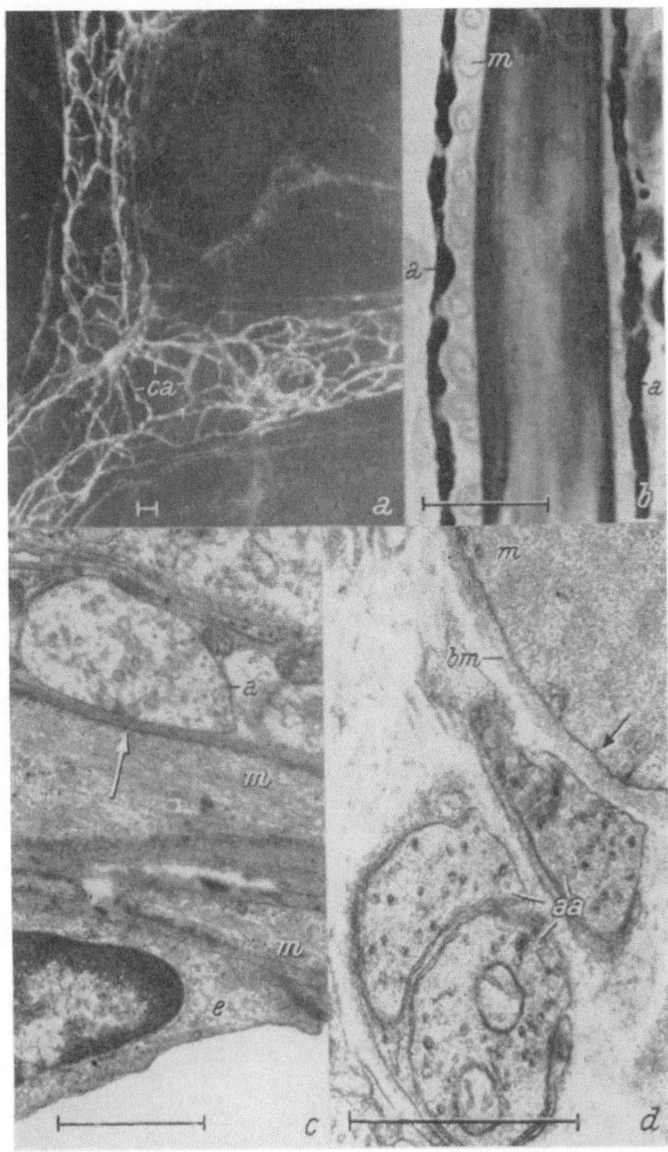

Abb. 7. Neurovegetative Plexus und neuro-muskuläre Synapsenstrukturen an arteriellen Gefäßen. a, Duodenum-Meerschweinchen; b, Kolon-Ratte; c, d, Duodenum-Ratte. Paraformaldehydbedampfung nach *Falck* (a), Osmiumsäure-Zinkjodid nach *Maillet* (b), Glutaraldehyd-Osmiumsäure (c), Kaliumpermanganat nach *Hökfelt* und *Jonsson* (d)

trachtet werden. Lichtmikroskopisch lassen sich in der Magen-Darmschleimhaut relativ dichte Nervengeflechte darstellen. Sie sind im Magen und Dünndarm (Abb. 8 a) stärker ausgebildet als im Dickdarm (Abb. 8 b). Mit die dichtesten Plexus enthalten die Duodenalzotten (*Stach*, 1972 c). Auch die Zuordnung der Axonbündel und Einzelaxone zum Drüsenepithel ist lichtmikroskopisch im Magen und Dünn-

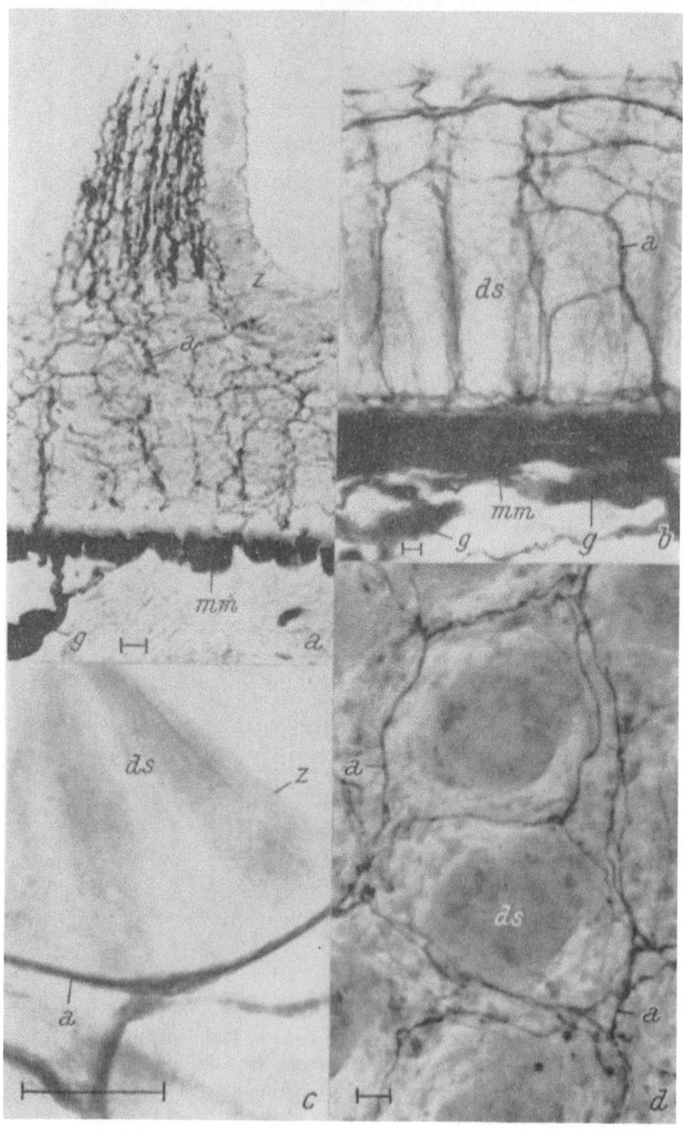

darm eindeutiger (Abb. 8 c) als im Dickdarm (Abb. 8 d). In Hunderten von Schnitten durch die Schleimhaut des Magen-Darm-Kanals haben wir bisher in keinem Fall Axone zwischen den Zylinderepithelzellen gefunden (*Stach*, 1969, 1973). Elektronenmikroskopische Untersuchungen (*Paley* und *Karlin*, 1959; *Pick*, 1967; *Ratzenhofer* und Mitarbeiter, 1969; *Stach*, 1973) stützen dieses negative Resultat. Es ist lediglich festzustellen, daß sich vesikuläre Axone der Epithelzellbasis nähern können. Häufig liegen noch Fortsätze von Muskel- und Bindegewebszellen dazwischen (Abb. 9 a). Die engsten (1000—500 Å) nur von Basalmembran gefüllten Spalten zwischen Axonen und basalen Abschnitten der Zylinderepithelzellen wurden von *Ratzenhofer* und Mitarbeiter (1969) sowie *Müller* und Mitarbeiter (1972) aus dem Magen und von *Stach* (1973) aus dem Duodenum demonstriert. Sie deuten auf eine efferente Innervation hin.

Wesentlich eindeutiger sind dagegen die neuro-epithelialen Beziehungen an den zum Darmkanal gehörenden Inselepithelzellen des Pankreas ausgeprägt. Obwohl bereits viele ältere Ergebnisse, unter anderem von *Castro* (1923), *Honjin* (1956) und *Coupland* (1958), darauf hindeuten, haben erst die elektronenmikroskopischen Untersuchungen der letzten Jahre, unter anderem von *Stahl* (1963), *Legg* (1967), *Watari* (1968), *Esterhuizen* und Mitarbeiter (1968), *Kobayashi* und *Fujita* (1969), *Klein* (1970) und *Stach* (1971), sichere Resultate erbracht.

Bereits lichtmikroskopisch ist sichtbar (Abb. 10 a), daß die Inseln über einen wesentlich stärkeren Innervationsapparat verfügen als die exokrinen Azini. Die elektronenmikroskopischen Bilder (Abb. 10 b, c) zeigen eindeutige Kontaktbeziehungen (Spaltweiten um 200 Å) von vesikulären Axonen zu A-, B- und D-Zellen. Sie sprechen nicht nur für eine bevorzugte Innervation des Inselapparates, sondern auch für eine eindeutige periphere neuro-hormonale Integration bei der Regulation von Stoffwechselvorgängen.

Die Qualität der Innervation stellt sich nach ultrahistochemischen und radiographischen Untersuchungen so dar, daß A- und B-Zellen sowohl cholinerg als auch adrenerg innerviert werden (*Esterhuizen* und Mitarbeiter, 1968). Eigene elektronenmikroskopische und ultrahistochemische (Abb. 10 d) Ergebnisse weisen denselben Tatbestand aus, so daß auch hier an einer spezifischen Zuordnung der Zelltypen zum sympathischen oder parasympathischen System (*Ferner*, 1952) gezweifelt werden muß.

Abb. 8. Plexus mucosus. a, Duodenum-Katze; b, Kolon-Katze; c, Duodenum-Meerschweinchen; d, Kolon-Ratte. Acetylcholinesterasereaktion nach *Gomori* (a, b), Osmiumsäure-Zinkjodid nach *Maillet* (c, d)

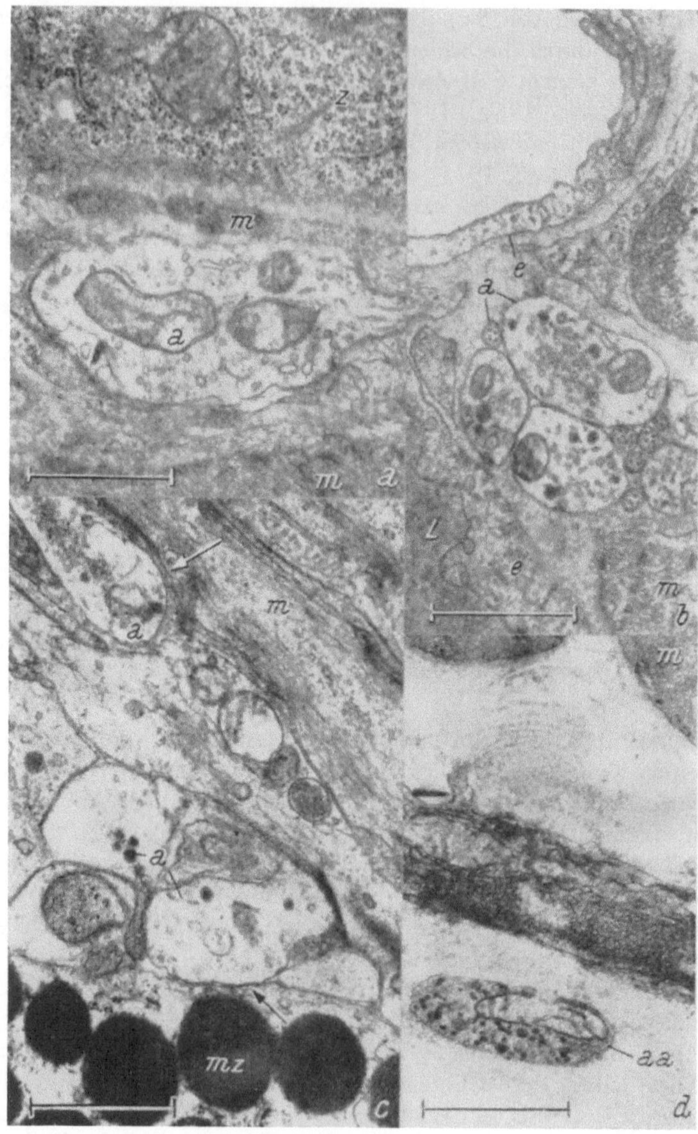

Abb. 9. Neuro-interstitielle Beziehungen im Schleimhaut- und Zottenstroma aus dem Duodenum der Ratte. Glutaraldehyd-Osmiumsäure (a, b, c), Kaliumpermanganat nach *Hökfelt* und *Jonsson* (d)

Abb. 10. Neuro-epitheliale Synapsenstrukturen aus Langerhanschen Inseln von Katze (a), Hund (b, c) und Goldhamster (d). Acetylcholinesterasereaktion nach *Gomori* (a), Glutaraldehyd-Osmiumsäure (b, c), Kaliumpermanganat nach *Hökfelt* und *Jonsson* (d)

4. *Neuro-interstitielle Beziehungen*

Es kann festgestellt werden, daß die Anzahl von partiell und vollständig aus den Schwannschen Zellen ausgefalteten Axonen, bezogen auf die Darmwand, von außen nach innen zunimmt. Besonders in den oberen Schleimhautschichten und in den Zotten werden häufig vollkommen nackte Axonbündel und Einzelaxone angetroffen, deren mit Vesikeln gefüllte Varikositäten unmittelbar an freie Bindegewebszellen bzw. den interstitiellen Raum grenzen (Abb. 9 b, c, d). An

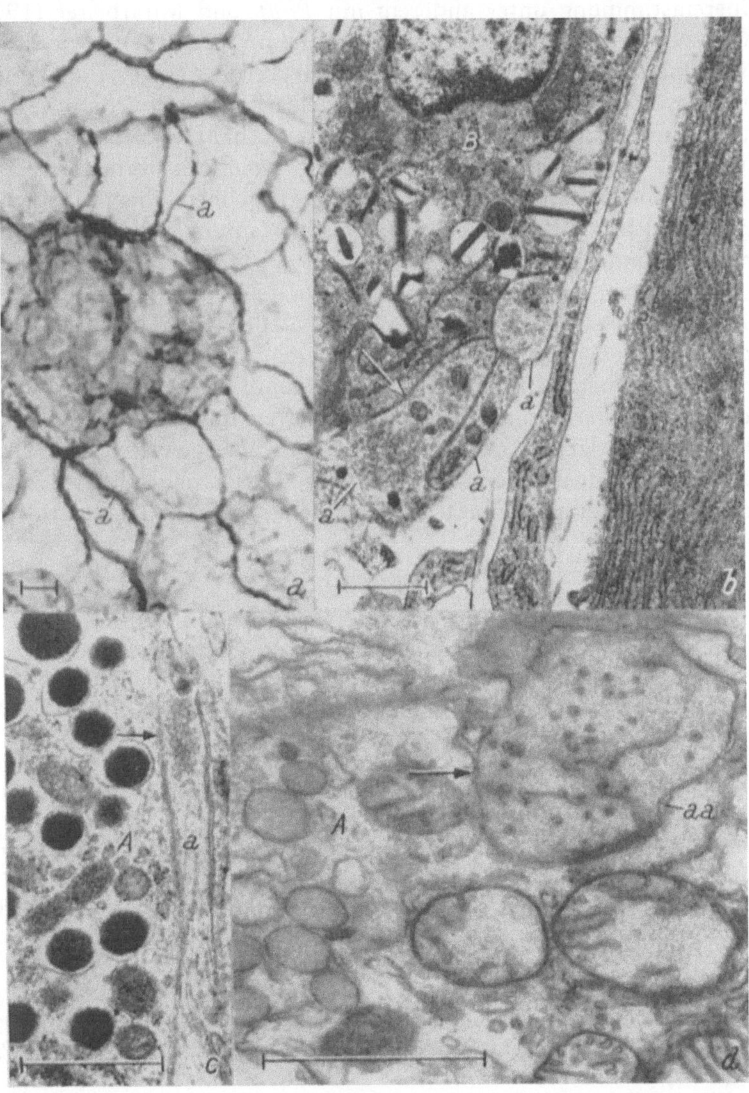

einigen Stellen ist sogar die Basalmembran stark reduziert oder fehlt vollkommen, z. B. im Kontaktbereich zu Muskel- und Bindegewebszellen (Abb. 6 c, 9 c), so daß die Axonmembran die einzige Grenzschicht darstellt. Neben nicht sicher differenzierbaren Bindegewebszellen, die häufig mit Achsenzylindern in Kontakt geraten, konnten auch sehr enge Beziehungen (Spaltweiten bis unter 200 Å) von Axonen zu Plasmazellen, Mastzellen und eosinophilen Leukozyten festgestellt werden. Insgesamt muß auf Grund derartiger Befunde in prinzipieller Übereinstimmung unter anderem mit *Papp* und Mitarbeiter (1962), *Ratzenhofer* und Mitarbeiter (1969), *Dermitzel* (1971) sowie *Müller* und *Ratzenhofer* (1972) angenommen werden (*Stach*, 1973), daß auch das Interstitium und damit die vielen Formen von Bindegewebszellen, insbesondere auch diejenigen, die hochaktive Substanzen produzieren, vom vegetativen Nervensystem beeinflußt werden.

Abschließend sollen die in der Vergangenheit aber auch gegenwärtig noch viel diskutierten Cajalschen interstitiellen Zellen Erwähnung finden.

Wir haben feststellen können (*Stach*, 1973), daß diese Elemente, die weder Nervenzellen, Schwannsche Zellen noch typische Bindegewebs- oder Muskelzellen sind, existieren und bevorzugt innerviert werden. In Übereinstimmung mit *Imaizumi* und *Hama* (1969) wurden nexus-artige Verbindungen von innervierten interstitiellen Zellen zu glatten Muskelzellen der Ringmuskulatur nachgewiesen. Aus ihrer bevorzugten Innervation kann auf eine spezifische, bisher allerdings noch nicht bekannte Funktion der interstitiellen Zellen geschlossen werden.

Literatur

Baumgarten, H. G., A.-F. Holstein und *Ch. Owman*: Auerbachs plexus of mammals and man: electron microscopic identification of three different types of neuronal processes in myenteric ganglia of the large intestine from rhesus monkeys, guinea pigs and man. Z. Zellforsch. *106*, 376—397 (1970).

Bennett, T., und *J. L. S. Cobb*: Studies on the avian gizzard: morphology and innervation of the smooth muscle. Z. Zellforsch. *96*, 173—185 (1969).

Botar, J., L. Battancs und *A. Becker*: Die Nervenzellen des Dünndarms. Anat. Anz. *93*, 138—149 (1942).

Bülbring, E.: The role of electrophysiology in the investigation of factors controlling intestinal motility. Rendic. R. Gastroenterol. *2*, 197—207 (1970).

Bullon, R. A.: Sobre la fina estructura del plexo de Auerbach del esofago, y sus relaciones con los conductores preganglionicos que tienen su origen en el nervio vago. Trab. Inst. Cajal Sec. fisiol. (Madrid) *37*, 215—258 (1945).

Bullon, R. A.: Contributión al conocimiento de la citoarquitectonia del plexo de Auerbach del recto. Trab. Laborat. Invest. biol. Univ. Madrid *36*, 253—272 (1947).

Castro, F. de: Contributión a la connaissance de l'innervation du pancréas. Trab. Labor. Rech. biol. Univ. Madrid *21*, 423—457 (1923).

Castro, F. de: Contributión al conocimiento de la innervación parasimpatica y simpatica del estomago. An. Acad. mac. med. (Madrid) *67*, 383—450 (1950).

Cauna, N.: The mode of termination of the sensory nerves and its significance. J. comp. Neurol. *113*, 169—210 (1959).

Costa, M., J.B. Furness und *G. Gabella*: Catecholamine containing nerve cells in the mammalian myenteric plexus. Histochemie *25*, 103—106 (1971).

Coupland, R. E.: The innervation of the pancreas of the rat, cat and rabbit as revealed by the cholinesterase technique. J. Anat. (London) *92*, 143—149 (1958).

Davenport, H. W.: Physiology of digestive tract, 2. Aufl. Chicago: Year Book Medical Publishers Inc. 1966.

Dermitzel, R.: Elektronenmikroskopische Untersuchungen über die Innervation der Pars pylorica des Mäusemagens. Z. mikrosk.-anat. Forsch. *84*, 225—256 (1971).

Dewey, M. M., und *L. Barr*: Intercellular connection between smooth muscle cells: the nexus. Science *137*, 670—672 (1962).

Dogiel, A. S.: Zwei Arten sympathischer Nervenzellen. Anat. Anz. *11*, 679—687 (1896).

Esterhuizen, A. C., T. L. B. Spriggs und *J. D. Lever*: Nature of islet-cell innervation in the cat pancreas. Diabetes *17*, 33—36 (1968).

Falck, B.: Observations on the possibilities of the cellular localization of monoamines by a fluorescence method. Acta physiol. Scand. *56*, Suppl. 197, 1—25 (1962).

Ferner, H.: Das Inselsystem des Pankreas. Stuttgart: Thieme. 1952.

Furness, J. B.: The origin and distribution of adrenergic nerve fibres in the guinea-pig colon. Histochemie *21*, 295—306 (1970).

Gabella, G., und *M. Costa*: Le fibre adrenergiche nel canale alimentare. Giorn. Accad. med. Torino *130*, 1—12 (1967).

Gomori, G.: Microscopic Histochemistry, Principles and Practice. Chicago: Univ. of Chicago Press. 1952.

Grillo, M. A.: Electron microscopy of sympathetic tissues. Pharmacol. Rev. *18*, 387—399 (1966).

Gunn, M.: Histological and histochemical observations on the myenteric and submucosal plexuses of mammals. J. Anat. (London) *102*, 223—239 (1968).

Gunn, M.: Cholinergic mechanisms in the gastrointestinal tract. J. Neurovisc. Relat. *32*, 224—240 (1971).

Hökfelt, T., und *G. Jonsson*: Studies on reaction and binding of monoamines after fixation and processing for electron microscopy with special reference to fixation with potassium permanganate. Histochemie *16*, 45—67 (1968).

Honjin, R.: The innervation of the pancreas of the mouse, with special reference to the structure of the peripheral extension of the vegetative nervous system. J. comp. Neurol. *104*, 331—371 (1956).

Honjin, R., *A. Takahashi* und *Y. Tasaki*: Electron microscopic studies of nerve endings in the mucous membrane of the human intestine. Okajimas Fol. anat. Jap. *40*, 409—427 (1965a).

Honjin, R., *A. Takahashi*, *S. Shimasaki* und *H. Maruyama*: Two types of synaptic nerve processes in the ganglia of Auerbach's plexus of mice, as revealed by electron microscopy (japanisch). J. Electronmicrosc. (Tokyo) *14*, 43—49 (1965b).

Iljina, W. J., und *B. J. Lawrentjew*: Zur Lehre von der Cytoarchitektonik des peripherischen autonomen Nervensystems. III. Ganglien des Rektums und ihre Beziehungen zu dem sakralen Parasympathicus. Z. mikrosk.-anat. Forsch. *30*, 530—542 (1932).

Imaizumi, M., und *K. Hama*: An electron microscopic study on the interstitial cells of the gizzard in the lovebird *(Uroloncha domestica)*. Z. Zellforsch. *97*, 351—357 (1969).

Jabonero, V.: Der anatomische Aufbau des peripheren neurovegetativen Systems. Acta neuroveg. (Wien), Suppl. 4. Wien: Springer. 1953.

Jabonero, V.: Studien über die Synapsen des peripheren vegetativen Nervensystems. Acta neuroveg. (Wien) *29*, 111—139 (1966).

Jacobowitz, D.: Histochemical studies of the autonomic innervation of the gut. J. Pharmacol. exp. Ther. *149*, 358—364 (1965).

Klein, C.: Ultrastructure des jonctions neuroglandulaires dans le pancreas endocrine d'un poisson teleostien, *Xiphophorus Helleri* H. C.R. Acad. Sci. (Paris) *271*, 1998—2000 (1970).

Kobayashi, Sh., und *Ts. Fujita*: Fine structure of mammalian and avian pancreatic islets with special reference to D cells and nervous elements. Z. Zellforsch. *100*, 340—363 (1969).

Lawrentjew, B. J.: Experimentell-morphologische Studien über den feineren Bau des autonomen Nervensystems. II. Über den Aufbau der Ganglien der Speiseröhre nebst einigen Bemerkungen über das Vorkommen und die Verteilung zweier Arten von Nervenzellen in dem autonomen Nervensystem. Z. mikrosk.-anat. Forsch. *18*, 233—262 (1929).

Lawrentjew, B. J.: Zur Lehre von der Zytoarchitektonik des peripherischen autonomen Nervensystems. I. Die Zytoarchitektonik der Ganglien des Verdauungskanals beim Hunde. Z. mikrosk.-anat. Forsch. *23*, 527—551 (1931).

Legg, P. G.: The fine structure and innervation of the beta and delta cells in the islet of Langerhans of the cat. Z. Zellforsch. *80*, 308—321 (1967).

Lever, J. D., *T. L. B. Spriggs* und *J. D. P. Graham*: A formol-fluorescence, fine structural and autoradiographic study of the vascular tree in the intact and sympathectomized pancreas of the cat. J. Anat. (London) *103*, 15—34 (1968).

Maillet, M.: Modifications de la technique de *Champy* au tetraoxyde d'osmium-iodure de potassium. Résultats de son application à l'étude des fibres nerveuses. Compt. rend. Soc. biol. (Paris) *153*, 939—941 (1959).

Marks, B. H., T. Samorajshi und *E. J. Webster*: Radioautographic localization of norepinephrine — H^3 in the tissues of mice. J. Pharmacol. exp. Ther. *138*, 376—381 (1962).

Milochin, A. A.: Die sensible Innervation der vegetativen Neuronen (Neue Vorstellungen über die Organisation des vegetativen Ganglions). (Russisch.) Leningrad: Nauka. 1967.

Morrison, A. R., und *R. E. Habel*: A quantitative study of the distribution of vagal nerve endings in the myenteric plexus of the ruminant stomach. J. comp. Neurol. *122*, 297—309 (1964).

Müller, O., und *M. Ratzenhofer*: Elektronenmikroskopische Untersuchungen der Lamina propria der Schleimhaut des Kaninchenmagens (Fundusteil). Z. mikrosk.-anat. Forsch. *85*, 1—22 (1972).

Norberg, K.-A.: Adrenergic innervation of the intestinal wall studied by fluorescence microscopy. Int. J. Neuropharm. *3*, 379—382 (1964).

Norberg, K.-A.: Transmitter histochemistry of the sympathetic adrenergic nervous system. Brain Res. *5*, 125—170 (1967).

Ono, M.: Electron microscopic observations on the ganglia of Auerbach's plexus and autonomic nerve endings in muscularis externa of mouse small intestine. Sapporo Med. J. *32*, 56—74 (1967).

Palay, S. L., und *L. J. Karlin*: An electron microscope study of the intestinal villus. I. The fasting animal. J. biophys. biochem. Cytol. *5*, 363—371 (1959).

Papp, M., P. Röhlich, I. Rusznyak und *I. Törö*: An electron microscopic study of the central *lacteal* in the intestinal villus of the cat. Z. Zellforsch. *57*, 475—486 (1962).

Pick, J.: Fine structure of nerve terminals in the human gut. Anat. Rec. *159*, 131—146 (1967).

Ratzenhofer, M., O. Müller und *H. Becker*: Zur Innervation der Drüsen- und Stromazellen im Kaninchenmagen. Mikroskopie (Wien) *25*, 283—296 (1969).

Richardson, K. C.: Electronmicroscopic observations on Auerbach's plexus in the rabbit, with special reference to the problem of smooth muscle innervation. Amer. J. Anat. *103*, 99—135 (1958).

Rogers, D. C., und *G. Burnstock*: Multiaxonal autonomic junctions in intestinal smooth muscle of the toad *(Bufo marinus)*. J. comp. Neurol. *126*, 625—652 (1966).

Schofield, G. C.: Experimental studies on the myenteric plexus in mammals. J. compl. Neurol. *119*, 159—185 (1962).

Sperelakis, N., und *M. Tarr*: Weak electrotonic interaction between neighboring visceral smooth muscle cells. Amer. J. Physiol. *208*, 737—747 (1965).

Stach, W.: Gibt es Methoden zur morphologischen Differenzierung adrenerger, cholinerger und histaminerger Nervenfasern? Acta histochem. (Jena) *18*, 377—383 (1964).

Stach, W.: Neurohistologische Untersuchungen an den Nervengeflechten der Dickdarmwand. Ein Beitrag zur Innervation des Magen-Darm-Kanals. Habil.-Schrift, Rostock (1969).

Stach, W.: Über die in der Dickdarmwand aszendierenden Nerven des Plexus pelvinus und die Grenze der vagalen und sakralparasympathischen Innervation. Z. mikrosk.-anat. Forsch. *84*, 65—90 (1971).

Stach, W.: Licht- und elektronenmikroskopische Untersuchungen zur Innervation der Bauchspeicheldrüse. In: IV. Society of Anatomists, Histologists and Embryologists in Bulgaria, Varna, 12.—16. 9. 1971. Summaries of reports, p. 25.

Stach, W.: Der Plexus entericus extremus des Dickdarmes und seine Beziehungen zu den interstitiellen Zellen (Cajal). Z. mikrosk.-anat. Forsch. *85*, 1—28 (1972a).

Stach, W.: Light and electron-microscopic investigations into the innervation of the musculature of the colon. Folia morph. (Warszawa). 1972b. (Im Druck.)

Stach, W.: Über die Nervengeflechte der Duodenalzotten. Licht- und elektronenmikroskopische Untersuchungen. Acta anat. (Basel) *85*, 216—231 (1973).

Stach, W., D. Gruber und *G. H. Schumacher*: Zur Innervation der Mundhöhle und des Schlundes. V. Mitteilung: Die Innervation der Blutgefäße. Dtsch. Zahn-, Mund- u. Kieferheilk. *55*, 120—126 (1970).

Stahl, M.: Elektronenmikroskopische Untersuchungen über die vegetative Innervation der Bauchspeicheldrüse. Z. mikrosk.-anat. Forsch. *70*, 62—102 (1963).

Taxi, J.: Contribution à l'étude des connexions des neurones moteurs du système nerveux autonome. Ann. Sci. Nat. Zool., 12 Ser. *7*, 413—674 (1965).

Taxi, J., und *B. Droz*: Localisation d'amines biogènes dans le système neurovegetatif périphérie. (Étude radioautographique en microscopie electronique après injection de noradrénaline — ^{3}H et de 5-hydroxytryptophane — ^{3}H.) In: Neurosecretion, IV. Internat. Symp. Neurosecr., Strasbourg (1966) (*Stutinsky, F.*, Hrsg.), S. 191—202. Berlin-Heidelberg-New York: Springer. 1967.

Thaemert, J. C.: Ultrastructural interrelationships of nerve processes and smooth muscle cells in three dimensions. J. Cell. Biol. *28*, 37—49 (1966).

Temesrèkàsi, D.: Die Synaptologie der Dünndarmgeflechte. Acta morphol. hung. *5*, 53—69 (1955).

Watari, N.: Fine structure of nervous elements in the pancreas of some vertebrates. Z. Zellforsch. *85*, 291—314 (1968).

Anschrift des Verfassers: Doz. Dr. sc. med. *W. Stach*, Anatomisches Institut der Universität Rostock, Gertrudenstraße 9, DDR-25 Rostock, Deutsche Demokratische Republik.

Discussion

Akert: My congratulations on your successful application of the zinciodideosmic acid method of Maillet. It would be useful to extend these studies to the electron microscopic level in order to gain information regarding shape, number and distribution of synaptic vesicles and the configuration of synaptic membrane complexes.

Lassmann: In my opinion the structures shown in the oesophagus are receptors.

Stach: This question cannot be answered at the time being. Whether the P-fibres are more numerous in lower animals, as suggested by Lassmann, cannot be decided, so far. We could demonstrate such fibres by means of electron microscopy in laboratory mammals.

Lassmann: Is there any evidence for a serotonergic synapse between axons of the Meissner's ganglion cells and the Auerbach plexus?

Stach: I could not observe any serotonergic neurones in the intestinal plexuses of mammals by means of the fluorescence technique.

Szentágothai: In my opinion, Dogiel II cells represent a very primitive type of neurones, similar to those in the diffuse nerve nets of lower invertebrates. Such cells are neither motor nor sensory in the strict sense of the word: their processes transmit impulses according to the momentary demands of the organism. One is inclined to believe that interstitial cells belong to an even more primitive "neuronoid" type, providing a contact between the true nervous elements and smooth muscle cells.

Stach: I should like to remind of the early Dogiel studies (1896) in which Type II cells were tentatively characterized as sensory cells, but, later (1898), also a secretomotor function was ascribed to them. In fact, there are two basic observations relevant in this respect: first, axons of Type I cells can be followed for many centimeters in Auerbach's or in Meissner's plexus and, second, axons of Type II cells can be traced only to the preterminal parts of the intramuscular plexus. Accordingly, the Jabonero (1953) view can still be sustained, *viz.*, Type I cells are associative in character, whereas Type II cells innervate the smooth musculature. The idea that Type II cells are primitive indifferent neurones is a very remarkable one since it permits a synthesis of earlier interpretations. On the other hand, the old story of the interstitial cells of Cajal is again revived. These cells are neither neurones, nor Schwann cells, nor typical connective tissue cells, nor muscle cells. The fact that close relationships ("nexuses") exist between interstitial cells and smooth muscle suggests a specific function.

Ungváry: I want to comment on the innervation of the islets of Langerhans (Fig. 10). Are there also any connections between axon terminals and exocrine pancreas cells?

Stach: I recall only a few synaptic structures on exocrine pancreatic cells — much less than on endocrine islet cells. Moreover, the synaptic gap is quite wide (∼1000 Å) in the exocrine pancreas.

Ungváry: Is it possible to ascribe a functional effect to an axon terminal upon an endocrine cell? An alternative possibility would be that the axon is related to the wall of a blood vessel, which is out of sight in an ultrathin section.

Stach: The synaptic innervation of endocrine (islet) cells is beyond doubt, since the contact between the nerve terminal and the endocrine cell is in the range of only 200 Å.

Journal of Neural Transmission, Suppl. XI, 103—124 (1974)

Dynamics of Transmissional Ultrastructures in Sympathetic Neurones of the Rat

J. Taxi

Laboratoire de Biologie animale, Université de Paris, Paris, France

With 15 Figures

Summary

The ultrastructural features of the segment proximal to a ligature of noradrenergic, sympathetic nerve fibres were studied in the sciatic nerve of the rat. The radioautographic method was used to identify these fibres on the basis of their property of specific uptake of ^{3}H-noradrenaline (NA) or an ^{3}H-precursor. It was observed that the labeled fibres exhibit various ultrastructural characteristics, in such a way that the storage of NA cannot be related to one definite type of cytoplasmic organelle.

The accumulation of ^{3}H-NA above a ligature as the result of a migration phenomenon from the perikaryon was tested for a period of 22 h: it appeared extremely low, negligible in normal conditions. Thus the migration cannot be accounted for supplying NA to nerve terminals. The differences in storage properties between the perikarya and the nerve terminals were emphasized. A local function is proposed for the small granulated vesicles of the perikaryon. An hypothesis was set up, according which the accumulation of NA proximal to a ligature may be understood as the result of local assembling of storage organelles from material, especially proteins manufactered in the perikaryon and stopped in their migration down to terminals.

In 1965, *Dahlström*, using the specific fluorescence method of Falck and Hillarp for the detection of catecholamines, described an accumulation of noradrenaline (NA) proximal to a ligature in certain fibres of the sciatic nerve of the rat, and set up a hypothesis according to which this accumulation is due to the migration of NA from the perikaryon to the periphery by a phenomenon of axonal transport, comparable to the axonal flow of *Weiss* and *Hiscoe* (1948), recently reviewed by *Weiss* (1969). After a serie of important works, mainly with biochemical techniques, *Dahlström* (1970), *Dahlström* and *Häggendal* (1970), *Häggendal* and *Dahlström* (1971) give a prominent role in the accumulation of NA to the flow of synthesis and/or storage structures.

Kappeler and *Mayor* (1967, 1969), *Geffen* and *Ostberg* (1969) have studied the ligated splenic nerves of the cat with the fluorescence method and the electron microscope. They described collections of numerous vesicles, 600—1000 Å in diameter, containing a dense core, in the region of the nerve fibres just proximal to the constriction. A large part of these vesicles lose their dense core under the action of reserpine. These authors considered that NA accumulation is the consequence of axonal transport of the vesicles.

On the other hand, *Geffen* and *Rush* (1968), and *Geffen et al.* (1969) emphasized that the life time of NA molecules is too short for allowing their transport in a vesicular pool from the perikaryon to the terminals. They suggested that NA accumulation is only the local consequence of the migration of dopamine-β-hydroxylase, which was first demonstrated by *Laduron* and *Belpaire* (1968), and that of storage protein as proved by immunohistochemistry (*Livett et al.*, 1968; *Geffen et al.*, 1969).

In this situation, and in spite of the limitations of the method (*Taxi* and *Droz*, 1969), *Sotelo* and I (1971) applied radioautography in electron microscopy after injection of ^{3}H-NA to one of the most studied materials: the sciatic nerve of the rat, in order to gain new information on the real meaning of the accumulation of NA proximal to a ligature, on the ultrastructure of nerve fibres able to store ^{3}H-NA, and on the question of the relationship of the perikaryon and the terminals regarding the storage of NA.

Before giving an account of our results, it is necessary to remind of our terminology. As has been well documented by *Richardson* (1962, 1964), *Wolfe et al.* (1962), *Taxi* and *Droz* (1966 a, b), *Van Orden et al.* (1966) and many others sympathetic axons fixed by osmium tetroxide or glutaraldehyde followed by osmium tetroxide

---

Fig. 1. Proximal segment of a rat sciatic nerve 22 h after ligation and IMAO injection. Noradrenaline-^{3}H was injected 30 min before fixation by osmium tetroxide

In an unmyelinated nerve fibre, composed of several axons sheathed in the same Schwann cell (S), only one axon is labeled (×21,200)

Fig. 2. Same nerve as in Fig. 1. The labeled fibre contains empty and granulated vesicles of various sizes (×26,800)

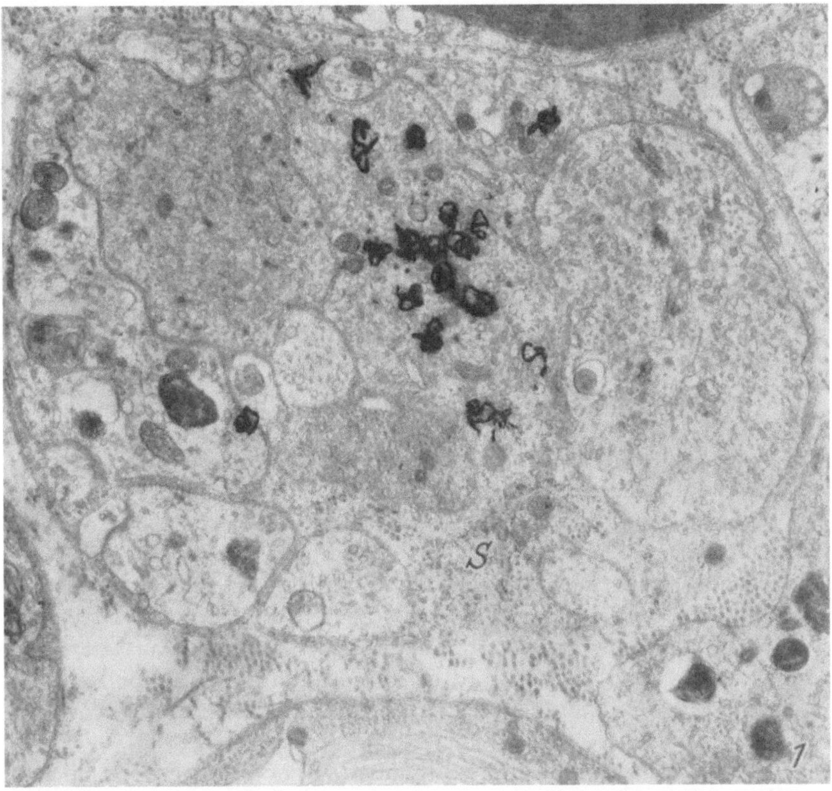

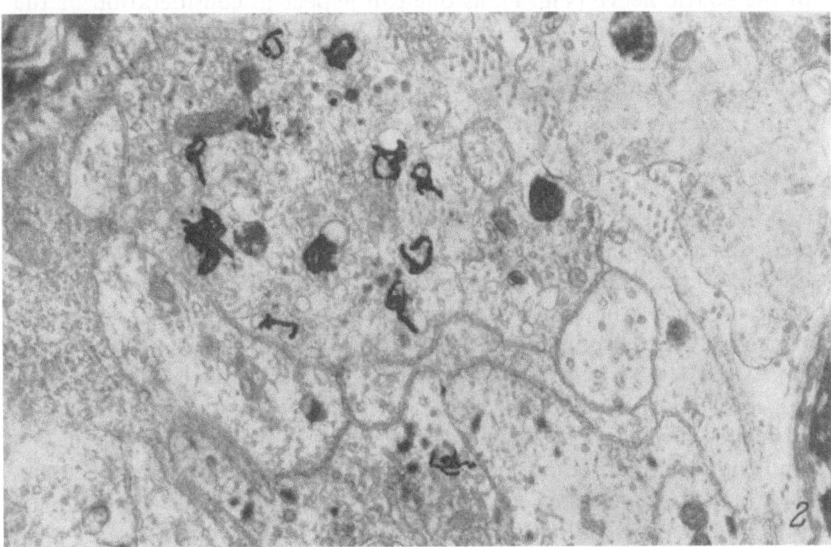
2

contain in their peripheral part numerous vesicles similar to synaptic vesicles, measuring 450 Å in diameter on an average, part of which is provided with a dense granule. The proportion of these "small granulated vesicles" (SGV) is largely increased after fixation by 3 % potassium permanganate (*Richardson*, 1966) or other fixation procedure (*Tranzer* and *Thoenen*, 1967). They are depleted by reserpine, like NA itself (*Pellegrino de Iraldi* and *De Robertis*, 1963) and appear the best morphological criterion for identifying normal sympathetic (noradrenergic) axons. The sympathetic terminals contain also a few larger, granulated vesicles (LGV), 800 Å in diameter, which do not appear modified by reserpine (*Taxi*, 1965).

Ultrastructure of Ligated Fibres Able to Store NA

A first type of experiments were carried out in view of recognizing noradrenergic fibres in the part of the sciatic nerve of the rat proximal to a constriction, and of establishing the ultrastructural features of these fibres.

In those experiments a ligature was made on the sciatic nerve. After an appointed delay, always 3 or 22 h, the rat received intravenously 5 mCi of ^{3}H-NA or ^{3}H-precursor. Half an hour later, the region of the ligature was fixed in 2 % osmium tetroxide buffered in phosphate 0.2 M and then prepared for radioautography according to the procedure described elsewhere (*Sotelo* and *Taxi*, 1973).

In such conditions, a limited number of nerve fibres were labeled in the sciatic nerve (Fig. 1), as one can expect in consideration of the few sympathetic fibres contained in this nerve. This attests the specificity of the labeling, already observed by *Lever et al.* (1970) in *in vitro* experiments. As a rule, the labeled fibres are enlarged, dilated, no matter the span of ligation. Of course the increase in diameter is progressive, although there are large individual variations. Within 24 h, numerous fibres reach a diameter 5 to 10 times the normal one, which is usually about 1 micron. It would be quite impossible to recognize the sympathetic fibres among all the enlarged, unmyelinated fibres without the labeling by radioautography, the situation being complicated by the fact that numerous myelinated fibres have lost or are losing their myelin sheath as soon as 24 h after ligation.

As the ultrastructure is concerned, the most typical feature of the enlarged labeled axons is the multitude of vesicular or tubular profiles contained by them (Figs. 1—4). As the microtubules and neurofilaments disappear almost completely from modified fibres proximal to a ligature, *Pellegrino de Iraldi* and *De Robertis* (1968) assumed that microtubules in these fibres are transformed *in situ* into vesicules

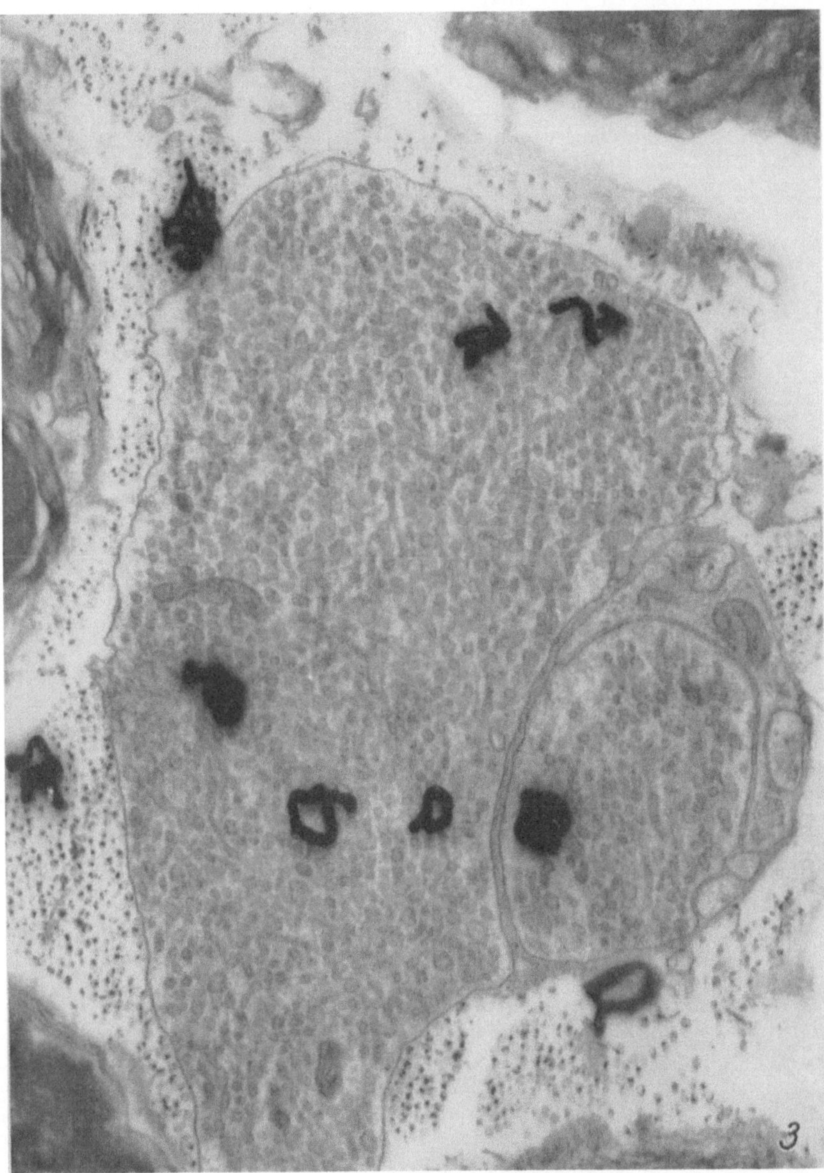

Fig. 3. Same experimental conditions as in Fig. 1, but fixation occurred only 3 h after ligation. The labeled axon contains almost exclusively empty vesicles or tubular profiles (×30,500)

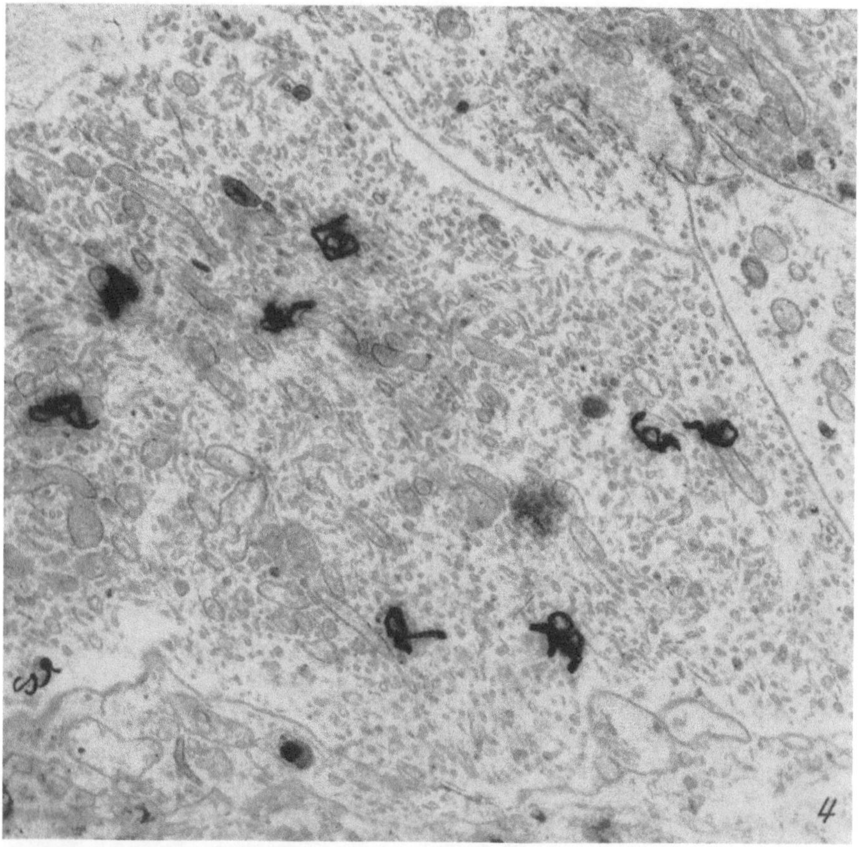

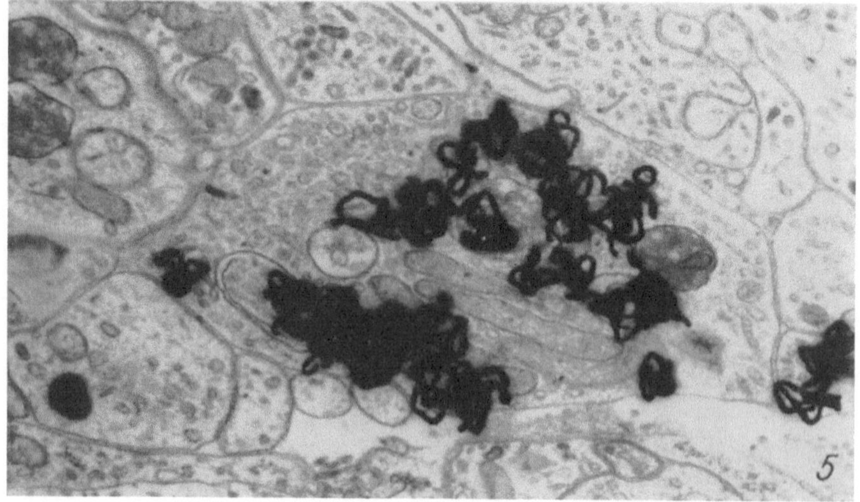
5

and tubules. These organelles are, however, bounded by an unit-membrane. For this reason they seem more related to the endoplasmic reticulum, which is always present in unmyelinated nerve fibres and perhaps able to grow rapidly in abnormal circumstances. In each fibre section, the numbers of tubules and of vesicles are in inverse relation, suggesting that vesicles may be formed from tubules by pinching off.

The ultrastructural features of labeled fibres are largely variable, and often not distinct from those of the numerous unlabeled ones. They may contain more or less vesicles, while sometimes tubules are numerous (Figs. 4, 7). When the vesicles are predominant they may appear either empty (Fig. 3), or a, partly, as LGV, rarely SGV (Fig. 2). Besides the vesicular and tubular formations, the labeled fibres may contain also lysosomes or sometimes mitochondria (Fig. 5), the labeling being overimposed on these organelles. In this case, the binding of ^{3}H-NA must be suspected to be an unspecific one at the level of the organelles, no matter the specificity of the uptake at the level of the plasmic membrane.

As it is well known that osmium tetroxide fails to preserve many SGV (*Richardson*, 1966), attempts were made to use radioautography after 3 % MnO_4K fixation according to *Hökfelt* (1968). Although it was established that a large part of ^{3}H-NA is lost during this type of fixation (*Devine* and *Laverty*, 1968; *Taxi*, 1968), a degree of labeling sufficient to identify the noradrenergic fibres was obtained. As expected, the number of granulated vesicles, especially SGV, is largely increased in certain labeled fibres after this fixation (Fig. 6). Other labeled fibres are, however, practically devoid of SGV or LGV, and thus the variety of the morphology of ligated noradrenergic fibres is confirmed, a part of them containing only empty vesicular or tubular profiles (cp. Fig. 6 and 7).

Moreover, controls made on the piece of nerve proximal to a crush of the cholinergic, preganglionic trunk of the superior cervical ganglion pointed out that SGV are also present in certain fibres. Thus, SGV, which must be considered as specific organelles of normal noradrenergic fibres, may have a different significance in fibres modified by a ligature, as far as the presence of some noradrenergic fibres in the preganglionic trunk could be totally precluded.

Fig. 4. Same experimental conditions as in Fig. 1, but without treatment by IMAO. Tubular profiles are the predominant organelles in the labeled fibre (×16,200)

Fig. 5. Sciatic nerve. Same experimental conditions as in Fig. 1, but dopamine-^{3}H was used instead of noradrenaline-^{3}H. The silver grains are superimposed on lysosomes and mitochondria (×29,000)

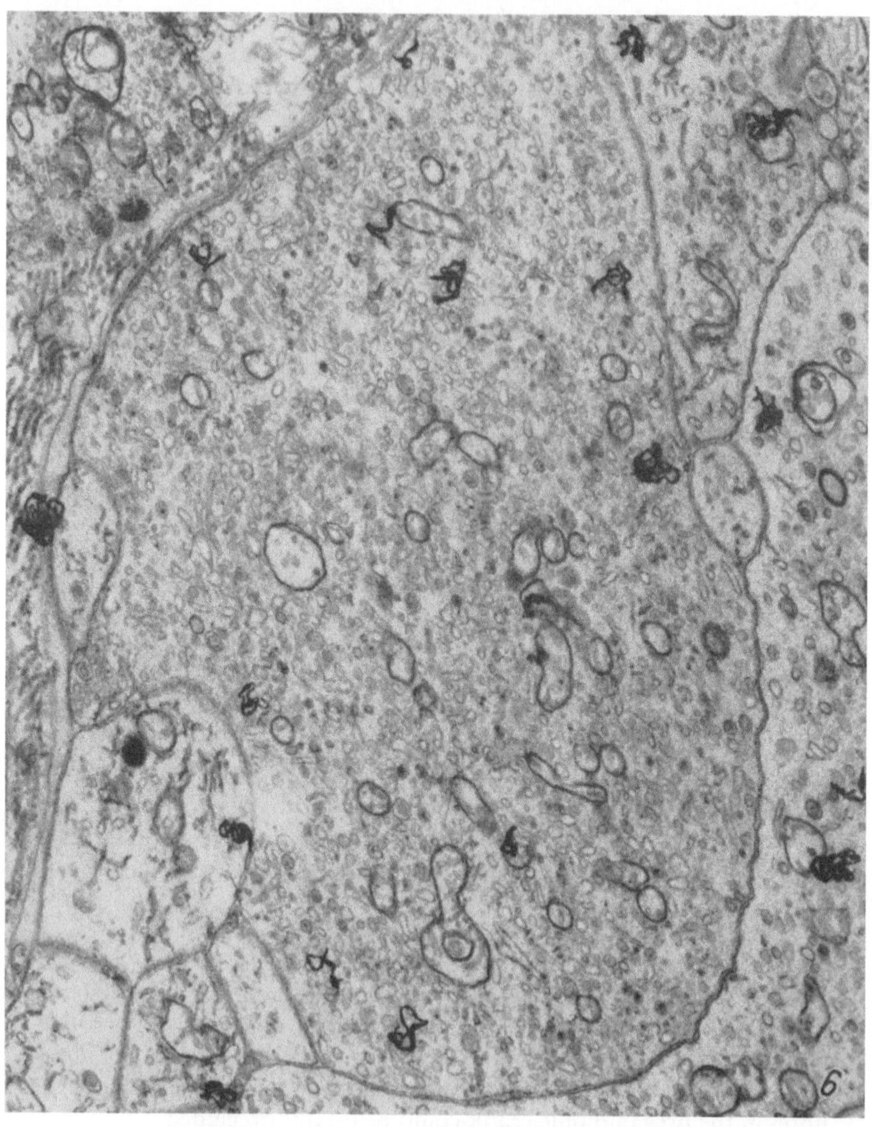

Fig. 6. Same conditions as in Fig. 1, but fixation was performed with 3% MnO_4K. The labeled fibre contains numerous granulated vesicles of various size (×20,050)

Fig. 7. Same nerve as in Fig. 6. In this case, the labeled fibre contains a large majority of empty vesicular or tubular profiles (×25,000)

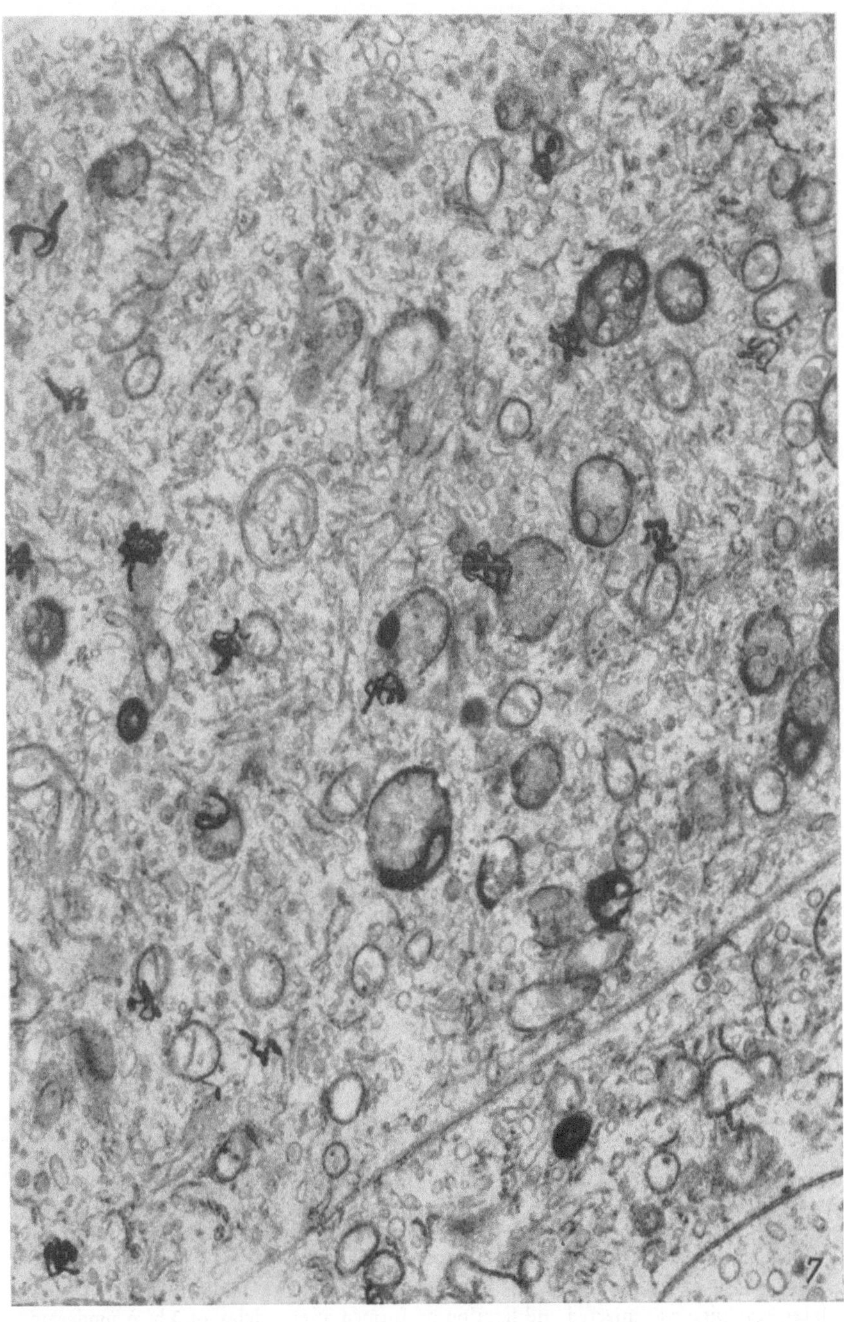
7

Effect of Drugs

The effect of a monoamine-oxydase inhibitor (IMAO) and that of reserpine on the labeling were tested. As IMAO, Catron (= PIH, JB 516, pheniprazine), 5 mg/kg i.p. injected or niamide (= nialamide), 200 mg/kg i. p. injected 30 min before the beginning of the experiment, were used. Comparison of precise quantitative data from experiments bearing on several animals appears beyond the possibilities of the method, due to the numerous factors which are very difficult to control (see *Taxi* and *Droz*, 1969). Thus, only rough estimations were made. In the presence of an IMAO, the labeling proved to be slightly increased, but it clearly appeared that the type of labeling is similar to that of the terminals, and not to that of the perikarya, which can be obtained with a significant density only in presence of an IMAO (*Taxi* and *Droz*, 1969).

Reserpine was i.p. injected at a dose of 5 mg/kg. In one type of experiment, reserpine was injected when the ligature has just been made; ^{3}H-NA was given 30 min before the time scheduled for removal and fixation of the ligated nerve was performed either 3$^{1}/_{2}$ or 22 h after making the ligature. In both cases the labeling was reduced, but not suppressed (Fig. 8). This result demonstrates a kind of reserpine-resistant uptake and storage in the ligated nerve. This very unusual type of storage might be due to unspecific binding phenomena, related to lysosomes or lysosomal components.

In another type of experiment, NA-^{3}H was administered to a rat, the sciatic of which had been ligated 19 h before. Reserpine was given 30 min later and fixation performed after a delay of 3 h, sufficient to obtain a large depletion of noradrenaline stores according to *Pellegrino de Iraldi* and *De Robertis* (1963). The labeling obtained was, indeed, poor. It is noticeable that LGV remain rather numerous in these fibres (Fig. 9).

If the same experiments were carried out on rats pretreated by an IMAO, the antagonistic action of reserpine and IMAO, already established biochemically (*Potter* and *Axelrod,* 1963) results in a nearly normal labeling.

Fig. 8. Sciatic nerve 20 h after ligation and injection of reserpine. Noradrenaline-^{3}H was injected 30 min before fixation. A moderate labeling is obtained on a fibre containing some LGV (×21,200)

Fig. 9. Sciatic nerve. Noradrenaline-^{3}H was administered 20 h after ligation; 30 min later, reserpine was injected and fixation performed after a delay of 3 h. A moderate labeling is obtained. Some LGV remain in the labeled fibre (×13,700)

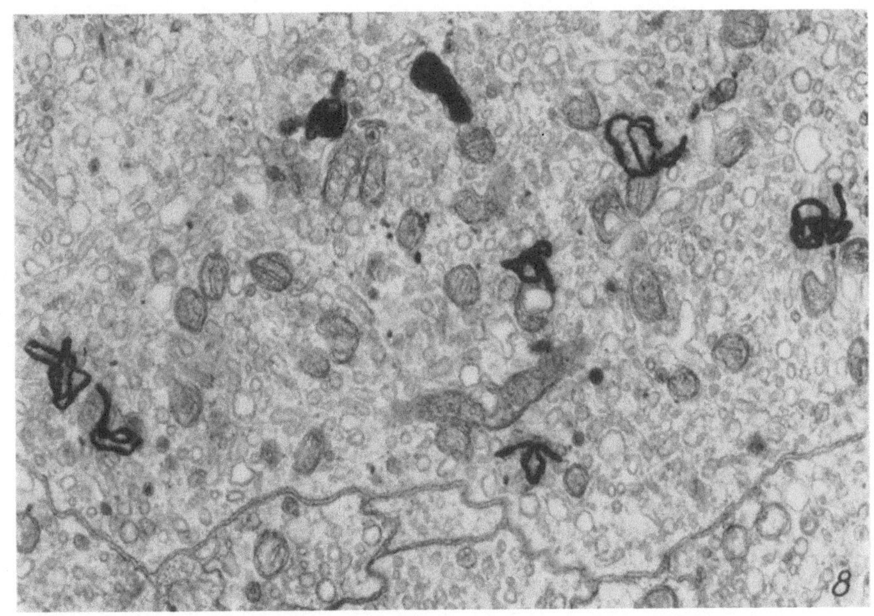
8

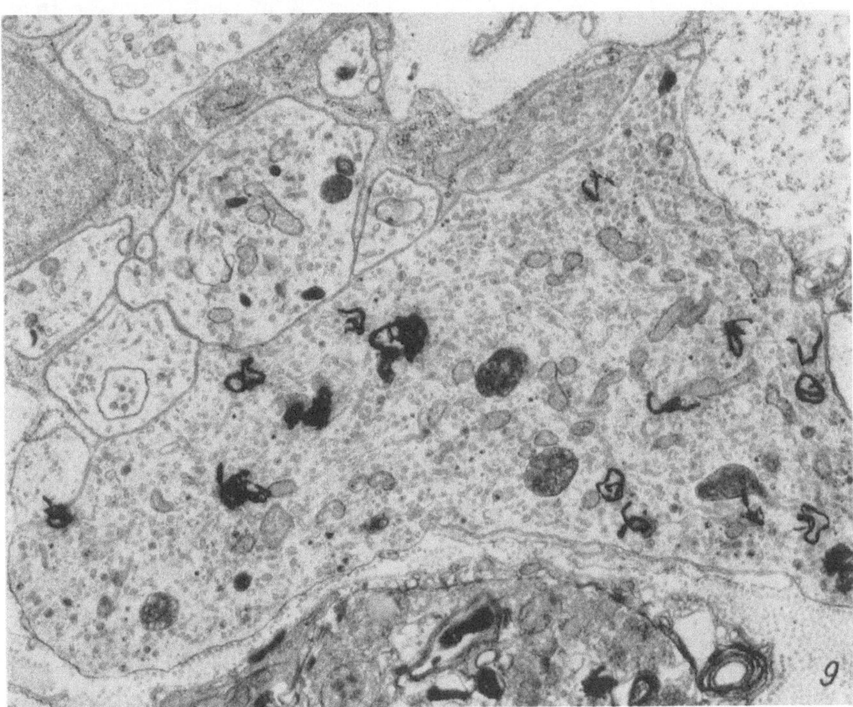
9

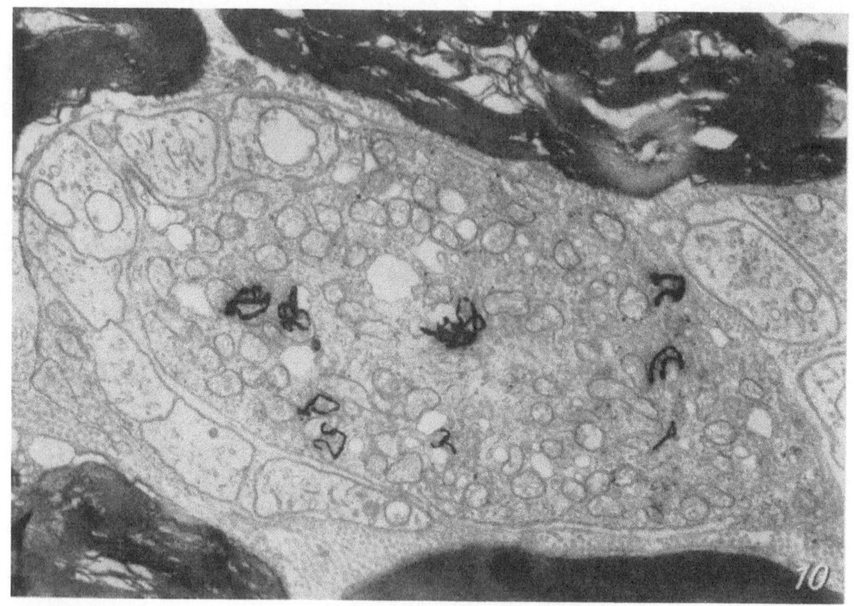
10

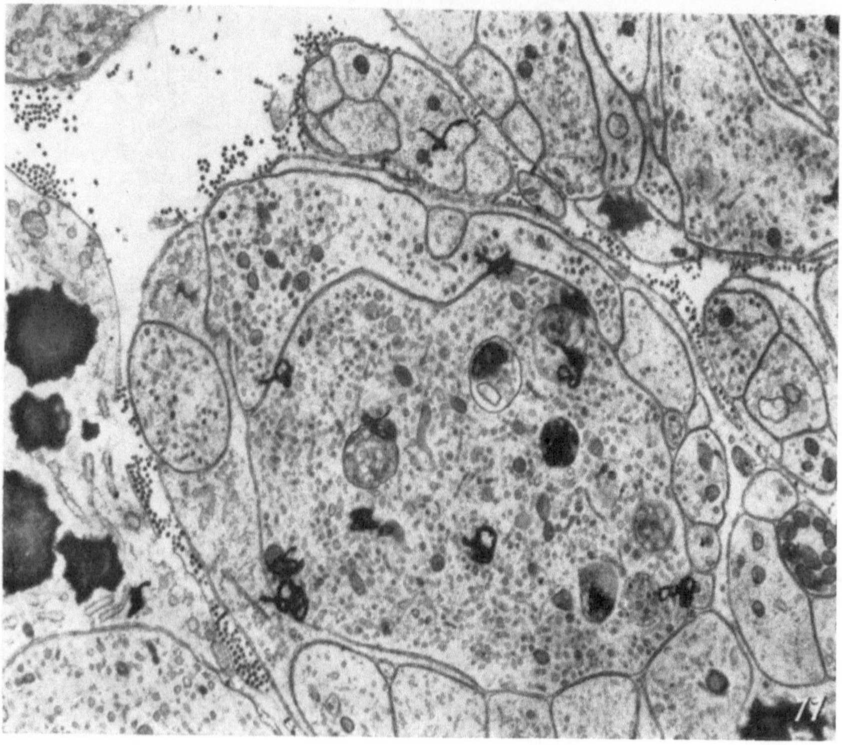

From this first type of experiments the following conclusions may be drawn: the labeling of nerve fibres proximal to a ligature is a highly specific phenomenon, limited to the modified, enlarged segment of certain fibres; it can be associated with a variable ultrastructure of the modified fibres in such a way that any correlation between the storage of ^{3}H-NA and a definite type of organelle can hardly be recognized, except for the endoplasmic reticulum in its various aspects. The storage process, as evaluated by radioautography, is sensitive to reserpine, the action of which is largely antagonized by IMAO.

Migration of ^{3}H-Noradrenaline

A second type of experiment was performed in order to evaluate the contribution of NA brought about by axonal transport to the local accumulation proximal to a ligature. First, ^{3}H-NA or ^{3}H-precursor was injected to a rat. This is rapidly taken up and stored by the sympathetic neurones (*Taxi* and *Droz*, 1966 a, b). After a delay of 1 to 2 h, necessary for assuring that the amine level in the blood reaches a value low enough to avoid interference between migration and local uptake phenomena, the nerve was ligated. Removal and fixation of the ligated part of the nerve were performed either 3 h or 22 h later. During this time, migration takes place.

In the absence of an IMAO, no or an extremely reduced labeling was obtained after 22 h of migration and but a moderate labeling was observed after 3 h of migration (Fig. 10).

If the experiment was performed on rats pretreated by an IMAO, labeling of middle intensity could be obtained after 22 h of migration (Fig. 11). The morphology of the labeled fibres varied in the same way as in the uptake experiments. The difference between the results of experiments performed in the presence or absence of an IMAO shows that, in normal circumstances, the quantity of amine transported along the axon is almost negligible in the accumulation proximal to a ligature, and more generally in the peripheral stores of NA, a conclusion which agrees with that of *Geffen et al.* (1971). As a moderate labeling can be obtained in the epiphysis cerebri 24 h after the injection of ^{3}H-NA, it seems likely to conclude that the NA

Fig. 10. Sciatic nerve. Labeling obtained after $3^{1}/_{2}$ h of migration of noradrenaline-^{3}H ($\times$ 16,200)

Fig. 11. Sciatic nerve. Labeling obtained after 22 h of migration after injection of dopamine-^{3}H and IMAO administration ($\times$16,200)

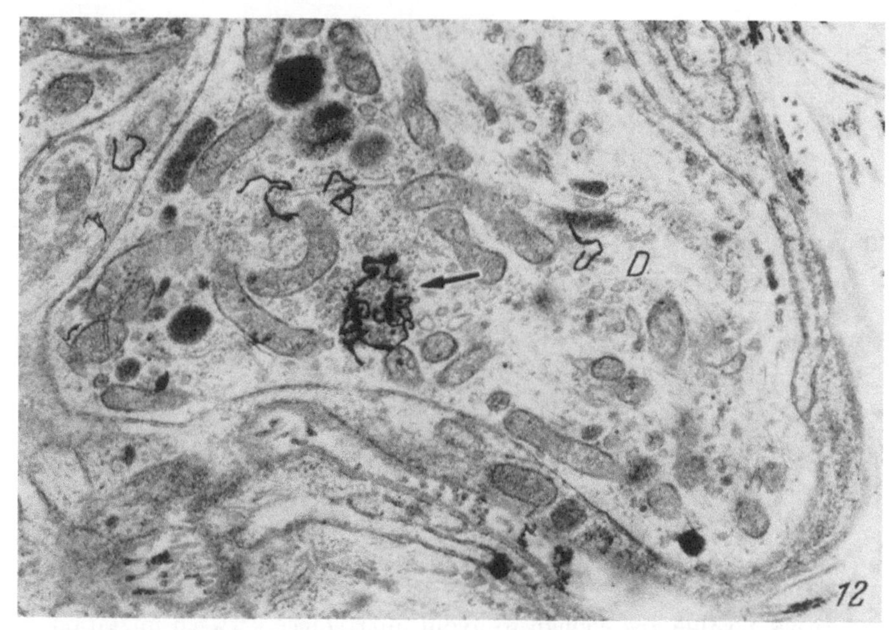
D
12

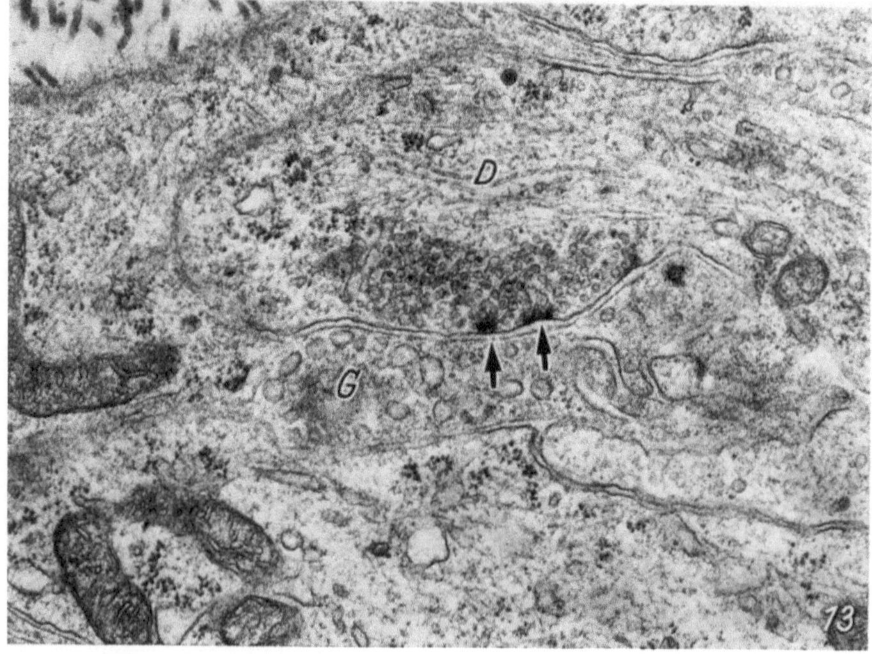
D
G
13

during its transport is not protected in the same way as in terminals. Thus the hypothesis of *Hökfelt* and *Dahlström* (1971) that dense-core vesicles, especially SGV, more numerous in the perikaryon than was previously accepted, move to the periphery carrying NA, seems difficult to reconcile with the MAO-sensitivity of ^{3}H-NA transported in the axon.

The above-mentioned hypothesis also raises the question of the role of the SGV in the perikaryon and dendrites of ganglionic neurones. First it should be remarked that at least a part of the SGV in the dendrites and the perikaryon of the sympathetic neurones of the rat have probably a specific function at this level, as first mentioned by *Taxi et al.* (1969). Radioautography after NA-^{3}H shows that the SGV have the same properties of NA storage as in the axon terminals (Fig. 12). Moreover, clusters of such vesicles are generally localized near the membrane; in many instances, there is a cytoplasmic differentiation in form of one or several dense patches attached to the neuronal membrane at this level (Figs. 13—15), giving an aspect similar to the presynaptic part of the active zones of synapses, as first described by *Gray* (1963). Since there are, obviously, no synapses at these sites, the neuronal surface being in contact with glia (Figs. 13—15) or sometimes with the interstitial medium of the ganglion (Fig. 14), such a disposition suggests that an output of NA can occur at this site, acting not in a synaptic system, but probably in a modulating one involving extra-synaptic receptors.

Secondly we want to emphasize that the storage of NA in the perikaryon exhibits other properties than that in the terminals. Labeling after injection of ^{3}H-NA requires the action of an IMAO (*Taxi* and *Droz*, 1969), except for the clusters of the SGV, as already mentioned. This sensitivity to MAO in the perikaryon, and also diffusion phenomena resulting, for instance, in labeling the nucleus of the ganglionic neuron and sometimes of glia, never labeled in the periphery, suggest that the bulk of ^{3}H-NA in the perikaryon is not bound in the same way as it is in the terminals. This situation may be probably related with the fact established by *Fisher* and *Snyder* (1965) that NA in the superior cervical ganglion of the rat is not bound to the microsomal fraction, but passes into the supernatant

---

Fig. 12. Rat superior cervical ganglion. Fixation 30 min after injection of noradrenaline-^{3}H. A cluster of vesicles, some of them having a dense core, is labeled (arrow) in a dendrite (D) (×25,000)

Fig. 13. Superior cervical ganglion. A cluster of vesicles, some of them having a dense core, in a dendrite (D). Notice the dense patches (arrows) attached to the neuronal membrane. G, neuroglia (×41,000)

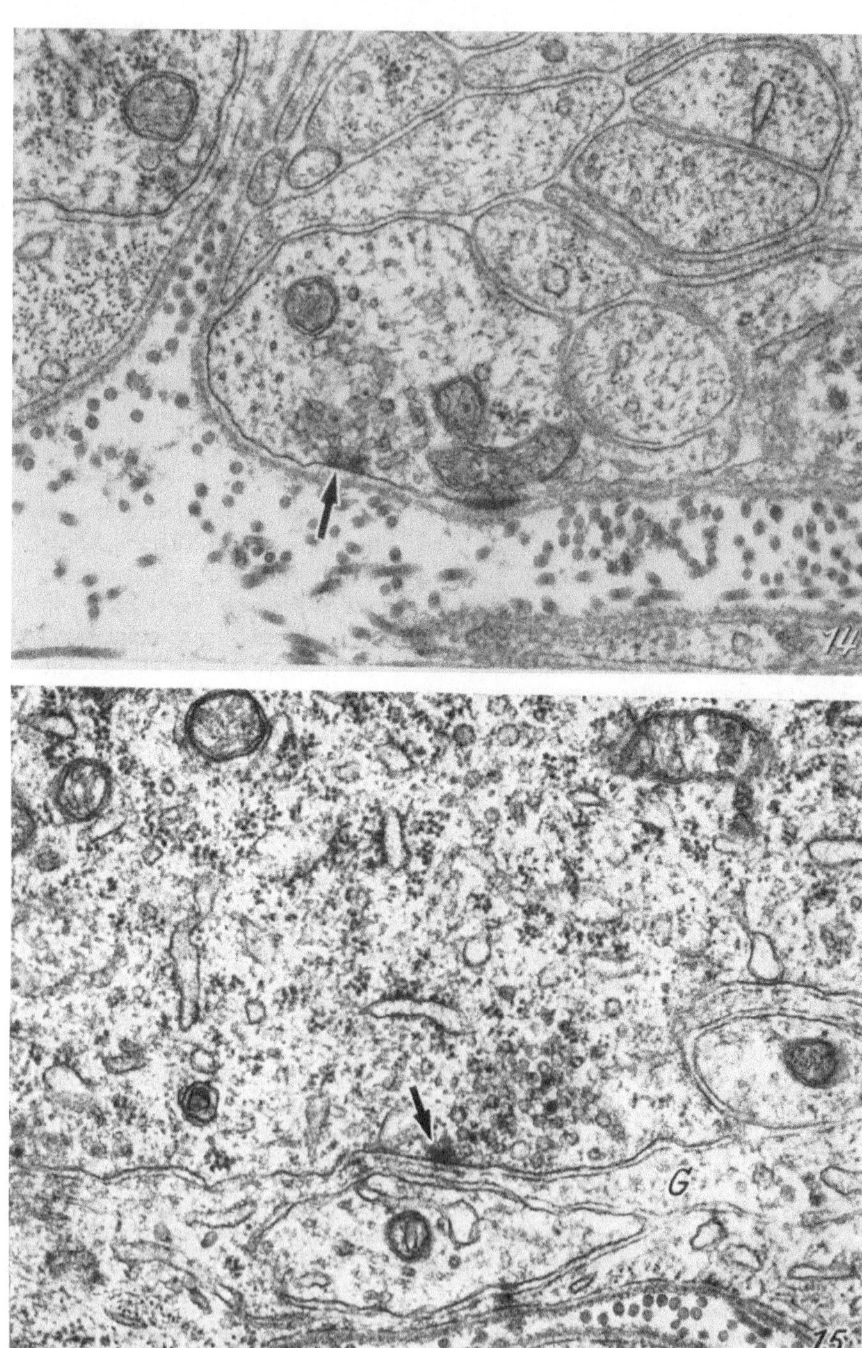
14
G
15

during homogenization. Nevertheless, fluorescence studies have shown that NA stores in the perikaryon and the terminals are both sensitive to reserpine. This implies that the two processes have something in common. The labeling of the perikaryon implies that NA is bound to something, probably the storage proteins — or chromogranins —, manufactured in the ergastoplasm. If NA would not be bound to something, it would be almost completely washed out by the fixative. Thus, it seems likely that storage of NA in the perikaryon is due to binding to specific storage proteins having not yet joined their definitive storage organelles, as they do in terminals.

In the axons, the concentration of NA is usually too low for giving a fluorescence reaction. Labeling after ^{3}H-NA injection is also generally lacking or too poor to be significant, although it has indeed been obtained in *in vitro* experiments (*Taxi,* 1967), in which local concentrations are certainly much higher than after intravenous administration.

An attractive hypothesis (*Kappeler* and *Mayor,* 1967; *Geffen* and *Ostberg,* 1969; *Dahlström,* 1970; *Häggendal* and *Dahlström,* 1971) is that the transport organelle for dopamine-β-hydroxylase and storage proteins is the LGV, rather numerous in sections of normal axons, while, on the contrary, the SGV are rare (at least 10 LGV on 1 SGV in the rat). It is, however, to be kept in mind that the organelle named LGV has probably several — at least two — functional meanings. This is suggested by the fact that they are insensitive to reserpine in normal terminals, which is also the case for at least part of the LGV accumulated proximal to a ligature (Figs. 8, 9). In normal fibres there are no morphological data in favour of a change of LGV into SGV, although there is a continuous range of granulated vesicle size in modified fibres proximal to a ligature. Certainly, LGV are not numerous in the perikaryon and the labeling obtained after radioautography is not related to them, as it is for the SGV at this level or at perphery.

For these reasons, an alternative hypothesis may be proposed concerning the morphological correlates of the transport of NA. In the perikarya, NA would be bound to storage proteins, but in the form of a hyaloplasmic component, that explains its sensibility to

---

Fig. 14. Superior cervical ganglion. In this case, the cluster of vesicles, with the dense patch attached to the membrane (arrow) is facing the interstitial medium (×41,000)

Fig. 15. Superior cervical ganglion. Cluster of vesicles and dense patch (arrow) in a perikaryon; G, neuroglia (×41,000)

MAO. Such an extravesicular pool was already taken in consideration, even in terminals, by *Van Orden et al.* (1967, 1970). In this hypothesis, events occurring proximal to a ligature may be understood as follows: the stopping of the axonal transport, mimicking to a certain extent the normal situation in the peripheral part of the axon, results in a rapid increase of local concentrations of dopamine-β-hydroxylase and storage proteins, occurring simultaneously with a tremendous outgrowth of the endoplasmic reticulum, which leads to the formation *in situ* of various storage organelles, partly atypical, as probably LGV and vesicles of a size intermediate between that of the LGV and SGV. New data will have to verify this hypothesis.

References

Dahlström, A.: Observations on the accumulation of noradrenaline in the proximal and distal parts of peripheral adrenergic nerves after compression. J. Anat. (London) *99*, 677—689 (1965).

Dahlström, A.: The effects of drugs on axonal transport of amine storage granules. In: New aspects of storage and release mechanisms of catecholamines (*Schümann, H. J.*, and *G. Kroneberg*, eds.), pp. 20—36. Berlin-Heidelberg-New York: Springer. 1970.

Dahlström, A., and *J. Häggendal*: Axonal transport of amine storage granules in sympathetic adrenergic neurons. Adv. in Bioch. Psychopharmacol. *2*, 65—93 (1970).

Devine, C. E., and *R. Laverty*: Fixation for electron microscopy and retention of ^{3}H-noradrenaline by tissues. Experientia *24*, 1156—1157 (1968).

Fisher, J. E., and *S. Snyder*: Disposition of norepinephrine-H^3 in sympathetic ganglia. J. Pharmacol. exp. Therap. *150*, 190—195 (1965).

Geffen, L. B., *L. Descarries*, and *B. Droz*: Intraaxonal migration of ^{3}H-norepinephrine injected into the coeliac ganglion of cats: radioautographic study of the proximal segment of constricted splenic nerves. Brain Res. *35*, 315—318 (1971).

Geffen, L. B., *B. G. Livett*, and *R. A. Rush*: Immunohistochemical localization of protein components of catecholamine storage vesicles. J. Physiol. (London) *204*, 593—605 (1969).

Geffen, L. B., and *A. Ostberg*: Distribution of granular vesicles in normal and constricted sympathetic neurons. J. Physiol. (London) *204*, 583 to 592 (1969).

Geffen, L. B., and *R. A. Rush*: Transport of noradrenaline in sympathetic nerves and the effect of nerve impulses on its contribution to transmitter stores. J. Neurochem. *15*, 925—930 (1968).

Gray, E. G.: Electron microscopy of presynaptic organelles of the spinal cord. J. Anat. (London) *97*, 101—106 (1963).

Häggendal, J., and *A. Dahlström*: The functional role of the amine storage granules of the sympatho-adrenal system. Memoirs Scc. Endocrinol. *19*, 651—667 (1971).

Hökfelt, T.: In vitro studies on central and peripheral monoamine neurons at the ultrastructural level. Z. Zellforsch. *91*, 1—14 (1968).

Hökfelt, T., and *A. Dahlström*: Effects of two mitosis inhibitors (colchicine and vinblastine) on the distribution and axonal transport of noradrenaline storage particles, studied by fluorescence and electron microscopy. Z. Zellforsch. *119*, 460—482 (1971).

Kappeler, K., and *D. Mayor*: The accumulation of noradrenaline in constricted sympathetic nerves as studied by fluorescence and electron microscopy. Proc. Roy. Soc. *B 167*, 282—292 (1967).

Kappeler, K., and *D. Mayor*: An electron microscopic study of the early changes proximal to a constriction in sympathetic nerves. Proc. Roy. Soc. *B 172*, 39—51 (1969).

Laduron, P., and *F. Belpaire*: Transport of noradrenaline and dopamine-hydroxylase in sympathetic nerves. Life Sci. *7*, 1—7 (1968).

Lever, J. D., *T. L. B. Spriggs*, *J. D. P. Graham*, and *C. Ivens*: The distribution of ^{3}H-noradrenaline and acetylcholinesterase (AChE) proximal to constrictions of hypogastric and splenic nerves in the cat. J. Anat. (London) *107*, 407—419 (1970).

Livett, B. G., *L. B. Geffen*, and *R. A. Rush*: Immunohistochemical evidence for the transport of dopamine-β-hydroxylase and a catecholamine binding protein in sympathetic nerves. Biochem. Pharmacol. *18*, 923 to 924 (1968).

Pellegrino de Iraldi, A., and *E. De Robertis*: Action of reserpine, iproniazid and pyrogallol on nerve endings of the pineal gland. Int. J. Neuropharmacol. 2, 231—239 (1963).

Pellegrino de Iraldi, A., and *E. De Robertis*: The neurotubular system of the axon and the origin of granulated and non granulated vesicles in regenerating nerves. Z. Zellforsch. *87*, 330—344 (1968).

Potter, L. T., and *J. Axelrod*: Studies on the storage of norepinephrine and the effect of drugs. J. Pharmacol. exp. Therap. *140*, 199—206 (1963).

Richardson, K. G.: The fine structure of autonomic nerve endings in smooth muscle of the rat vas deferens. J. Anat. (London) *96*, 427—442 (1962).

Richardson, K. G.: The finc structure of the albino rabbit iris, with special reference to the identification of adrenergic and cholinergic nerves and nerve endings in its intrinsic muscles. Amer. J. Anat. *114*, 173—205 (1964).

Richardson, K. G.: Electron microscopic identification of autonomic nerve endings. Nature *210*, 756 (1966).

Sotelo, C., and *J. Taxi*: Axonal flow of catecholamines in constricted sciatic nerves of the rat. An autoradiographic study. Anat. Rec. *169*, 435 (1971).

Sotelo, C., and *J. Taxi*: On the axonal migration of catecholamines in constricted sciatic nerve of the rat. A radioautographic study. Z. Zellforsch. *138*, 345—370 (1973).

Taxi, J.: Contribution à l'étude des connexions des neurones moteurs du système nerveux autonome. Ann. Sci. Nat., Zool., 12ème ser., VII, 413—674 (1965).

Taxi, J.: Identification des fibres nerveuses adrénergiques dans quelques muscles lisses de Mammifères par la méthode radioautographique utilisée en microscopie électronique. Bull. Ass. Anat., 52ème Réunion (Paris-Orsay), 1132—1139 (1967).

Taxi, J.: Sur la fixation et la signification du contenu dense des vesicles des fibres adrénergiques étudiées au microscope électronique. C.R. Acad. Bulg. Sci. *21*, 1229—1231 (1968).

Taxi, J., and *B. Droz*: Étude de l'incorporation de noradrénaline-^{3}H (NA-^{3}H) et de 5-hydroxytryptophane-^{3}H (5-HTP-^{3}H) dans les fibres nerveuses du canal déférent et de l'intestin. C.R. Acad. Sci. (Paris) *263*, 1237—1240 (1966a).

Taxi, J., and *B. Droz*: Étude de l'incorporation de noradrénaline-^{3}H (NA-^{3}H) et de 5-hydroxytryptophane-^{3}H (5-HTP-^{3}H) dans l'épiphyse et dans le ganglion cervical supérieur. C.R. Acad. Sci. (Paris) *263*, 1326—1329 (1966b).

Taxi, J., and *B. Droz*: Radioautographic study of the accumulation of some biogenic amines in the autonomic nervous system. In: Cellular dynamics of the neuron (*Barondes, S. H.*, ed.). New York: Academic Press. 1969.

Taxi, J., *J. Gautron*, and *P. L'Hermite*: Données ultrastructurales sur une éventuelle modulation adrénergique de l'activité du ganglion cervical supérieur du Rat. C.R. Acad. Sci. (Paris) *269*, 1281—1284 (1969).

Van Orden III, L. S., *K. G. Bensch*, and *N. J. Giarman*: Histochemical and functional relationships of catecholamines in adrenergic nerve endings. II. Extravesicular norepinephrine. J. Pharmacol. exper. Therap. *155*, 428—439 (1967).

Van Orden III, L. S., *F. E. Bloom*, *R. J. Barnett*, and *N. J. Giarman*: Histochemical and functional relationships of catecholamines in adrenergic nerve endings. I. Participation of granular vesicles. J. Pharmacol. exper. Therap. *154*, 185—199 (1966).

Van Orden III, L. S., *J. P. Burke*, *M. Cuper*, and *F. V. Lodoen*: Localization of depletion-sensitive and depletion-resistant norepinephrine storage sites in autonomic ganglia. J. Pharmacol. exp. Therap. *174*, 56—71 (1970).

Weiss, P.: Neuronal dynamics and neuroplasmic ("axonal") flow. In: Cellular dynamics of the neuron (*Barondes, H. S.*, ed.). New York: Academic Press. 1969.

Weiss, P., and *H. B. Hiscoe*: Experiments on the mechanism of nerve growth. J. exp. Zool. *107*, 315—396 (1948).

Wolfe, E. D., *L. T. Potter*, *K. C. Richardson*, and *J. Axelrod*: Localizing tritiated norepinephrine in sympathetic axons by electron microscopic autoradiography. Science *138*, 440—442 (1962).

Author's address: Prof. Dr. *J. Taxi*, Laboratoire de Biologie animale, Université de Paris-VI, 12 Rue Cuvier, F-75005 Paris 5ème, France.

Discussion

Häggendal: 1. The mixture of granules with different diameters seems to be at variance with the results of other authors, as discussed in my paper. What do you think is the reason for this discrepancy? 2. Do you think that in different tissues or species the adrenergic system differs so much in such an important point? Is this not due to different techniques used? 3. In some main aspects I don't think that our views differ materially. Since the granules appear to be formed in a more or less mature form in the perikarya, their maturation might take place during their axonal transport down to the nerve terminals. In fact, Dr. Dahlström and I have some recent results showing an increased NA loading of the granules when arrested above a constriction. Quantitatively, however, this NA increase seems to be so small that it does not markedly influence the calculated life-span for the granules with respect to their capacity to store endogenous NA. Such a process of maturation of the granules seems to be in agreement with the factor Dr. Dahlström and I have discussed: the effect of time on different capacities of the granules. 4. To which extent do you think the precipitate found, *e.g.* in the lysosomes, is not due to ^{3}H-NA itself, but to its metabolites?

Taxi: (1) The discrepancy between the results you mention and those we obtained may be explained at least in two ways, which are not exclusive one for the other:

(a) as to difference in species, the results of other authors were obtained on the splenic nerve of the cat. There are many examples in the literature and also in my personal experience of great differences in the morphology of the vesicular content of the sympathetic fibres of various animal species, although fixed in the same way;

(b) the modified zone proximal to a ligature has a length of about 1 mm in the sciatic nerve of the rat. As the sections for radioautography are of about 10^{-4} mm in average, it is probable that, in spite of our attention, we cut slightly different levels from one experiment to another. Moreover, it is well-known that there are large individual variations in the delay of appearance of modifications in different fibres of the same nerve. On the other hand, authors using only a morphological approach have no criteria to identify the fibres able to retain NA when they are not modified in the way which they expect.

(2) Like you, I was surprised by the differences between the results obtained by other authors and our own. But I don't think this is a matter of technique, as we generally used fixation by osmium tetroxide, as do, for instance, Kappeler and Mayor. Also using the 3% MnO_4K fixation according to Hökfelt, which can be considered the best fixative for preserving the vesicular contents, we obtained the same variety of aspects.

(3) I think that, indeed, our points of view are not very different. The difference is on the subject of the appearance of granulated vesicles. It appears to me likely that the material for granules, especially proteins, are manufactured in the perikaryon, but in normal conditions the assembling

takes place mainly at the periphery, or eventually proximal to a ligature when there is one. The increase of NA loading you observed at this point may correspond to the fact that the storage is more perfect in relation to the formation of storage organelles at this level.

(4) It cannot be excluded that the labeling of lysosomes is due to some metabolites. However, such a labeling may be obtained a short time after the injecion of NA-^{3}H, when the metabolites are still low in quantity.

Csillik: Could you possibly comment on the significance of labeling of lysosomes?

Taxi: The labeling of lysosomes seems to me an example of unspecific binding of NA inside the nerve fibres, the input in which is highly specific, due to the specificity of the phenomenon of active transport through the neuronal membrane. This kind of binding is one of the difficulties of the radioautographic method, and more generally of using the uptake of transmitter in the study of its cytological localization.

Salánki: Besides the green fluorescence, there was a very strong yellow fluorescence to be seen in one of your pictures. Do you think this was also due to NA or rather to something else, *e.g.* serotonin?

Taxi: I have a limited experience with the fluorescence method, but, as far as I know, Swedish authors, masters in this field, have postulated that, in case of high concentrations of NA, the normal green fluorescence may turn into a yellow one. For instance, the so-called "small intensely fluorescent cells" of the sympathetic ganglia of the rat, which were demonstrated to contain dopamine in high concentration, regularly show a yellow fluorescence.

Journal of Neural Transmission, Suppl. XI, 125—133 (1974)

The Effect of Electric Stimulation on the Axoplasmic Transport of Newly Synthetized Neuronal Proteins

J. Kiss, Eszter Láng, and **J. Hámori**

Second and First Department of Anatomy, Semmelweis University Medical School, Budapest, Hungary

With 4 Plates

Summary

1. One hour of electric stimulation of the preganglionic oculomotor nerve results in an enhanced speed of fast axonal transport of the newly synthetized proteins, as revealed by light and electron microscope autoradiography of labelled and transported proteins within the synaptic endings (calices) of the stimulated axons present in the ciliary ganglion of the chick.

2. In the ciliary ganglion cells the synthesis of proteins is also markedly activated by electric stimulation of the preganglionic fibres.

Introduction

It is now generally established (*Weiss* and *Hiscoe*, 1948) that neuronal proteins are synthetized almost exclusively in the perykaryon, *i.e.* in the nerve cell soma. The newly synthetized proteins are then carried by the (1) fast and (2) slow axonal transport to the axonal ending (*Droz* and *Koenig*, 1971). A simple though indirect method to study the kinetics of axonal protein transport is the autoradiographic demonstration of ^{3}H-labelled newly synthetized proteins in the axon terminal (*Droz*, 1967). Because of their exceptionally large size, the caliciform endings (*Lenhossék*, 1911; *Szentágothai et al.*, 1954) of the Edinger-Westphal neurones present in the ciliary ganglion of the chick are especially useful in the study of axonal protein transport. Indeed, *Koenig* and *Droz* (1971 a) observed that a maximum of labelled protein accumulation in these terminals occurs about 18 hours after the injection of the labelled amino acid. Considering a 6 mm distance of the axonal terminal from the neuronal soma in this particular case, the speed of the fast transport in the parasympathetic oculomotor axons will be somewhat above 50 mm/day.

On the basis of the data mentioned, we have attempted to investigate whether the speed of the fast axonal protein transport is constant or is effected and therefore related to the bioelectric processes going on in the axonal membrane during the functioning of the neuron.

Table 1. *Diagram of the Parasympathetic Oculomotor Pathway and the Locus of Electric Stimulation*

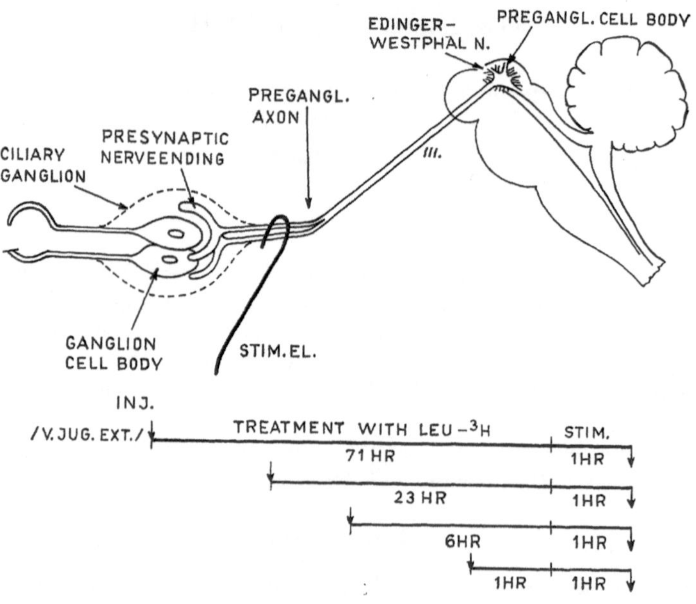

Material and Methods

10—12 days old chicken were injected through the jugular vein with 1.5 mCi ^{3}H-Leucin in physiological concentration; 1, 6, 23 and 72 hours after the injection (3 animals for each group), the ganglia at both sides were exposed, and the left preganglionic oculomotor trunc stimulated electrically for one hour at 3×16 Cps frequency and 8 Volt (Table 1). The right ciliary ganglia served as controls. The ciliary ganglia were fixed in buffered formaldehyde containing inactive leucin, and postfixed in osmic acid. The tissue pieces were embedded in Araldite, and 1 μ thick and thin sections were cut on an LKB mikrotome. Then the sections treated for further radioautographic procedures according to the methods of *Kopriwa* and *Leblond* (1962) and of *Granboulan* and *Granboulan* (1965), respectively. The stained thin sections were examined with a JEM-100/B electron microscope. Thick sections were stained with toluidine-blue.

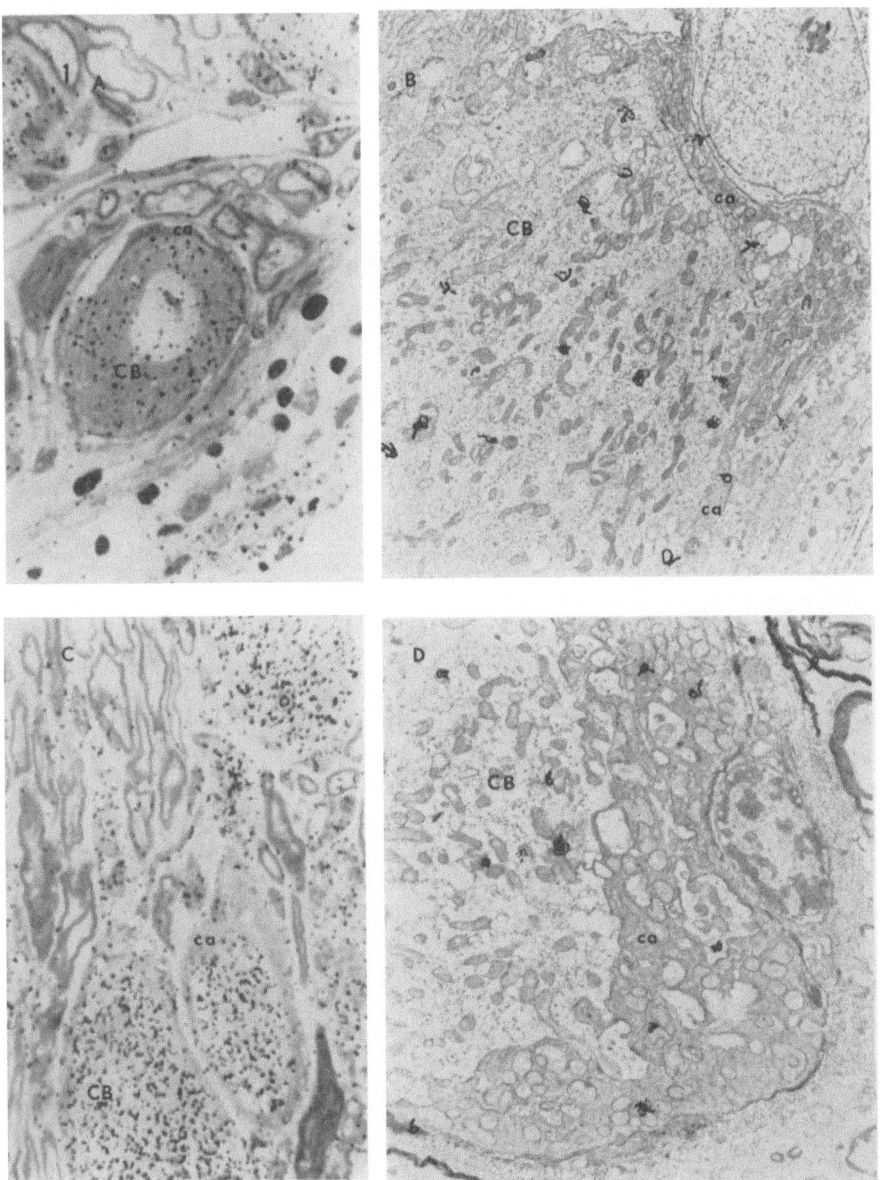

Plate 1. Ciliary ganglion of the chick: light and electron microscopic autoradiograms 2 hours after intravenous injection of ^{3}H-leucin.

A and B: control, non-stimulated. The labelling of the presynaptic calyx (ca) is light, the postsynaptic ganglion cell body (CB) shows a moderate activity. C and D: preganglionic stimulation. Both the calyx and ganglion cell body are heavily labelled as compared to the control

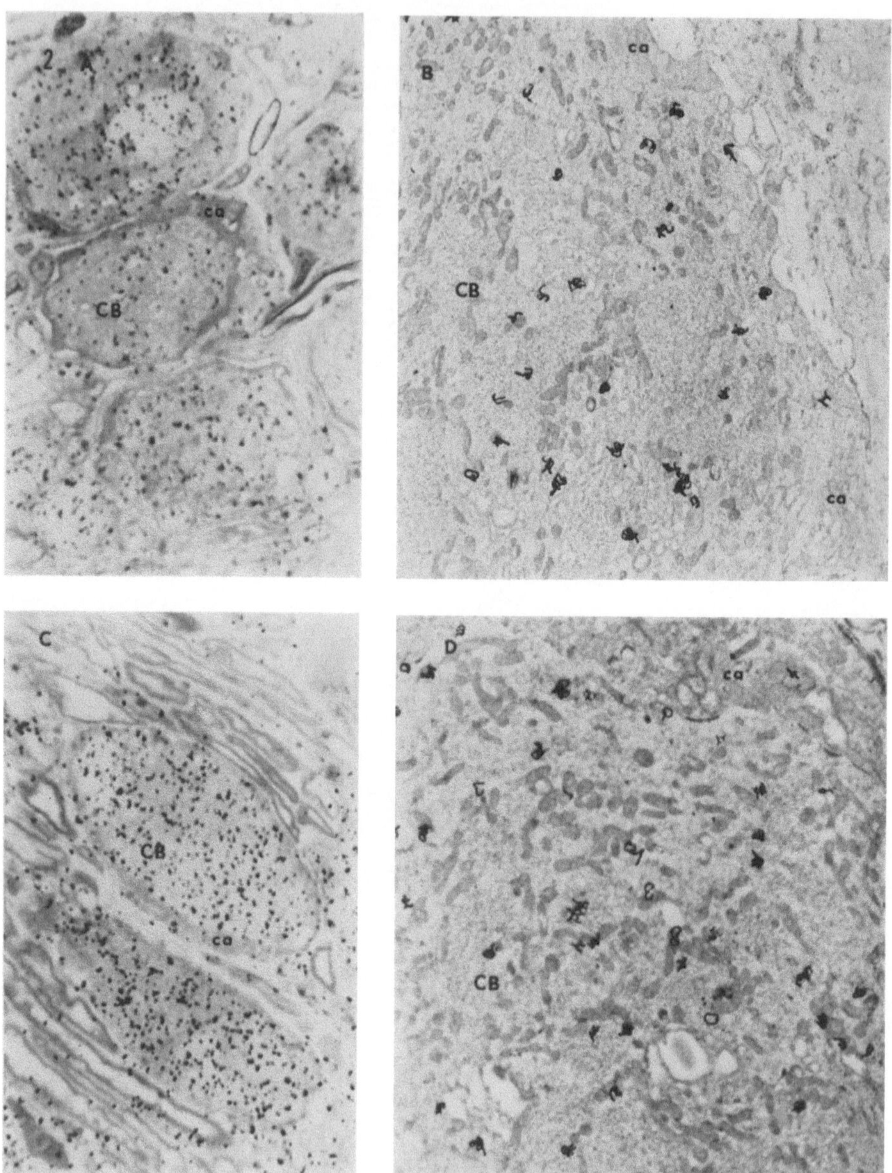

Plate 2. Ciliary ganglion cell body 2 hours after intravenous injection of ^{3}H-leucin. A and B, control; C and D, stimulated. The stimulated perikaryon (CB) is more heavily labelled than in the controls

Results

2 hours after the leucin injection, a strong labelling of the presynaptic calyx can be observed at the stimulated side (Fig. 1). The radioactivity of the calices of the corresponding control (non-stimulated) side is, as shown in Table 2, 2 1/2 times weaker than at the stimulated side. Using higher magnification it can also be seen (Fig. 3) that the grains are localized mainly above the area rich in

Table 2. *Concentration of Radioactivity in Nerve Endings (Calices) and Perikarya of Ganglion Cells in Ciliary Ganglia, Following Leu-^{3}H Administration and Electrical Stimulation or without Stimulation*

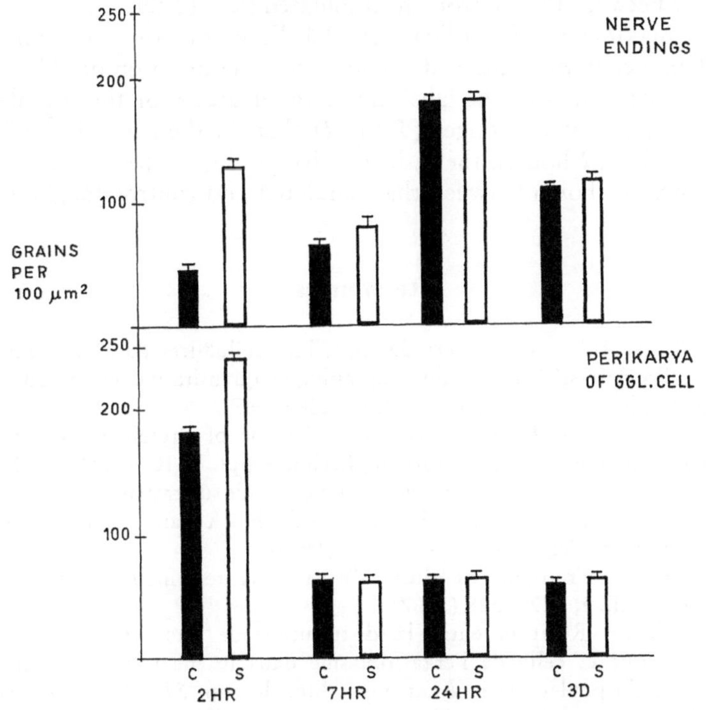

		Silver Grains per 100 μm²			
		2 hours	7 hours	24 hours	3 days
Nerve endings	Control	45.8 ± 10.2	57.6 ± 8.8	178.4 ± 6.2	112.2 ± 7.4
	Stim.	126.0 ± 12.2	74.6 ± 10.4	182.7 ± 5.4	120.0 ± 8.2
Perikarya of Ggl. Cells	Control	179.2 ± 8.6	63.2 ± 5.1	60.2 ± 3.8	58.1 ± 6.5
	Stim.	236.8 ± 6.4	59.8 ± 4.4	62.2 ± 6.2	60.3 ± 7.2

synaptic vesicles as well as close or above the presynaptic membrane. Labelled mitochondria can also be occasionally observed.

7 hours after injection, the labelling of the calices is stronger than at 2 hours, and although the stimulated calices still show somewhat stronger radioactivity than the control ones, the difference is no more statistically significant.

A maximum of radioactive labelling of the calices is observed *24 hours* following the ^{3}H-leucin injection. No qualitative difference between the stimulated and control side could be detected at that time (Table 2).

72 hours after injection (Fig. 4), the labelling of the calices becomes already weaker than at 24 hours. There is no difference in activity between the control and stimulated sides (Table 2).

A maximum of radioactive labelling of the postsynaptic ganglionic cells was observed *2 hours* after the injection of ^{3}H-leucin (Fig. 2). At that time the labelling of the neurones of the stimulated side is significantly stronger (Table 2) than in the control ganglion. Later (7, 24, 72 hours), the radioactivity gradually decreases, and no differences is shown between the stimulated and control ganglion cell bodies.

References

Droz, B.: Synthèse et transfert des protéines cellulaires dans les neurones ganglionnaires; étude radioautographique quantitative en microscopie éléctronique. J. Microscopie *6*, 201—228 (1967).

Droz, B., and *H. L. Koenig*: Dynamic Condition of Protein in Axons and Axon Terminals. Acta neuropath., Berlin, Suppl. V, 109—118 (1971).

Granboulan, N., and *P. Granboulan*: Cytochémie ultrastructurale du nucléole. II. Étude des sites de synthèse du RNA dans le nucléole et le noyau. Exp. Cell. Res. *38*, 604—619 (1965).

Kerkut, G. A.: Transport of glutamate to nerve terminals. Neurosc. Res. Progr. Bull. *5*, 322—325 (1967).

Koenig, H. L.: Relation entre la distribution de l'activité acétylcholinesterasique et celle de l'ergastoplasme dans les neurones du ganglion ciliaire du poulet. Arch. Anat. micr. Morph. exp. *54*, 937—964 (1965).

Koenig, H. L., and *B. Droz*: Transports axonaux de protéines aux terminaisons nerveuses du ganglion ciliaire du poulet, après injection intraventriculaire cérébrale de leucine-^{3}H. C.R. Acad. Sci. (Paris) *272*, 2812—2815 (1971a).

Koenig, H. L., and *B. Droz*: Effect of nerve section on protein metabolism of ganglion cells and preganglionic nerve endings. Acta neuropath., Berlin, Suppl. V, 119—125 (1971b).

Kopriwa, B. M., and *C. P. Leblond*: Improvements in the coating technique of radioautography. J. Histochem. Cytochem. *10*, 269—284 (1962).

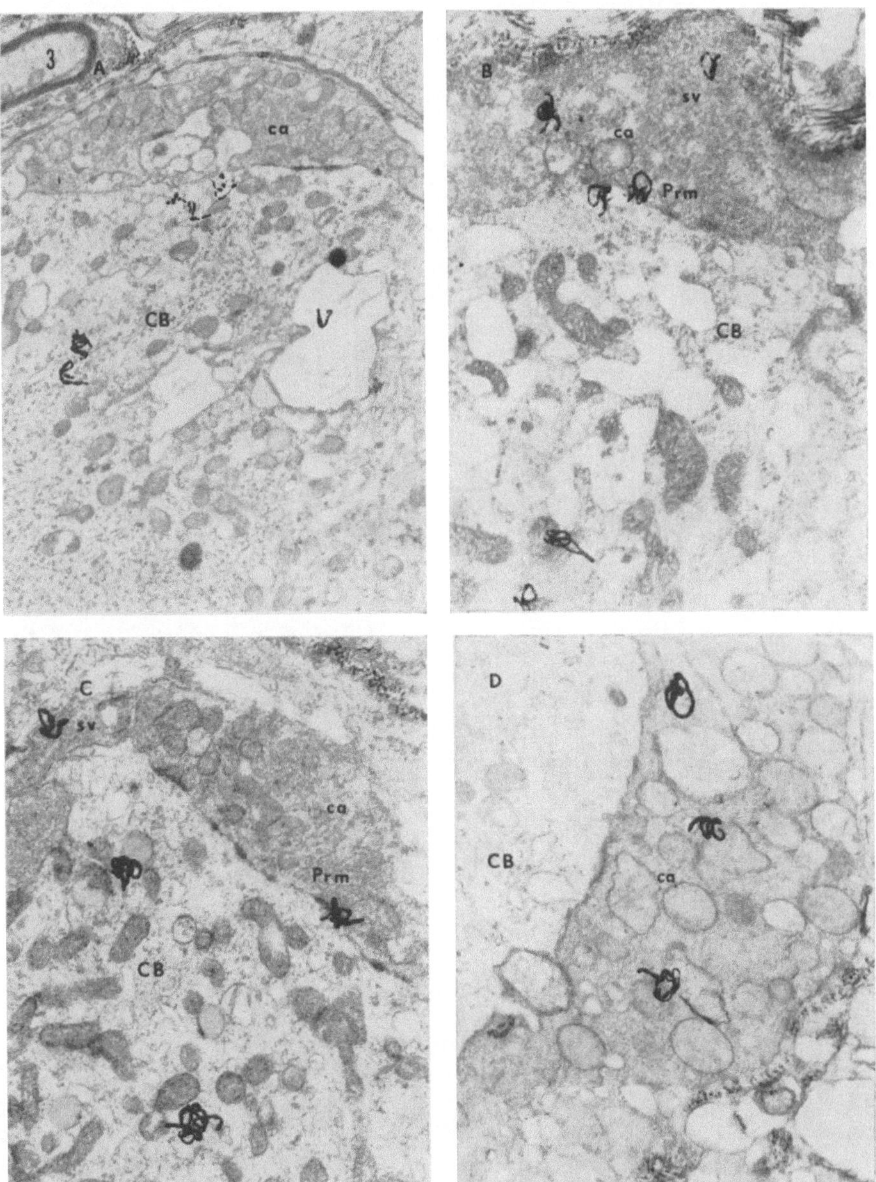

Plate 3. EM-autoradiograms of ciliary calices 2 hours (A and B) and 72 hours (C and D) after intravenous injection of ^{3}H-leucin. A, non-stimulated. No labelling of the calyx; B, stimulated. Heavy labelling of the calyx; C, non-stimulated; D, stimulated. A similar activity of the presynaptic calices is shown. The labelled proteins are confined mostly to synaptic vesicles (sv), occasionally to mitochondria and possibly presynaptic membrane (Prm)

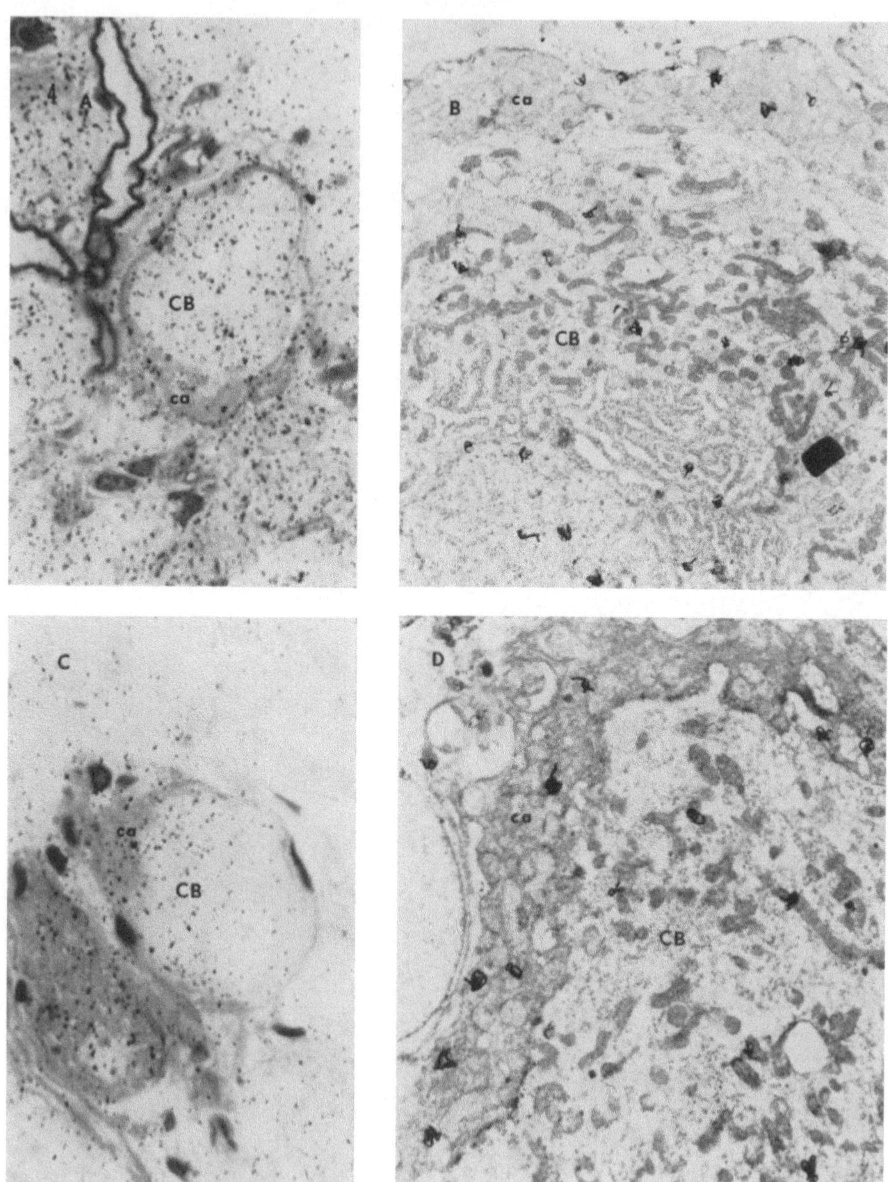

Plate 4. Ciliary ganglion of the chick 24 hours after intravenous injection of ^{3}H-leucin. The radioactivity of the calices (ca) and ganglion cell bodies is essential on the same level in both the control (A, B) and stimulated (C, D) side. Legends the same as in former plates

Lenhossék, M.: Das Ganglion ciliare der Vögel. Arch. für mikr. Anat. *76*, 745 (1911).

Szentágothai, J., A. Donhoffer, and *K. Rajkovits*: Die Lokalisation des Cholinesterase in der intraneuronalen Synapse. Acta histochem. *1*, 272—281 (1954).

Weiss, P. A., and *H. B. Hiscoe*: Experiments on the mechanisme of nerve growth. J. Exp. Zool. *107*, 315 (1948).

Author's address: Dr. *J. Kiss,* Second Department of Anatomy, Semmelweis University Medical School, Budapest, Hungary.

Discussion

Csillik: Do your results imply that stimulation of the presynaptic axon modifies the metabolism of the postsynaptic neuron?

Kiss: Koenig and *Droz* (1971 b) have shown that postganglionic axotomy effects both, the protein synthesis in the perikaryon and the fast axonal transport in the preganglionic axon. *Kerkut* (1967), on the other hand, demonstrated a change in the speed of axonal glutamate transport in the ischiadic nerve following electric stimulation of the CNS. Our present results clearly demonstrate, that *direct* electric stimulation of the axon effects, in fact increases, the speed of the *fast* axonal transport, at least in the first few hours after stimulation.

The increase of synthetic activity in the postsynaptic ganglion cells following the stimulation of the preganglionic fibres indicates that bioelectric changes not only effect the synaptic activity of the stimulated axonal process but also the protein synthesis of the postsynaptic neurones in the ciliary ganglion.

Journal of Neural Transmission, Suppl. XI, 135—161 (1974)

Some Aspects of the Release of the Adrenergic Transmitter

J. Häggendal

Department of Pharmacology, University of Göteborg, Göteborg, Sweden

With 8 Figures

Summary

The release of noradrenaline (NA) from the sympathetic nerve endings (varicosities) is discussed in relation to the economy of the amine storage granules, their content of NA, and the rate of contribution of new granules to the nerve terminals via intra-axonal transport.

I. Different models for the involvement of amine storage granules in release are considered mainly on the basis of quantitative data on the amount of NA released per nerve impulse. The conclusion is drawn that granules in all probability can be used several times for transmitter release at nerve activity and that newly synthesized NA plays an important role at this release. Also the transmitter "quantum" seems to be far less than the total content of NA in a single granule.

II. The marked increase of the NA amount released from the sympathetic nerve endings at nerve activity after α-receptor blockade has been studied. Various previous explanations for this phenomenon appear at present to be unlikely. The current view is that the results are better explained as due to an increase in the amount of transmitter released per nerve impulse from the nerve endings. This indicates that the amount of transmitter released per nerve impulse is not constant but may vary due to regulation via local feed-back mechanisms. In consequence α-receptor blockers are unsuitable as tools for quantitative studies on the physiological, "true" release of transmitter at nerve activity.

III. Studies on the recovery of different functions in the nerve terminals after a large dose of reserpine indicate that granules formed in the cell bodies and recently downtransported to the nerve terminals via intra-axonal transport are probably capable not only of storing endogenous NA but also of uptake-retention of small amounts of ^{3}H-NA and of preferential release of the transmitter at nerve activity. These last-mentioned functions appear to be lost rather soon after the arrival of the new granules to the nerve terminals. The capacity to store endogenous NA seems, however, to be a long-lasting capacity.

This functional heterogenicity of the nerve terminal amine storage granules is discussed in relation to morphological and biochemical differences in granules from the axons and from the nerve terminals. A factor of importance for the agening of the granules may be release of granuler proteins during nerve activity.

The results support the view (section I) that the granules are not fully consumed at transmitter release following a normal nerve impulse, but that release occurs via partial exocytosis which may cause a change of properties of the storage particles.

The intention of this article is not to give a review of the broad field of transmitter release, as several reviews have been published during the last years, *e.g. Carlsson* (1965), *Iversen* (1967), *Kopin* (1968), *Bloom* and *Giarman* (1968), *Andén et al.* (1969), *Geffen* and *Livett* (1971), *Shore* (1972) and *De Potter et al.* (1972). The present paper will, instead, present some quantitative aspects of the release of transmitter, especially in relation to the economy of amine storage granules. Our present views on the quantitative release of adrenergic transmitter and the importance of intra-axonal transport of amine storage granules will be discussed. Experiments on the quantitative release of endogenous noradrenaline (NA) were mostly performed together with Dr. Björn Folkow, on the release of ^{3}H-NA together with Drs. Börje Johansson, Jan Jonason and Bengt Ljung, and on axonal transport of NA-containing granules together with Dr. Annica Dahlström.

For practical reasons we have mainly worked with the NA-containing sympathetic neurones in the peripheral nervous system (PNS). However, the monoamine-containing neurones in the central nervous system (CNS) have many features in common with the peripheral NA-containing neurones. Morphologically, the NA-containing neurones have the same appearance in the PNS and CNS. Thus, the axons are split up into a typical network of nerve terminals, their varicosities containing amine storage granules. Probably, the basic mechanisms for the release of NA and the role of amine storage granules for nerve terminal function are also similar in the PNS and CNS, even if quantitative variations may occur, *e.g.* with respect to the rate of intra-axonal transport.

The axons are of importance, not only for the propagation of the nerve impulses, but also for supplying the nerve terminals with necessary compounds, *e.g.* amine storage granules, which are formed in the cell bodies and transported to the nerve terminals by intra-axonal transport.

I. Some Different Possibilities for Involvement of Amine Storage Granules in the Release of NA

The release of transmitter is generally thought to occur in small packets, so-called "quanta" (cf. *Katz,* 1962; *Eccles,* 1964).

In the cholinergic neuron system subthreshold miniature end-plate potentials occur at random during nerve impulse rest. They are considered to be due to the release of multimolecular packets of acetylcholine (ACh). The vesicles present in the presynaptic nerve endings were suggested to be the structural correlates of the quanta of transmitter, particularly since both the miniature end-plate potentials and the vesicles are often remarkably constant in size. When a nerve impulse arrives at the motor end-plate, the random release of ACh packets is markedly increased so that a large number of packets are almost simultaneously discharged at the motor end-plate thus producing an excitatory end-plate potential (cf. *Katz,* 1962; *Eccles,* 1964). According to the theory, this will correspond to the release of a large number of ACh-containing vesicles.

With respect to adrenergic neurons, subtreshold miniature potentials, occurring at random, have also been observed at the adrenergic neuro-effector junctions (*Burnstock* and *Holman,* 1961). Also, in the presynaptic adrenergic nerve terminals a large number of structures, dense-core vesicles, have been observed electronmicroscopically, corresponding to the so called "amine storage granules".

Based on analogous findings in the two neuron systems, the quantum of released NA has been suggested to correspond to the NA content within one amine granule. Whilst it may be accepted that the subthreshold miniature potentials are caused by the release of a certain amount of the transmitter in "quanta", there exist several possibilities for the relation of these "quanta" to vesicles or granules.

In the NA-containing sympathetic neurons, the presence of normally functioning amine-storage granules in the nerve terminals seems to be a prerequisite for the release of transmitter under normal conditions (for ref. see *e.g.*, *Carlsson,* 1965, and *Andén et al.*, 1969). Furthermore, in recent years the sympathetic nerve endings have been shown to release not only transmitter during nerve activity, but also granular proteins: DA-β-hydroxylase (DβH) and chromogranin A (*e.g., Geffen et al.*, 1969; *De Potter et al.*, 1969). Provided that granular proteins are normally also released under physiological conditions, the release may occur in at least four different ways, schematically illustrated in figure 1.

Comments on possibilities 1 and 2: — (where the whole granule is released into the synaptic junction, and the granule releases its total content of transmitter and protein into the synaptic "gap", respectively): If the total granular content of NA and protein is released, the granule will be consumed and cannot be reused. Probably, the granules cannot be formed in the nerve terminals but are formed in the cell body (for ref. see *Dahlström*, 1967, 1971). Then more or less mature, they are transported to the nerve terminals at a fast rate of 5—10 mm/h (*Dahlström* and *Häggendal*, 1966 b; *Dahlström, Häggendal* and *Larsson*, to be published; *Kapeller* and *Mayor*, 1967; *Livett et al.*, 1968). Once in the nerve terminals, they appear to have the ability to store endogenous NA for several weeks, *i.e.*, their life-span is about 3—5 weeks (*Dahlström* and *Häggendal*, 1966 b). The daily contribution of amine storage granules by intra-axonal transport to the nerve terminal area may be about 3 to 5 % of the total number of granules in the terminals (see also *Geffen* and *Rush*, 1968), provided

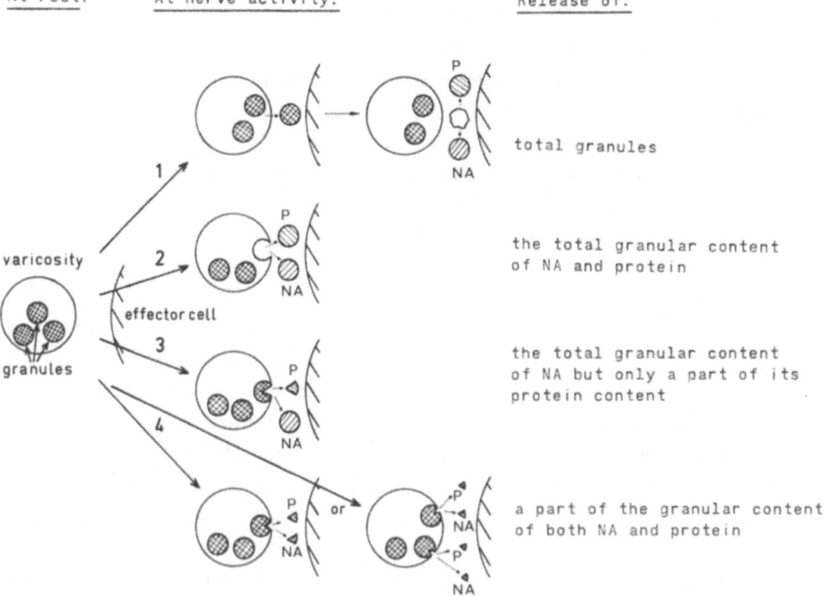

Fig. 1. Schematic illustration of four possible mechanisms for the release of transmitter in the adrenergic neuron upon nerve activity. P, protein (Dopamine-β-hydroxylase and chromogranin A); NA, noradrenaline. In possibility 4, the released part of NA is probably larger than the released fraction of protein

that the NA content per granule is about the same in the axonal granules as in an average nerve terminal granule.

The nerve terminal granules have been suggested to contain up to 10 to 100 times the NA content of axonal granules (*Stjärne*, 1968; *Lagercrantz*, 1971; *De Potter et al.*, 1972). However, these suggestions are mainly based on the ratio between NA and granular proteins in NA-storage particles isolated from axons and nerve terminals, respectively. The proportionally higher amounts of NA in relation to protein, present in the nerve-terminal particles, may equally well be explained by a loss of protein during nerve activity, together with an efficient *de novo* synthesis of NA. Furthermore, *Chubb et al.* (1970) have demonstrated by gradient centrifugation that NA in the nerve terminals is present in two types of particles having different densities, one of which contains almost no DβH. This will be further discussed when dealing with functional heterogeneity of amine storage granules (section III).

The number of granules per varicosity seems to vary widely; thus, between 45 and 862 dense-core vesicles per varicosity were found in the dilator muscle of the rat iris as revealed in serial sections (*Hökfelt*, 1969). The mean value was 318 (plus 6 granules with a diameter above 750 Å). Even if the number of granules per varicosity varies in different tissues and different species, it seems likely that this figure of about 350 granules per varicosity gives an approximate figure for the order of magnitude of the average number of granules present in an adrenergic varicosity.

The average daily contribution by intra-axonal transport to a varicosity will thus be about 5 % of 350 granules, ~ 20 granules, which is about one granule per hour.

During nerve impulse rest, the individual varicosity seems to fire spontaneously one quantum every 60th to 90th second to give a subtreshold miniature potential, as indicated from the studies on vas deferens by *Burnstock* and *Holman* (1962). Whether the quantum corresponds to the release of one whole granule (possibility 1) or to the total content of one granule (possibility 2) is obviously inconsistent with the economy of formation and transport to the nerve terminals of new granules.

The sympathetic tone in many adrenergically innervated tissues at rest seems to be about 0.5 imp/sec (*Folkow*, 1952). Also, if total granules are released only during nerve activity — one granule/nerve impulse and varicosity — it seems evident that the axonal down-transport of new granules is too small to maintain even the normal sympathetic tone. 0.5 imp/sec means 43,200 imp/24 h which would correspond to the release of 43,200 granules/24 h if one granule was released per impulse. Partly based on the figure of 26 to 40 h for the turnover of DβH, *De Potter* and *Chubb* (1971) suggested that the granules are lost during nerve activity. According to this suggestion, about 70 % of the granules in a varicosity would have to be renewed every day by axonal flow, equivalent to about 250 out of 350 granules per varicosity per 24 h. In order to meet with the speculated loss of 43,200 granules per 24 h, only one varicosity out of about 200 (43,200/250) will release one granule per nerve impulse. As will be discussed below, it is unlikely that the neuronal innervation of the adrenergic effector cell is so poor. (The use of DβH as a marker of the amine-storage granules will be discussed in section III.)

Comments on possibilities 3 and 4: — (the total content of transmitter and only a part of the protein content are released, and only parts of the contents of both transmitter and proteins are released, respectively).

If the granules can be re-used and reloaded with newly synthetized NA, many of the difficulties discussed above are overcome. Such a local synthesis is well known to occur in the nerve terminals (for ref., see, *e.g.*, *Andén et al.*, 1969). This may occur also in a granule that has released all or part of its NA content, but only part of its content of protein, *e.g.* DβH.

The 3rd and 4th possibilities for release may be discussed from another point of view — the quantity of NA released per varicosity and nerve impulse into the "synaptic gap", or nerve terminal-effector cell junction:

The average number of granules per varicosity seems, as discussed above, to be only a few hundred, about 350. Almost all the NA in, *e.g.*, skeletal muscle seems to be present in the adrenergic vasoconstictor nerve terminals (*Fuxe* and *Sedvall*, 1965), where almost all the NA appears to be bound to the granules in the varicosities (*Carlsson*, 1965).

If the NA content of only one granule is released, as much as 1/350 of the NA present in an average varicosity would be released. According to the concept of quantal release, all the granules contain about the same amount of transmitter. If several quanta, corresponding to the NA content of several granules, are needed for causing a suprathreshold potential of the effector cells, the fractional release will be higher than 1/350. For the cholinergic motor end-plate, as many as 100—300 quanta simultaneously discharged may be needed for producing an action potential (for ref. see *Eccles*, 1964). In the adrenergic nerve terminal-effector cell junction, the corresponding number of quanta required is uncertain. If it is as low as 5 to 10, for example 7, it means that 7/350 = 1/50 of the NA content in the varicosity will be released by one nerve impulse, to yield an effector cell response.

The fractional release of transmitter from different tissues (per imp released NA/total tissue NA) has been studied with various techniques. The decrease of the NA content in the tissue has been followed during sympathetic stimulation with or without blockers of NA synthesis by, *e.g.*, *Almgren* (1971). He obtained a figure of about 1/20,000 for the fractional release of NA from the salivary gland when its synthesis was blocked. A more direct way is to measure the NA released into the perfusing blood or nutritive fluids during nerve stimulation. From the vasoconstrictor fibres in cat hind limb, the fractional release was found to be 1/50,000, taking into account the mechanisms for inactivation of released NA at the nerve terminal-effector cell level (*Folkow et al.*, 1967).

In the following discussion the fractional release of NA will be assumed to be about 1/50,000 per nerve impulse.

The magnitude of this figure is obviously inconsistent with the magnitude of the theoretical figures of 1/350 for fractional release if the NA content in one granule is released per nerve impulse, or 1/50 if several "quanta" (calculated for a number of 7) of NA granules are released.

There seems to be at least two ways for explaining this discrepancy: (a) either only a fraction of NA, less than the NA content of one granule, is released from almost every varicosity per nerve impulse, or (b) the total NA content in one or several granules is released, but each time only from a small number of the varicosities that are present in the tissue.

If the first possibility (a) is valid, this implies that the NA amount released will be less than 1 % of the NA content in *one* granule. If the amount of NA released by 1 nerve impulse is the sum of several NA quanta (*e.g*, 7 quanta), then a quantum is much less (only about 0.1 %) of the NA content in one granule.

If the second possibility (b) is valid, *i.e.*, release from only some varicosities (corresponding to possibility 3 in figure 1), only one out of about 150 ($\frac{50,000}{350} = 143$) varicosities will release transmitter at every nerve impulse in order to satisfy the fractional release experimentally found. According to the hypothesis for quantal release, however, the NA content in one granule corresponds only to one quantum (of which several are needed for giving an activation of the effector cell), in which case more varicosities are needed. Thus, if *e.g.*, 7 quanta are needed, only 1 varicosity out of about 1000 will release transmitter at each nerve impulse.

Many adrenergically innervated tissues, such as the nictitating membrane, vas deferens and many blood vessels seem to respond swiftly during nerve stimulation. The nerve impulse flow is probably rather low under resting physiological conditions and an impulse frequency of one imp/sec produces about 50 % of the maximal response in many adrenergic effector units (cf. *Folkow*, 1952). Even if smooth muscle cells are electrically coupled, forming "functional units" (cf. *Ljung*, 1970), the number of varicosities available, about 8000, appears to be rather low within such a functional unit containing about 3000 to 4000 cells (cf. *Folkow* and *Häggendal*, 1970). If only one varicosity out of 150—1000 releases NA at an impulse frequency of 1 imp/sec, this would imply that only about 8—60 ($\frac{8000}{1000}$, $\frac{8000}{150}$) of the varicosities in the functional unit will release NA. This would give a very ineffective nervous control of the tissue, which at this frequency (1 imp/sec) shows almost 50 % of its maximal response.

Furthermore, release of the NA content of one granule (containing some 40,000 molecules of NA [*Folkow* and *Häggendal*, 1970]) will create an enormously high concentration of NA in the synaptic gap. If the gap is suggested to have an average width of 1000 Å and a square area of 1—2 μ^2, the concentration of NA can be calculated to be 50 to 100 μg/ml. Such high NA concentrations are unlikely to be needed, particularly when considering that such a high NA concentration would give only a subthreshold potential. In this connection it may be observed that exposure of *all* junction gaps to a sustained NA concentration of only 1 to 3 μg/ml induces 50 % of the maximal response in organ baths (*Ljung*, 1969).

In some blood vessels, the distance between the varicosity and the smooth muscle cell appears to be longer than 1000 Å, *e.g.*, in the rabbit pulmonary artery (*Bevan et al.*, 1969). However, this tissue seems to have a more sluggish response upon nerve stimulation than most other adrenergically innervated tissues (cf. *Bevan*, 1962). These other tissues, including not only iris and vas deferens but also the precapillaries, appear to have a rather short pre- to post-synaptic distance, some of them as low as 200 Å, between the varicosities and smooth muscle cells. In the beautiful electron microscopical pictures shown by Dr. Stach at this Symposium (1972), it appears as if every smooth muscle cell in the precapillaries is distinctly innervated. This may suggest a direct nervous control of these smooth muscle cells.

When all the data and considerations on NA release are taken together, it seems rather unlikely that a quantum of NA could correspond to the total NA content in one granule (for further discussion see *Folkow et al.*, 1967; *Folkow* and *Häggendal*, 1970).

From the quantitative studies presented above the first three alternatives presented in figure 1 seem unlikely. More probable is that only a *small* fraction of the NA content in *one* or *several* granules is released during nerve activity, and that a quantum of transmitter is much less than the NA content in one granule. Every granule may thus be used several times during its life span. Synthesis of NA from DA may occur in the granules until most of the DβH in the granules has been lost by repeated exocytosis. Due to *de novo* synthesis, the fraction of NA released from the granule per nerve impulse may be larger than the fraction of protein released. The release of only a fraction of the NA content from one or several granules during nerve activity also seems to be in good agreement with our present view on a functional heterogeneity of the amine storage granules in the nerve terminals, which will be discussed later.

In this discussion, the fractional release has been assumed to be 1/50,000. Several authors have, however, found a higher figure (*e.g.*, *Bevan et al.*, 1969; *Stjärne et al.*, 1969; *Langer* and *Vogt*, 1971). These results will be discussed below. However, much of the above discussion concerning different mechanisms for release seems to be valid even if the fractional release is, *e.g.*, 10 times larger than 1/50,000.

In most of the experiments where a high fractional release was found an α-receptor blocking drug, often phenoxybenzamine (PBZ), has been used to facilitate the NA overflow. The effect of such drugs will now be discussed with respect to a constant or variable quantitative release of NA from the nerve terminals during nerve activity.

II. Possibilities for a Variable Release of NA from Nerve Terminals during Nerve Activity

The amount of transmitter released into a synaptic gap per nerve impulse was often assumed to be constant. The facilitating effect of α-receptor blockers have thus been thought to be due to a blocking effect of the drug on one or several of the mechanisms that inactivate released NA at the nerve terminal-effector cell level. Therefore, these drugs have been used in order to disclose the "true quantitative release".

Figure 2 illustrates the five ways for inactivation of released NA that have been proposed: (1) binding to the receptors; (2) local enzymatic destruction; (3) re-uptake into the nerve terminals; (4) extra-neuronal uptake; (5) overflow into the blood stream.

Ad (1). Binding to the receptors does not seem to be of any quantitative importance. The mechanism by which the NA-receptor

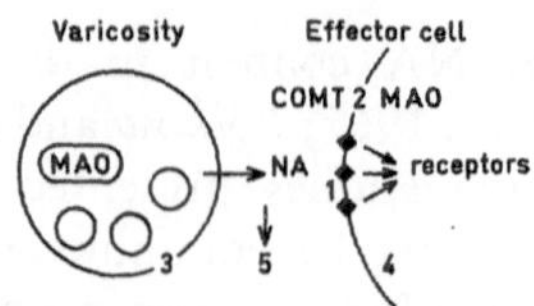

Fig. 2. Schematic illustration of the five mechanisms for inactivation of released noradrenaline at the nerve terminal-effector cell level. Small circles in the varicosity indicate amine storage granules. NA, noradrenaline released from the varicosity into the synaptic junction. This NA is inactivated by the different inactivating mechanisms indicated by figures. MAO, monoamine-oxidase; COMT, catechol-O-methyl transferase.

1 – Binding to the receptors; 2 – Local enzymatic destruction by COMT and MAO; 3 – Active mechanisms for re-uptake into the nerve terminals ("The amine membrane pump"); 4 – Extra neuronal uptake; 5 – Diffusion into the blood stream

coupling is broken is uncertain. However, after coupling either the NA is set free (unmetabolized), in which case the NA will appear free in the gap between the effector cell and the nerve terminal where the other inactivating mechanisms can act, or the NA is metabolized. In the latter case, catechol-O-methyl transferase (COMT) has been discussed to be of particular importance. Metabolized NA will contribute to the sum of metabolites found in the perfusion solution, since the NA metabolites are probably not accumulated in the tissue under physiological conditions. The NA that was bound to the receptors will in either case appear for the 4 other ways for inactivation.

Ad (2). In most experiments the sum of NA metabolites seems to be rather low during stimulation, being about 30 % or less of the total release of NA and its metabolites (*e.g.*, *Rosell et al.*, 1963; see *Häggendal et al.*, 1970, also for references). Furthermore, inhibitors of monoamine oxidase (MAO) or COMT do not appear to increase markedly the NA output (*e.g.*, *Brown*, 1965; *Folkow et al.*, 1967; *Farnebo* and *Hamberger*, 1971).

Ad (3). Membrane pump blockers (without α-receptor blocking effects) will increase only moderately the NA overflow into the perfusion solution in most experiments where the released NA is likely to diffuse easily (*e.g.*, in organ baths or at high blood flow) (*Häggendal et al.*, 1970; *Farnebo* and *Hamberger*, 1971).

Ad (4). A physiologically occurring accumulation of NA in extraneuronal tissue has apparently not been demonstrated. If, however, such an uptake occurs, the NA will probably be metabolized (cf. *Iversen*, 1971). In this case the NA is likely to contribute to the sum of metabolites in the overflow (see ad 2). Normetanephrine, which inhibits extraneuronal uptake only increases very slightly the NA overflow (*Farnebo* and *Hamberger*, 1971).

Ad (5). The magnitude of the blood flow seems to be a factor of importance. When the blood flow is strongly reduced, the overflow of NA is small (*Rosell et al.*, 1963; *Carlsson et al.*, 1964). During moderate or high blood flow, or when diffusion can occur easily, *e.g.*, when using thin isolated tissue preparations, the *overflow* of NA seems to be the most important mechanism for local inactivation of NA released during nerve activity.

The facilitating effect of PBZ on the NA overflow is thus difficult to explain merely as a blockade of one or several of the other inactivating mechanisms.

These difficulties for explaining the effect of PBZ are further stressed by some recent results dealing with the different inactivating mechanisms (1—5 in Figure 2).

Ad (1). Following apparently total α-receptor blockade obtained after PBZ administration (no or a decreased effector cell response) the overflow of NA and its metabolites will continue to increase with increasing concentrations of PBZ in the perfusion solutions (*Häggendal et al.*, 1972). If receptor binding was important for inactivation, one would expect a stabilization of the overflow of NA plus metabolites when total receptor blockade was reached. This was not the case. Therefore, the PBZ effect is not explained by hindering NA-receptor coupling.

Ad (2). The sum of ^{3}H-NA plus its metabolites increased after PBZ (*Farnebo* and *Hamberger*, 1970; *Häggendal et al.*, 1972). This indicates that the increased NA overflow is not due to only blocking of the local metabolism.

Ad (3). After an apparently optimal treatment with blockers of the "membrane pump", the addition of PBZ to the preparation markedly increased the NA overflow (*Geffen*, 1965; *Häggendal*, 1970; *Farnebo* and *Hamberger*, 1971). Therefore, it is difficult to explain the effect of PBZ on the NA overflow in terms of membrane pump inhibition.

Ad (5). During stimulation after administration of PBZ, the blood flow is not reduced since vasoconstriction is prevented by the α-receptor blockade. The maintained blood flow has thus been thought to facilitate the NA output. However, when by other means (*e.g.* by continuous muscle exercise by means of motor nerve stimulation) the blood flow is kept at the same level as after PBZ, the NA overflow is much lower than after PBZ (*e.g. Häggendal*, 1970).

Considering the difficulties of explaining the increased NA overflow after administration of PBZ and some other α-receptor blockers in terms of a blockade of the local inactivating mechanisms, it was suggested that *the amount of NA released per nerve impulse* into the synaptic gap is *not constant* but may *vary* (*Häggendal*, 1969, 1970). Such an increased overflow of NA after administration of α-receptor blockers can then be observed and interpreted as an increased release, both in the peripheral and central nervous system (*e.g. Farnebo* and *Hamberger*, 1970, 1971; *Starke et al.*, 1971; *Langer et al.*, 1972).

In agreement with the idea of an increased release of transmitter after administration of α-receptor blockers, an increased release of DβH has also been demonstrated after PBZ and phentolamine (*De Potter et al.*, 1971; *Johnson et al.*, 1971).

A variable release of NA during nerve activity may be regulated by some local feed-back mechanism. Several possibilities for such a mechanism have been discussed (*Häggendal*, 1969, 1970; *Hedqvist*, 1969 a, b; *Farnebo* and *Hamberger*, 1970, 1971; *Hedqvist*, 1970; *Wennmalm*, 1971). Modulation of the NA release could be mediated via some substance, *e.g.* prostaglandin (*Hedqvist*, 1969 a, b; *Wennmalm*, 1971). Another possible mechanism is inhibition of NA release via α-receptor-like structures on the nerve terminals (*Farnebo* and *Hamberger*, 1971; *Langer et al.*, 1972). In agreement with this hypothesis, α-receptor blockers increased the NA overflow from mouse atrium which is supposed to contain only β-receptors in the effector cells, while β-receptor blockers were without any marked effect (*Farnebo* and *Hamberger*, 1971). However, a low concentration of β-receptor blockers appears to decrease the NA output (*Åblad*, 1972, personal communication). The situation at present appears to be rather complex when one is trying to explain the mechanisms behind the variable release of transmitter at nerve activity after α-receptor blockers, and more studies are necessary.

Since there appears to exist a variable release of transmitter during nerve activity it is difficult to speak of the "true quantitative release of transmitter", particularly after using drugs such as PBZ, often administered in high concentrations.

As mentioned in the introduction, the granules (vesicles) are suggested to be the morphological correlates of the transmitter quanta since both are remarkably constant in size. The following paragraph will, however, discuss the heterogeneity of amine storage granules in the nerve terminals.

III. Functional Heterogeneity of Amine Storage Granules in the Adrenergic Nerve Terminals

The amine storage granules are complex structures containing not only NA and ATP, but also proteins, both "chromogranin A" and DβH. The granules are probably formed, more or less to maturity, in the cell bodies. Subsequently, they are transported to the nerve terminals by a specific mechanism, probably connected to the neurotubules (for a review see *e.g. Dahlström*, 1971).

The rate of this transport seems to be rapid in the PNS. In the sciatic nerve of the rat it is, for instance, about 5 to 10 mm/h. On the assumption (*Dahlström* and *Häggendal*, 1966 a) that the endogenous NA content per granule when accumulated above a constriction of the axons is about the same as the average NA content in the granules in the nerve terminals, it can be postulated that the daily contribution of granules to the nerve terminals is probably low (see also *Geffen* and *Rush*, 1968). In other words, the life-span, *i.e.* the time-period during which the individual granules are able to store endogenous NA, seems to be rather long, being in the order of several weeks. Even though recent results indicate that the granules arrested above a constriction of the axon can somewhat increase their NA content (*Dahlström* and *Häggendal*, in preparation), it still seems that the granules are capable of storing endogenous NA for some weeks in the nerve terminals. The likelihood of a long life-span gains support from studies on the recovery of endogenous NA levels after giving reserpine in a large dose, as will be discussed below.

In the nerve terminals, the granules play an important role in such functions as (a) the storage of endogenous NA, (b) uptake-storage of exogenous amines, *e.g.* ^{3}H-NA, (c) synthesis of NA from DA and also (d) release of transmitter upon nerve activity. After a large dose of reserpine (*e.g.* 10 mg/kg i.p. to rats) the transmission soon fails, the amine storage granules are deprived of their NA content, their ability to take up and store exogenous amines, *e.g.* ^{3}H-NA when given i.v., is lost, and the whole neuron is empty of NA.

The times for onset of recovery and for full recovery of these different functions are given schematically in figure 3, in which data from different sources are assembled.

Using the histochemical fluorescence technique of Hillarp and Falck and biochemical techniques, the recovery of endogenous NA is seen first in the cell bodies, about 12 to 14 h after the reserpine injection. Thereafter, by using nerve crushes, NA can be seen in the axon 15 h after administration of reserpine at a level close to the cell bodies, and after 18 h at a level 1.5—2 cm more distally (*Dahlström*, 1967). The endogenous NA levels in the nerve terminals start to recover between 24 and 36 h after injecting reserpine (*Häggendal* and *Dahlström*, 1971 a, b). The time lag between the onset of recovery in the cell bodies, in different parts of the axons, and in the nerve terminals seems to be in agreement with the formation in the cell bodies and axonal downtransport of newly formed, functioning amine storage granules.

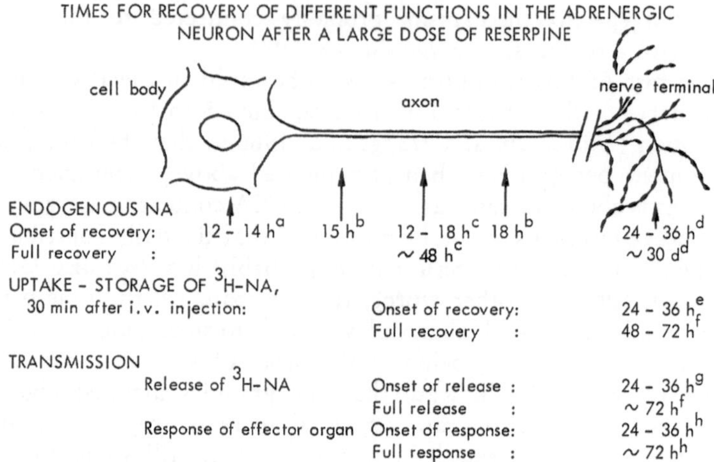

Fig. 3. Schematic illustration of the recovery of different functions in a long adrenergic neuron after a large dose of reserpine (10 mg/kg i.p. to rats). The different recoveries were studied in *in vivo* preparations. Recovery of endogenous noradrenaline (NA) was studied with histochemical and/or biochemical techniques. Small letters indicate the following references:
[a] *Dahlström et al.* (1965), *Norberg* (1965), *Dahlström* (1967); [b] *Dahlström* (1967); [c] *Dahlström* and *Häggendal* (1969); [d] *Häggendal* and *Dahlström* (1971 a); [e] *Andén et al.* (1964), *Iversen et al.* (1965), *Andén* and *Henning* (1966), *Häggendal* and *Dahlström* (1970, 1971 b, 1972), *Almgren* and *Lundborg* (1971); [f] *Almgren* and *Lundborg* (1971), *Häggendal* and *Dahlström* (1971 a, 1972); [g] *Almgren* and *Lundborg* (1971); [h] *Häggendal* and *Lindqvist* (1963, 1964), *Andén et al.* (1964), *Andén* and *Henning* (1966), *Almgren* and *Lundborg* (1971)

The production and axonal downtransport of new functioning amine granules is normal at about 48 h after reserpine injection (Fig. 4) as indicated by the accumulation of endogenous NA above a 6 h crush or ligation of the rat sciatic nerve. The amount of NA found accumulated above the crush is likely to be proportional to the approximate number of newly formed functioning amine granules, transported from the cell body into the part of the nerve above the crush. Therefore, in the following dicussion variations in the accumulated NA will be considered as variations in the amount of transported granules. After normalization, the axonal downtransport of NA-containing granules shows a period of overshooting during the first week, followed by a somewhat reduced granule transport. Normal accumulation was again found after about 3 weeks.

The recovery of the endogenous NA levels in the nerve terminals occurs slowly (Fig. 5). The pattern seems to be a function of the axonal downtransport of new granules. Thus, during the first week there is a rapid increase in the NA levels, followed by a period of less rapid increase. Such a recovery curve will be obtained if the granules accumulated in the nerve terminals can store NA for a rather long time. Also, the tendency to decreased NA levels in the tissues 4—5 weeks after reserpine injection can be explained if the granules that are downtransported in supranormal amounts during the first week are loosing their capacity to store endogenous NA after a rather long but limited life-span. At this time the axonal downtransport of granules is only normal, and not sufficient to compensate for the simultaneous loss of a large number of old granules.

In our studies the NA is a marker of the granules. The downtransported NA probably contributes very little to the turnover of NA in the nerve terminals. Instead the turnover is probably due to NA losses, *e.g.* by release, balanced mainly by a local new synthesis of NA and, to some extent, by re-uptake of released transmitter. The downtransported granules should be looked upon as factories for NA synthesis with both storage- and transmitter-releasing capacities.

The results presented indicate that the reserpine-blocked granules in the nerve terminals do not recover to any marked extent, *i.e.*, they are irreversibly damaged after a large dose of reserpine. This view is in agreement with recent results obtained by *Norn* and *Shore* (1971).

The results on recovery of endogenous NA in nerve terminals support the view that the new amine storage granules which reach the nerve terminals have a rather long life-span with respect to the *capacity to store endogenous NA.*

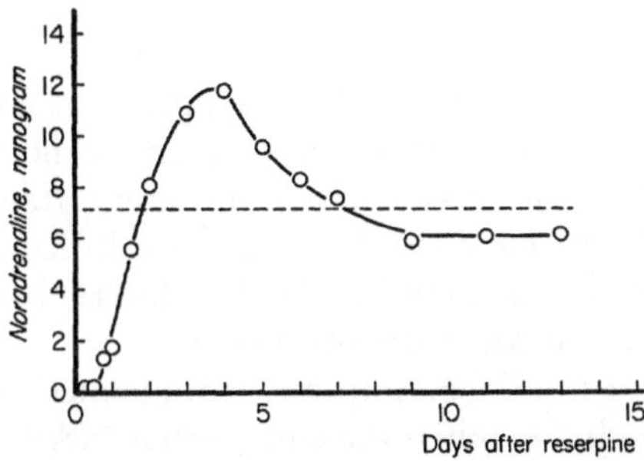

Fig. 4. Amounts of noradrenaline (NA) transported to the 1 cm part of the rat sciatic nerve just above a 6 h ligation (lig.), made at different time intervals after reserpine (10 mg/kg i.p.). NA is probably located in amine storage granules transported to this part of the nerve during 6 h. The curve is based on figure 1 in *Dahlström* and *Häggendal* (1969) and obtained by subtraction of the amounts of the amine in 1 cm unligated nerves from those found proximal to a 6 h ligation, at different times after reserpine injection. The hatched line shows the normal 6 h accumulation in unreserpinized rat. (From *Häggendal* and *Dahlström*, 1971 a, by courtesy of the J. Pharm. Pharmacol.)

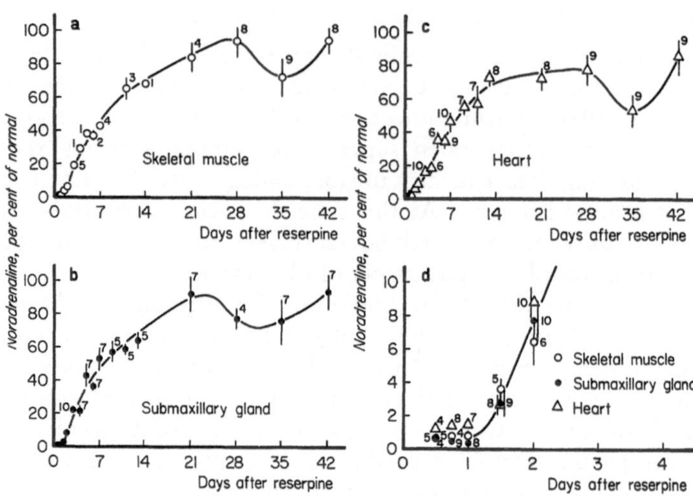

Fig. 5. The course of recovery of endogenous noradrenaline (NA) after reserpine treatment (10 mg/kg i.p.) in different tissues of the rat: (a) gastrocnemic muscle, (b) submaxillary gland, (c) heart. The NA values are expressed as percentages of the values in control rats, killed and assayed in parallel. The values were expressed per organ and not per g of tissue, to compensate for weight reductions in reserpine-treated rats. The vertical bars indicate the s.e.m., and the small figures indicate the number of experiments. (d) The NA levels in heart, salivary gland, and gastrocnemic muscle during the early period after the reserpine injection. In all three tissues the NA starts to increase between 24 and 36 h after reserpine (details from a—c). (From *Häggendal* and *Dahlström*, 1970. By courtesy of the J. Pharm. Pharmacol.)

The pattern of recovery of ^{3}H-NA uptake storage in the nerve terminals is quite different from that of endogenous NA recovery. ^{3}H-NA (2.5 μg/kg) was given i.v. at different time intervals after reserpine, and 30 min later the animals were killed and assayed for ^{3}H-NA in the salivary glands. Twelve hours before death preganglionic denervation was performed (Fig. 6).

The recovery of the ^{3}H-NA uptake-storage mechanism commences between 24 and 36 h, *i.e.* when the endogenous NA levels also start to recover. This is the time interval when newly formed functioning granules may be calculated to have reached the nerve terminals. Thereafter, however, the recovery of the uptake-storage capacity of ^{3}H-NA is very rapid, reaching normal levels after about 2—3 days. A tendency to overshoot during the first week then occurs, followed by a significantly reduced uptake-storage capacity. This curve is very similar to that for axonal downtransport of newly formed granules after administration of reserpine. In the two curves both the rise, the maximum and, most important, the decline, are strikingly congruent. This indicates that the newly downtransported granules are mainly responsible for the uptake-storage of ^{3}H-NA and that the granules *soon lose this capacity*. If this loss occurs monoexponentially, the half-life of this capacity can be calculated to be as short as about half a day. Figure 7 shows the theoretical curves for different hypothetical T 1/2 values; the curve based on T 1/2 = 12 h shows the best congruence with the empirical curve.

The results imply that the new granules appear to have different half-lives for different functions; a short half-life for the uptake-storage of ^{3}H-NA but a long-lasting one for the capacity to store endogenous NA.

There is also a striking parallel between the recovery of the uptake-storage capacity of ^{3}H-NA after reserpine and the recovery of nerve transmission (Fig. 3). The two recoveries start between 24 and 36 h after reserpine injection and both, normal transmission and normal ^{3}H-NA uptake-storage capacity, are found after about 3 days (*Andén* and *Henning*, 1966; *Almgren* and *Lundborg*, 1971). Since ^{3}H-NA recovery seems to be mainly due to the presence of

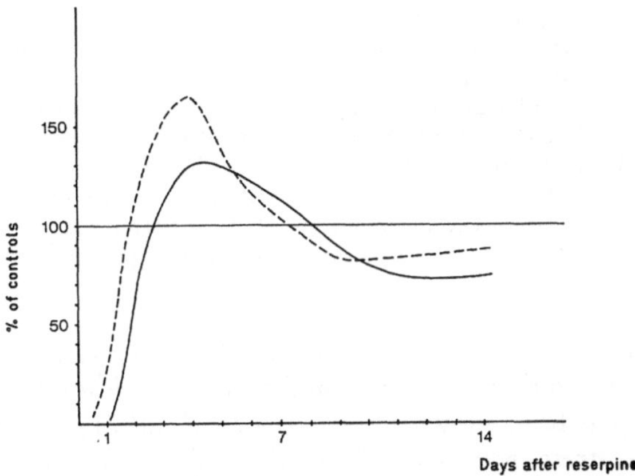

Fig. 6. The recovery curve for ^{3}H-NA uptake-retention capacity in decentralized rat salivary glands after reserpine treatment, shown together with the curve for noradrenaline accumulation in 6 h ligated rat sciatic nerves at different times after reserpine treatment (hatched line). This curve is based on results given in *Dahlström* and *Häggendal* (1969) and probably indicates the relative amounts of new amine storage granules that are transported distalward in the axons to the terminals per unit of time after reserpine treatment. The ^{3}H-NA activity per gland was measured 30 min after the i.v. injection of 2,5 μg/kg of ^{3}H-NA. (For details see *Häggendal* and *Dahlström*, 1972. By courtesy of the J. Pharm. Pharmacol.)

young amine storage granules, the release of transmitter from nerve terminals may also be associated mainly with the young amine granules (*Dahlström* and *Häggendal,* 1972; *Häggendal* and *Dahlström,* 1972).

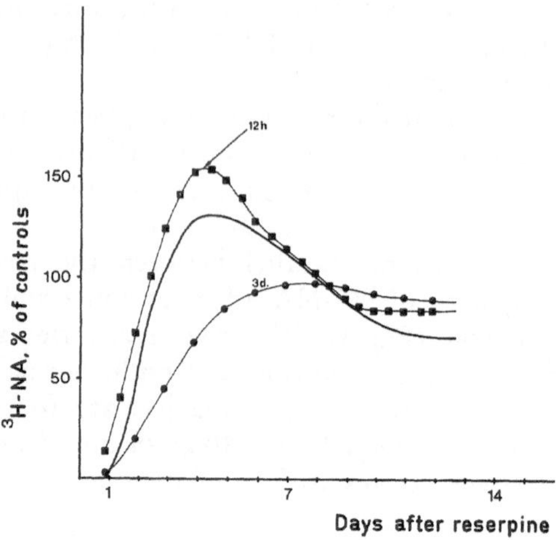

Fig. 7. Theoretical curves for the recovery of the capacity of uptake-retention of ^{3}H-NA in rat adrenergic nerve terminals after reserpine treatment. The curves were produced on the assumption that this capacity declines mono-exponentially with theoretical halv lives (T 1/2) for the capacity of the young amine granules to take up and store ^{3}H-NA of 12 h (■—■) and 3 d (●—●). The empirical curve (from Fig. 6) for ^{3}H-NA uptake-retention capacity is also indicated (——). Both the theoretical and the empirical curves are expressed in percentages of the control, *i.e.*, the ^{3}H-NA content in decentralized glands of non-reserpinized rats. (*Dahlström* and *Häggendal,* 1972. By courtesy of the Acta physiol pol.)

Furthermore, ^{3}H-NA taken up into the nerve terminals is preferentially released shortly after administration (*Chidsey* and *Harrison,* 1963). This has been shown by *Thierry et al.* (1971) to occur in the CNS when stress was applied to the animal. In the periphery, Dr. Dahlström and I found that significantly more ^{3}H-NA was released from the salivary glands upon sympathetic stimulation 20 min after ^{3}H-NA administration than 12 h later, although the total ^{3}H-NA content was not significantly changed before stimulation (Fig. 8).

Newly synthetized NA in the nerve terminals has been shown to be preferentially released upon nerve activity (*e.g. Kopin et al.,* 1968; see also *Wennmalm,* 1971 for ref.). Since the newly downtransported granules are likely to contain more DβH than the granules which have already been present in the nerve terminals for some time (see below), the local synthesis of NA is also likely to occur predominantly in the young amine storage granules.

Based on the results discussed above, we have formulated the following working hypothesis. It is likely that the amine storage granules, downtransported to the nerve terminals, have different half-lives or life-spans for different functions. The population of amine storage granules in the nerve terminals thus seems to be heterogeneous from a functional point of view. Young amine storage granules may have capacities to synthesize and store endogenous NA, to take up and store ^{3}H-NA, and to release transmitter during nerve activity. Most of these functions may decline rather quickly, whilst the capacity to *store endogenous NA* seems to be more long-lasting (*Dahlström* and *Häggendal*, 1966 a, 1972; *Häggendal* and *Dahlström*, 1971 b).

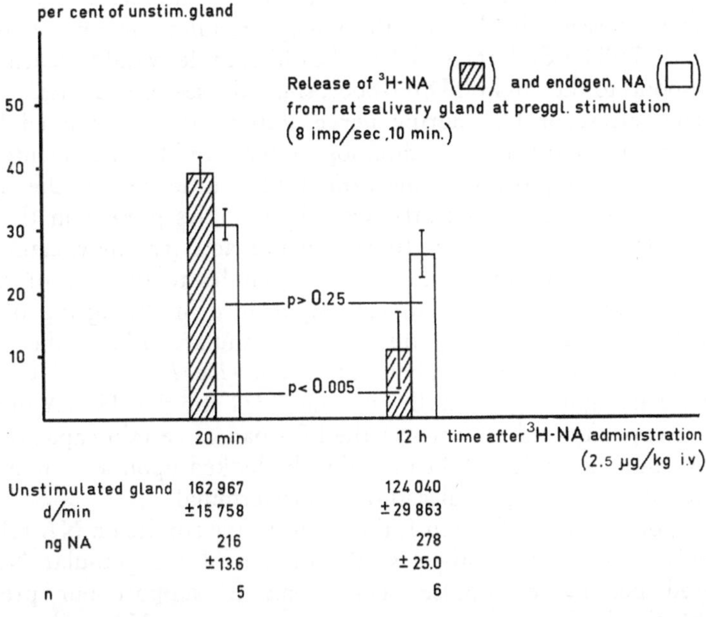

Fig. 8. Decrease of ^{3}H-NA and endogenous noradrenaline (NA) in rat salivary gland during preganglionic unilateral stimulation (8 imp/sec, for 10 min) started 20 min or 12 h after the ^{3}H-NA administration (2.5 μg/kg) i.v.). The amounts of released ^{3}H-NA and endogenous NA (difference between stimulated side and control side) are given in per cent of the respective contents in the contralateral unstimulated gland. The control values (unstimulated side) for ^{3}H-NA activity and NA content are indicated. Vertical bars indicate s.e.m. and n indicates number of observations

Thus, an ageing of the amine storage granules may be discussed, in agreement with the results of other investigators:

1. The granules in the axons, when studied electronmicroscopically are generally larger — having a diameter of 800—1000 Å — than the majority of the nerve terminal granules which are 400—500 Å in diameter (*Geffen* and *Ostberg*, 1969; *Hökfelt*, 1969; *Kapeller* and *Mayor*, 1969). Only a few per cent of the granules in the nerve terminals appear to be of the large type.
2. Sucrose gradient studies have demonstrated "light" and "heavy" particles in the nerve terminals, but only "heavy" particles in the axons (*Roth et al.*, 1968; *Chubb et al.*, 1970; *De Potter*, 1971).
3. The heavy particles were found to contain more DβH in relation to NA content than the "light" nerve terminal particles (*De Potter*, 1971; *Lagercrantz*, 1971).
4. The "heavy" particles have been identified electronmicroscopically as large dense cored vesicles (*Bisby* and *Fillenz*, 1970).

Thus, it seems likely that the young granules correspond to the "heavy", DβH-rich, large vesicles. The old granules would accordingly correspond to the "light", DβH-poor, small dense-core vesicles.

The release of NA during nerve activity is accompanied by a release of granular proteins, "chromogranin A" and DβH, as mentioned earlier. Probably, the proteins cannot be synthesized in the nerve terminals, but are incorporated into the granules present in the cell bodies. The loss of protein (during nerve activity) may cause the decrease in volume of the large, new granules to the size of small granules and may be a factor of importance in the ageing of the granules. Interestingly, the life-span of granular DβH in the nerve terminals is rather short, about 26—40 h (*De Potter* and *Chubb*, 1971). This figure agrees well with the T 1/2 for the ^{3}H-NA uptake storage capacity and may reflect the life-span of certain capacities of *young* amine granules. DβH may thus be looked upon as a mainly a marker of the young granules in the nerve terminals.

As pointed out in Section I, the quantitative studies on NA release strongly indicate that only a small fraction of the granular NA is released per nerve impulse. This seems to support our present hypothesis that young amine granules can release NA and protein several times by partial exocytosis until they are eventually changed into "old" granules.

In the nerve terminals, the transmitter has often been discussed as being present in at least two different pools: one small, easily releasable pool, and one large pool, less readily available for release (cf. *Carlsson*, 1965; *Iversen*, 1967). In view of the hypothesis presented on different functions of young and old granules, possibly the young amine granules mainly constitute the small, easily available pool whilst the old granules mainly constitute the large, less readily available pool.

Our views on a functional heterogeneity of amine storage granules, indicating that mainly young granules are functionally active, stresses the importance of a steady supply by intra-axonal transport of new amine storage granules to the nerve terminals for the normal function of adrenergic neurones.

Acknowledgements

The studies performed in the author's laboratory have been supported by the Swedish Medical Research Council (No. 14X-166).

References

Almgren, O.: Influence of synthesis and membrane pump inhibition on the nerve impulse induced disappearance of noradrenaline from rat salivary glands. Acta physiol. scand. *83*, 515—526 (1971).

Almgren, O., and *P. Lundborg*: Correlation of the recovery of the granular uptake-storage mechanism and the nerve impulse induced release of (^{3}H) noradrenaline after reserpine. J. Pharm. Pharmacol. *23*, 671—677 (1971).

Andén, N.-E., and *M. Henning*: Adrenergic nerve function, noradrenaline levels and noradrenaline uptake in nictitating membrane after reserpine treatment. Acta physiol. scand. *67*, 498—504 (1966).

Andén, N.-E., *A. Carlsson*, and *J. Häggendal*: Adrenergic mechanisms. Ann. Rev. Pharmacol. *9*, 119—134 (1969).

Andén, N.-E., *T. Magnusson*, and *B. Waldeck*: Correlation between noradrenaline uptake and adrenergic nerve function after reserpine treatment. Life Sci. *3*, 19—25 (1964).

Bevan, J.: Some characteristics of the isolated sympathetic nerve-pulmonary artery preparation of the rabbit. J. Pharmacol. exp. Ther. *137*, 213 to 218 (1962).

Bevan, J., *G. Chesher*, and *C. Su*: Release of adrenergic transmitter from terminal nerve plexus in artery. Agents and Actions *1*, 20—26 (1969).

Bisby, M. A., and *M. Fillenz*: Isolation of two types of vesicles containing endogenous noradrenaline in sympathetic nerve terminals. J. Physiol. (London) *210*, 49—50 (1970).

Bloom, F. E., and *N. J. Giarman*: Physiologic and pharmacologic considerations of biogenic amines in the nervous system. Ann. Rev. Pharmacol. *8*, 229—258 (1968).

Brown, G. L.: The release and fate of the transmitter liberated by adrenergic nerves. Proc. Roy. Soc. B. *162*, 1—19 (1965).

Burnstock, G., and *M. E. Holman*: Spontaneous potentials at sympathetic nerve endings in smooth muscle. J. Physiol. (London) *155*, 115—133 (1961).

Burnstock, G., and *M. E. Holman*: Spontaneous potentials at sympathetic nerve endings in smooth muscle. J. Physiol. (London) *160*, 446—460 (1962).

Carlsson, A.: Drugs which block the storage of 5-hydroxytryptamine and related amines. In: Handbuch der exp. Pharmacol. (*Eichler, O.*, and *A. Farah*, eds.), Vol. 19, pp. 529—592. Berlin-Heidelberg-New York: Springer. 1965.

Carlsson, A., *B. Folkow*, and *J. Häggendal*: Some factors influencing the release of noradrenaline into the blood following sympathetic stimulation. Life Sci. *3*, 1335—1341 (1964).

Chidsei, C. A., and *D. C. Harrison*: Studies on the distribution of exogenous norepinephrine in the sympathetic neurotransmitter stores. J. Pharmacol. Exp.-Ther. *140*, 217—223 (1963).

Chubb, I. W., *W. P. De Potter*, and *A. F. De Schaepdryver*: Evidence for Two Types of Noradrenaline Storage Particles in dog spleen. Nature (London) *228*, 1203 (1970).

Dahlström, A.: The effect of reserpine and tetrabenazine on the accumulation of noradrenaline in the rat sciatic nerve after ligation. Acta physiol. scand. *69*, 167—179 (1967).

Dahlström, A.: Axoplasmic transport with particular respect to adrenergic neurons. Phil. Trans. Roy. Soc. (London) *B 261*, 325—358 (1971).

Dahlström, A., *K. Fuxe*, and *N.-Å. Hillarp*: Site of action of reserpine. Acta pharmacol. toxicol. *22*, 277—292 (1965).

Dahlström, A., and *J. Häggendal*: Recovery of noradrenaline levels after reserpine compared with the life-span of amine storage granules in rat and rabbit. J. Pharm. Pharmacol. *18*, 750—751 (1966a).

Dahlström, A., and *J. Häggendal*: Studies on the transport and life-span of amine storage granules in a peripheral adrenergic neuron system. Acta physiol. scand. *67*, 278—288 (1966b).

Dahlström, A., and *J. Häggendal*: Recovery of noradrenaline in adrenergic axons of rat sciatic nerves after reserpine treatment. J. Pharm. Pharmacol. *21*, 633—638 (1969).

Dahlström, A., and *J. Häggendal*: Axonal transport of amine storage granules in sympathetic adrenergic neurons. In: Adv. Biochem. Psychopharmacol. (*Costa, E.*, and *E. Giacobini*, eds.), Vol. 2, pp. 65—93. New York: Haven Press. 1970.

Dahlström, A., and *J. Häggendal*: On the possible relation between different pools of adrenergic transmitter and heterogeneity of amine storage granules in nerve terminals. Acta physiol. Pol. *23*, suppl. 4, 67—79 (1972).

De Potter, W. P.: Noradrenaline storage particles in splenic nerve. Phil. Trans. Roy. Soc. (London) *B 261*, 313—317 (1971).

De Potter, W. P., and *I. W. Chubb*: The turnover rate of noradrenergic vesicles. Biochem. J. *125*, 375—376 (1971).

De Potter, W. P., I. W. Chubb, A. Put, and *A. F. De Schaepdryver*: Facilitation of the release of noradrenaline and dopamine-β-hydroxylase at low stimulation frequencies by α-blocking agents. Arch. int. Pharmac. Ther. *193*, 191—197 (1971).

De Potter, W. P., I. W. Chubb, and *A. F. De Schaepdryver*: Pharmacological aspects of peripheral noradrenergic transmission. Arch. int. Pharmacodyn. *196*, Suppl., 258—287 (1972).

De Potter, W. P., A. F. De Schaepdryver, E. J. Moerman, and *A. D. Smith*: Evidence for the release of vesicle-proteins together with noradrenaline upon stimulation of the splenic nerve. J. Physiol. (London) *204*, 102 to 104 (1969).

Eccles, J. C.: The Physiology of Synapses. Berlin-Göttingen-Heidelberg-New York: Springer. 1964.

Farnebo, L.-O., and *B. Hamberger*: Effects of desipramine, phentolamine and phenoxybenzamine on the release of noradrenaline from isolated tissues. J. Pharm. Pharmacol. *22*, 855—857 (1970).

Farnebo, L.-O., and *B. Hamberger*: Drug-induced changes in the release of ^{3}H-noradrenaline from field stimulated rat iris. Br. J. Pharmac. *43*, 97—106 (1971).

Folkow, B.: Impulse frequency in sympathetic vasomotor fibres correlated to the release and elimination of the transmitter. Acta physiol. scand. *25*, 49—76 (1952).

Folkow, B., and *J. Häggendal*: Some aspects of the quantal release of the adrenergic transmitter. In: New Aspects of Storage and Release Mechanisms of Catecholamines. Bayer Symp. II, 1969, pp. 91—97. Berlin-Heidelberg-New York: Springer. 1970.

Folkow, B., J. Häggendal, and *B. Lisander*: Extent of release and elimination of noradrenaline at peripheral adrenergic nerve terminals. Acta physiol. scand. Suppl. *307*, 1—38 (1967).

Fuxe, K., and *G. Sedvall*: The distribution of adrenergic nerve fibres to the blood vessels in skeletal muscle. Acta physiol. scand. *64*, 75—86 (1965).

Geffen, L. B.: The effect of desmethylionipramine upon the overflow of sympathetic transmitter from cat's spleen. J. Physiol. (London) *181*, 69—70 P (1965).

Geffen, L. B., and *B. G. Livett*: Synaptic vesicles in sympathetic neurons. Physiol. Rev. *51*, 98—157 (1971).

Geffen, L. B., B. G. Livett, and *R. A. Rush*: Immunological localization of chromogranins in sheep sympathetic neurons, and their release by nerve impulses. J. Physiol. (London) *204*, 58—59 (1969).

Geffen, L. B., and *A. Ostberg*: Distribution of granular vesicles in normal and constricted sympathetic neurons. J. Physiol. (London) *204*, 583 to 592 (1969).

Geffen, L. B., and *R. A. Rush*: Transport of noradrenaline in sympathetic nerves and the effect of nerve impulses on its contribution to transmitter stores. J. Neurochem. *15*, 925—930 (1968).

Häggendal, J.: On release of transmitter from adrenergic nerve terminals at nerve activity. Acta physiol. scand. Suppl. *330*, 29 (1969).

Häggendal, J.: Some further aspects on the release of the adrenergic transmitter. In: New Aspects of Storage and Release Mechanisms of Catecholamines. Bayer Symp. II, 1969, pp. 100—109. Berlin-Heidelberg-New York: Springer. 1970.

Häggendal, J., and *A. Dahlström*: Uptake and retention of ^{3}H noradrenaline in adrenergic nerve terminals after reserpine and axotony. Europ. J. Pharmacol. *10*, 411—415 (1970).

Häggendal, J., and *A. Dahlström*: The recovery of noradrenaline in adrenergic nerve terminals of rat after reserpine treatment. J. Pharm. Pharmacol. *23*, 81—89 (1971 a).

Häggendal, J., and *A. Dahlström*: The functional role of the amine storage granules of the sympatho-adrenal system. In: Mem. Soc. Endocrinol. (*Heller, H.*, and *K. Lederis*, eds.) *19*, 651—669 (1971 b).

Häggendal, J., and *A. Dahlström*: The recovery of the capacity for uptake-retention of ^{3}H-noradrenaline in rat adrenergic nerve terminals after reserpine. J. Pharm. Pharmacol. *24*, 565—574 (1972).

Häggendal, J., *B. Johansson*, *J. Jonason*, and *B. Ljung*: Correlation between noradrenaline release and effector response to nerve stimulation in rat portal vein *in vitro*. Acta physiol. scand., Suppl. *349*, 17—32 (1970).

Häggendal, J., *B. Johansson*, *J. Jonason*, and *B. Ljung*: Effects of phenoxybenzamine on transmitter release and effector response in the isolated portal vein. J. Pharm. Pharmac. *24*, 161—164 (1972).

Häggendal, J., and *M. Lindqvist*: Behaviour of and monoamine levels during long-term administration of reserpine to rabbits. Acta physiol. scand. *57*, 431—436 (1963).

Häggendal, J., and *M. Lindqvist*: Disclosure of labile monoamine fractions in brain and their correlation to behaviour. Acta physiol. scand. *60*, 351—357 (1964).

Hedqvist, P.: Modulating effect of prostaglandin E_2 on noradrenaline release from the isolated cat spleen. Acta physiol. scand. *75*, 511—512 (1969 a).

Hedqvist, P.: Antagonism between prostaglandin E_2 and phenoxybenzamine on noradrenaline release from the cat spleen. Acta physiol. scand. *76*, 383—384 (1969 b).

Hökfelt, T.: Distribution of noradrenaline storage particles in peripheral adrenergic neurons as revealed by electron microscopy. Acta physiol. scand. *76*, 427—440 (1969).

Iversen, L. L.: The uptake and storage of noradrenaline in sympathetic adrenergic nerves. London: Cambridge University Press. 1967.

Iversen, L. L.: Role of transmitter uptake mechanisms in synaptic neurotransmission. Br. J. Pharmac. *41*, 571—591 (1971).

Iversen, L. L., *J. Glowinski*, and *J. Axelrod*: Uptake and storage of ^{3}H-norepinephrine in reserpine-pretreated rat heart. J. Pharmac.-exp. Ther. *150*, 173—183 (1965).

Johnson, D. G., *N. B. Thoa*, *R. Weinshilboum*, *J. Axelrod*, and *I. J. Kopin*: Enhanced release of dopamine-β-hydroxylase from sympathetic nerves by calcium and phenoxybenzamine and its reversal by prostaglandins. Proc. Nat. Acad. Sci. U. S. A. *68*, No. 9, 2227—2230 (1971).

Kapeller, K., and *D. Mayor*: The accumulation of noradrenaline in constricted sympathetic nerves as studied by fluorescence and electron microscopy. Proc. Roy. Soc. (London) *B 167*, 282—292 (1969).

Katz, B.: The transmission of impulses from nerve to muscle and the subcellular unit of synaptic action. Proc. Roy. Soc. (London) *B 155*, 455—479 (1962).

Kopin, I. J.: False adrenergic transmitters. Ann. Rev. Pharmacol. *8*, 377 to 394 (1968).

Kopin, I. J., *G. R. Breese*, *K. R. Krauss*, and *V. K. Weise*: Selective release of newly synthesized norepinephrine from the cat spleen during sympathetic nerve stimulation. J. Pharmacol. exp. Ther. *161*, 271—278 (1968).

Lagercrantz, H.: Isolation and characterization of sympathetic nerve trunk vesicles. Thesis. Acta physiol. scand., Suppl. *366*, 1—44 (1971).

Langer, S. Z., *M. A. Enero*, *E. Adler-Graschinsky*, and *F. J. E. Stefano*: The role of the alfa receptor in the regulation of transmitter overflow elicited by stimulation. Abstracts of volunteers papers. Fifth International Congress on Pharmacology, 1972, nr. 802, p. 134.

Langer, S. Z., and *M. Vogt*: Noradrenaline release from isolated muscles of the nictitating membrane of the cat. J. Physiol. (London) *214*, 159—171 (1971).

Livett, B. G., *L. B. Geffen*, and *L. Austin*: Proximo-distal transport of ^{14}C-noradrenaline and protein in sympathetic nerves. J. Neurochem. *15*, 931—940 (1968).

Ljung, B.: Local transmitter concentrations in vascular smooth muscle during vasoconstrictor nerve activity. Acta physiol. scand. *77*, 212—222 (1969).

Ljung, B.: Nervous and myogenic mechanisms in the control of a vascular neuroeffector system. Thesis, Göteborg (1970).

Norberg, K.-A.: Drug-induced changes in monoamine levels in the sympathetic adrenergic ganglion cells and terminals. A histochemical study. Acta physiol. scand. *65*, 221—234 (1965).

Norn, S., and *P. A. Shore*: Further studies on the nature of persistent reserpine binding: evidence for reversible and irreversible binding. Biochem. Pharmacol. *20*, 1291—1295 (1971).

Rosell, S., *I. J. Kopin*, and *J. Axelrod*: Fate of ^{3}H-norepinephrine in skeletal muscle before and following sympathetic stimulation. Amer. J. Physiol. *205*, 317—321 (1963).

Roth, R. H., *L. Stjärne*, *F. E. Bloom*, and *N. J. Giarman*: Light and heavy norepinephrine storage particles in the rat heart and in bovine splenic nerve. J. Pharmac. exp. Ther. *162*, 203—212 (1968).

Shore, P. A.: Transport and storage of biogenic amines. Ann. Rev, Pharmacol. *12*, 209—226 (1972).

Stach, W.: Morphologie und Histochemie von Synapsen im Bereich des Magen-Darmkanals. Symposium on Neurovegetative Transmission Mechanisms. Tihany, 1972. This volume, p. 79—101.

Starke, K., H. Montel, and *J. Wagner*: Effect of phentolamine on noradrenaline uptake and release. Naunyn-Schmiedebergs Arch. Pharmak. *271*, 181—192 (1971).

Stjärne, L.: *In*: Ciba Foundation, Study Group no. 33. Adrenergic Neurotransmission. (*Wolstenholme, G. E. W.*, and *M. O'Connor*, eds.), pp. 113—115. London: Churchill. 1968.

Stjärne, L., P. Hedqvist, and *S. Bygdeman*: Neurotransmitter quantum released from sympathetic nerves in cat's sceletal muscle. Life Sci. *8*, 189—196 (1969).

Thierry, A. M., G. Blanc, and J. Glowinski: Effect of stress on the disposition of catecholamines localized in various intraneuronal storage forms in the brain stem of the rat. J. Neurochem. *18*, 449—461 (1971).

Wennmalm, Å.: Studies on mechanisms controlling the secretion of neurotransmitters in the rabbit heart. Thesis. Acta physiol. scand., Suppl. *365*, 1—36 (1971).

Author's address: Dr. *J. Häggendal*, Department of Pharmacology, University of Göteborg, Fack, S-400 33 Göteborg 33, Sweden.

Discussion

Réthelyi: The rate of disappearance of monoamines in the spinal cord below the transection as described by you is more rapid than that reported in an earlier publication (Andén, Häggendal, Magnusson and Rosengren, 1964).

Häggendal: This is due to experimental parameters. On the first and second postoperative days, there is already a decrease of NA in the one centimeter part just below the transection, but the rest of the cord distally has an almost unchanged NA content. Thus the NA content of the whole spinal cord below the level of transection is not as markedly decreased until several days after the operation. The fluorescence microscopical picture shows degenerating NA terminals in parallel to the decrease in NA content. I do not know whether anybody succeeded in demonstrating electron microscopically monoamine-containing granulated (dense-core) vesicles in the spinal gray matter.

Zs. Nagy: I should like to comment on the recovery of NA-neurones after reserpine treatment. I observed a similar pattern of restitution in dopaminergic neurones after reserpine depletion in the fresh water mussel *Anodonta cygnea.*

Häggendal: A similar mechanism of recovery can be postulated in mammalian dopaminergic neurones.

Ariëns Kappers: The term "synaptic gap", often used in connection with adrenergic systems, is not entirely correct, since, very often, the distance between the nerve terminal and the effector cell is rather large. Real synaptic contacts between adrenergic terminals and effector cells occur only seldom.

Häggendal: I agree to that, even though in many regions of the vascular bed the distance between the varicosity of the axon terminal and the smooth muscle cell is rather short. In other regions, *e.g.* in the lung, the distance may be much longer. Therefore, instead of the term "synaptic gap", it would indeed be better to use another expression.

Wollemann: Why is dopamine-β-hydroxylase released from the granules together with noradrenaline when there exists a stable pool and the granules persist?

Häggendal: I think that the release of dopamine-β-hydroxylase and chromogranin A together with noradrenaline, is an indication of a partial exocytosis at NA release. Presumably it is only the instable or soluble form of proteins that are released. Whether these proteins, when released, have any effects on the nerve terminals or effector cells, is at present unknown.

Journal of Neural Transmission, Suppl. XI, 163—180 (1974)

Pharmacology of the Adrenergic Mechanism

A. K. Pfeifer †

Institute for Experimental Medicine, Hungarian Academy of Sciences, Budapest, Hungary

With 6 Figures

Summary

The change of susceptibility to seizures was chosen as a model to investigate transmission mechanisms in the central nervous system. Catecholamine depletors, such as reserpine, prenylamine and guanethidine increase the susceptibility to seizures, given after administration of a monoamine-oxidase inhibitor, however, they are anticonvulsive. α-Methyl-m-tyrosine and α-methyl-m-tyramine which decrease brain catecholamine levels did not influence the susceptibility to seizures, possibly these methyl analogues act as false transmitters in this respect. P-Cl-amphetamine inhibited the noradrenaline and dopamine depletion produced by α-methyl-m-tyrosine, but did not change the subcellular distribution of metaraminol. P-Cl-amphetamine and some other amphetamine derivatives exerted anticonvulsive effect and inhibited the reserpine induced facilitation of convulsions. The anticonvulsive effect of p-Cl-amphetamine seemed to be independent of its serotonin depleting action. Studying the interaction of amphetamine derivatives and catecholamine depletors or receptor inhibitors the results suggested, that the central nervous system excitation and the ability to decrease susceptibility to seizures caused by amphetamine and its derivatives are independent effects. The subcellular distribution of amphetamine and p-Cl-amphetamine showed that this drugs are present in a relatively high concentration in the fraction containing synaptic vesicles. It is suggested that amphetamine and some of its derivatives are able to substitute cerebral catecholamines functionally and some effects of these drugs (e. g. the anticonvulsive effect) may be direct ones.

Drugs are considered as valuable tools to get information concerning several functional mechanisms in the organism. As it is

Table 1. *The Effect of Amine Depletors on the Pentetrazol Convulsion Threshold and Brain Amines Level in Mice*

Treatment	Time (hours)	Convulsion threshold ml/10 g pentetrazol mean ± S.D.	γ/g mean ± S.D.		
			NA	DA	5-HT
	—	0.199 ± 0.036 (25)	0.436 ± 0.061 (12)	0.624 ± 0.266 (18)	0.600 ± 0.185 (10)
Reserpine 2.5 mg/kg i.p.	2	0.133 ± 0.015 (10)	0.103 ± 0.074 (10)	0.294 ± 0.021 (4)	0.151 ± 0.035 (4)
Prenylamine 50 mg/kg s.c.	2	0.151 ± 0.03 (15)	0.105 ± 0.098 (9)	0.298 ± 0.155 (9)	0.618 ± 0.142 (4)
Guanethidine 5 mg/kg i.p.	2	0.162 ± 0.02 (20)	0.256 ± 0.136 (6)	0.620 ± 0.160 (4)	0.648 ± 0.025 (4)
MMT 50 mg/kg i.p.	2	0.212 ± 0.041 (10)	0.210 ± 0.140 (4)	0.261 ± 0.141 (4)	0.570 ± 0.210 (5)
MMT-ine 100 mg/kg i.p.	2	0.215 ± 0.06 (10)	0.230 ± 0.106 (5)	0.282 ± 0.106 (4)	0.585 ± 0.139 (5)

( ) Number of experiments.

impossible to recount all the work done on the pharmacology of adrenergic mechanisms, this paper presents some of our results dealing with the adrenergic mechanism in the central nervous system.

The change of susceptibility to seizures is a good model—according to our opinion—to investigate transmission mechanisms. A drug which facilitates convulsions facilitates the discharge of a stimulus while a drug which inhibits or diminishes convulsions impedes such a discharge.

We established that a low catecholamine (CA) level in the brain facilitates the susceptibility to seizures, whereas this is decreased by a high CA level.

It is well known that reserpine, which depletes brain noradrenaline (NA), dopamine (DA) and serotonin (5-HT) facilitates convulsions brought about by pentetrazol or by electrical stimulation.

Table 1 shows that CA depletors, such as prenylamine and guanethidine, increase the susceptibility to pentetrazol convulsions in mice. The convulsions were induced by slow intravenous infusion and the convulsion threshold was determined by the volume of pentetrazol solution which caused tonic extensor seizures. Prenylamine and guanethidine decreased the convulsion threshold as did reserpine. These compounds also decreased brain NA and DA content, but they did not influence the 5-HT content. α-Methyl-m-tyrosine (MMT) and α-methyl-m-tyramine (MMT-ine) decreased brain catecholamine levels without changing the 5-HT content and they did not influence the susceptibility to seizures. This phenomenon will be discussed later.

When the amine depletors were given after administration of the monoamine-oxidase (MAO) inhibitor tranylcypromine, all the CA depletors became anticonvulsive; NA and DA levels are about normal or sligthly higher, while the 5-HT contents were as high as in mice given tranylcypromine alone (Table 2). Tranylcypromine *per se* did not influence the convulsive threshold besides high NA, DA and 5-HT levels. One can assume that, when amine depletors are administered after tranylcypromine, the free CA content is increased and responsible for the anticonvulsant effect. I would like to emphasize that MMT and MMT-ine had also an anticonvulsant effect after pretreatment with the MAO inhibitor though they did not increase the susceptibility to seizures when given alone.

MMT and MMT-ine do not produce sedation in spite of the large CA depletion occurring. MMT shows the same metabolic pathway as does DOPA (Fig. 1). *Carlsson* and *Lindqvist* (1962) suppose that MMT-ine and metaraminol act as false transmitters; therefore, sedation does not take place at a low brain CA level. This theory is supported by the observation of the authors mentioned that the brain

Table 2. *The Effect of Both Tranylcypromine and Amine Depletors on the Pentetrazol Convulsion Threshold and Brain Amines Level in Mice*

Treatment	Time (hours)	Convulsion threshold ml/10 g pentetrazol mean ± S.D.	γ/g mean ± S.D. NA	DA	5-HT
—	—	0.201 ± 0.05 (15)	0.319 ± 0.112 (8)	0.554 ± 0.165 (11)	0.634 ± 0.176 (20)
5 mg/kg Tranylcipromine s.c.	3	0.199 ± 0.06 (30)	0.406 ± 0.149 (9)	0.712 ± 0.022 (9)	1.332 ± 0.407 (12)
5 mg/kg Tranylcipromine s.c. 2.5 mg/kg Reserpine i.p.	3 2	0.372 ± 0.110 (15)	0.349 ± 0.132 (5)	0.583 ± 0.236 (6)	1.423 ± 0.385 (6)
5 mg/kg Tranylcipromine s.c. 50 mg/kg Prenylamine s.c.	3 2	0.403 ± 0.132 (18)	0.246 ± 0.081 (5)	0.595 ± 0.235 (9)	1.249 ± 0.379 (6)
5 mg/kg Tranylcipromine s.c. 50 mg/kg MMT i.p.	3 2	0.348 ± 0.09 (10)	0.431 ± 0.158 (6)	0.647 ± 0.124 (8)	1.568 ± 0.276 (6)
5 mg/kg Tranylcipromine s.c. 100 mg/kg MMT-ine i.p.	3 2	0.374 ± 0.08 (10)	0.402 ± 0.160 (6)	0.424 ± 0.207 (6)	1.572 ± 0.226 (6)
5 mg/kg Tranylcipromine s.c. 5 mg/kg Guanethidine i.p.	3 2	0.231 ± 0.061 (20)	0.424 ± 0.153 (20)	0.651 ± 0.180 (10)	1.120 ± 0.252 (3)

MMT-ine and metaraminol levels are about as high as the amount of missing CA (*Andén*, 1964).

We investigated whether the methyl analogues could also act as false transmitters regarding the susceptibility to seizures.

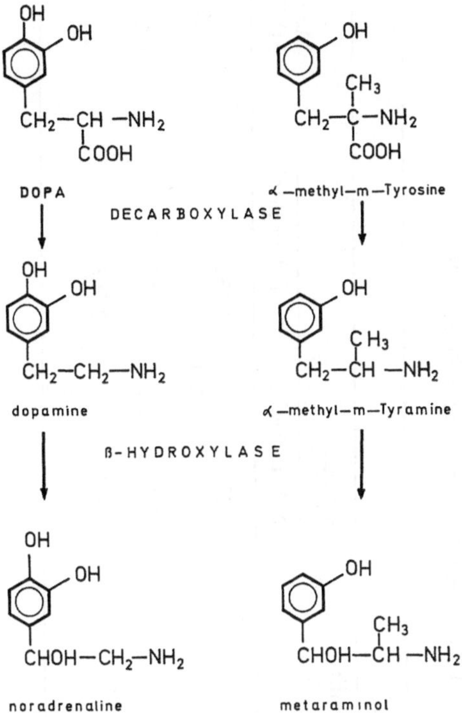

Fig. 1. The synthesis of noradrenaline and metaraminol

MMT does not influence the susceptibility to pentetrazol convulsion in mice after 2 hours (Table 3). Two hours after the administration of MMT, the amount of NA plus DA depleted is considerable and the level of the methyl-analogues in the brain is very high. After 4 and 18 hours the increase of susceptibility to seizures is significant. After 4 hours the amount of NA and DA lacking is larger than after two hours while the MMT-ine+metaraminol levels are slightly lower. After 18 hours the CA depletion is the same as after two hours but the brain content of methyl-analogues is about $^1/_4$ of the amount measured after two hours. These results support our theory that a low brain CA content facilitates the discharge of a stimulus. They also suggest that the methyl-analogues act as false transmitters in the case of susceptibility to seizures.

Table 3. *The Effect of MMT on the Pentetrazol Convulsion Threshold, and on the NA and DA, MMT-ine and MA Level of the Brain in Mice*

	50 mg/kg MMT i.p.				
	Increase of susceptibility to seizures	Missing			
Time (hours)	%	NA mμMol/g		DA mμMol/g	Metaraminol + MMT-ine mμMol/g
2	17	1.254	3.704	2.450	8.19
4	35	1.923	6.314	4.418	6.20
18	32	1.526	3.624	2.098	2.41

Thus it can be supposed that CA plays a role in the change of susceptibility to seizures. We investigated this question from another side using the so-called specific 5-HT depletor p-Chloroamphetamine. Since the early studies by *Pletscher et al.* (1964), many investigators confirmed that the chlorinated amphetamines cause a selective and long-lasting depletion of cerebral 5-HT. (*Fuller et al.*, 1965; *Pfeifer* and *Galambos*, 1967; *Miller et al.*, 1970).

P-chloroamphetamine exerts anticonvulsive effects both in mice and rats. It is remarkable that the substance depletes brain 5-HT only in rats, but not in mice. The anticonvulsive effect is long-lasting in both species, but does not parallel the 5-HT depletion in rats (Fig. 2). The anticonvulsive effect eases after 8 hours. After 18 hours the brain

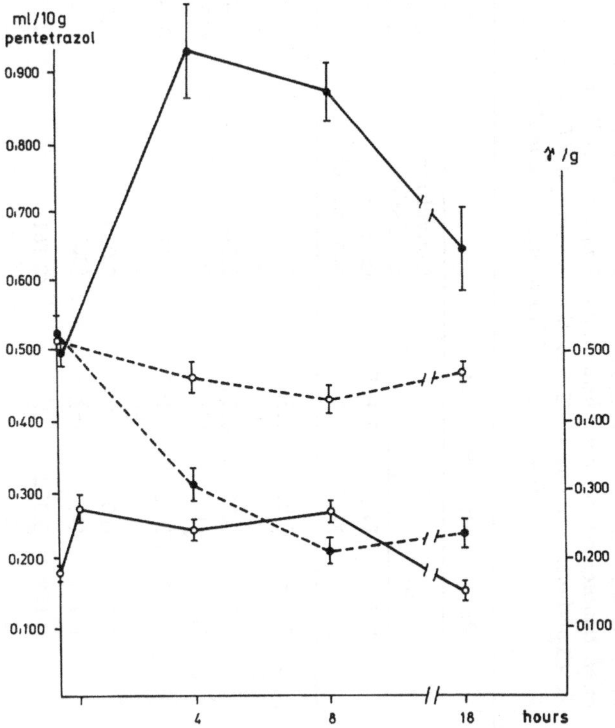

Fig. 2. The effect of 10 mg/kg p-chloroamphetamine on the pentetrazol convulsion threshold and on brain 5-HT level.
O——O convulsion threshold in mice,
•——• convulsion threshold in rats,
O-----O 5-HT level in mice,
•-----• 5-HT level in rats

Table 4. *The Effect of Reserpine and MMT on the Pentetrazol Convulsion Threshold and on the Brain Amines Content in Control Mice and in Mice Pretreated with p-Chloroamphetamine; p-Chloroamphetamine (10 mg/kg i.p.) Was Administered 4 Hours before the Experiments. Reserpine (2.5 mg/kg i.p.) and MMT (50 mg/kg i.p.) Were Given Two Hours after p-Chloroamphetamine Administration*

Treatment	Pretreatment: 0.9% NaCl				Pretreatment: p-Cl-amphetamine			
	Brain			Convulsion threshold	Brain			Convulsion threshold
	NA γ/g	DA γ/g	5-HT γ/g	ml/10 g pentetrazol	NA γ/g	DA γ/g	5-HT γ/g	ml/10 g pentetrazol
—	0.445 ± 0.08 (13)	0.997 ± 0.20 (13)	0.565 ± 0.06 (4)	0.177 ± 0.044 (50)	0.514 ± 0.10 (13)	1.044 ± 0.19 (12)	0.551 ± 0.10 (4)	0.237* ± 0.073 (20)
Reserpine	0.087 ± 0.03 (5)	0.134 ± 0.04 (5)	0.199 ± 0.02 (4)	0.118 ± 0.029 (20)	0.149 ± 0.07 (5)	0.178 ± 0.09 (5)	0.152 ± 0.02 (4)	0.304* ± 0.081 (20)
MMT	0.179 ± 0.05 (8)	0.332 ± 0.07 (8)		0.161 ± 0.043 (20)	0.376* ± 0.14 (7)	0.598* ± 0.04 (8)		0.331* ± 0.057 (15)

5-HT level is as low as when the anticonvulsive effect is maximal, but the convulsive threshold does not differ from that in the control. This fact suggests that the anticonvulsive action is not related to the alteration of the brain 5-HT level; it is rather a *per se* effect of the molecule.

How does p-Cl-amphetamine influence the effects of reserpine and MMT on the susceptibility to seizures and on the contents of monoamines of the brain?

The anticonvulsive effect of p-Cl-amphetamine takes place in the presence of reserpine in spite of NA, DA and 5-HT levels being as low as in the brain of mice treated with reserpine only (Table 4). On the other hand, p-Cl-amphetamine diminished significantly the NA and DA depletion produced by MMT. The anticonvulsive effect can also be demonstrated in this case. If we accept that the low CA level is due to the convulsion-facilitating effect of reserpine, we may suppose that p-Cl-amphetamine is able to replace the CA-s functionally. The CA depleting effect of MMT is explained by the theory that the methyl-analogues of the physiologic mediators formed dispose CA from their storages. So we supposed that p-Cl-amphetamine inhibits the binding of the methyl-analogues to the storages sites.

We investigated the subcellular distribution of ^{3}H-metaraminol (MA) administered intraventricularly in control rats and in rats which were pretreated with p-chloro-amphetamine. Table 5 shows the

Table 5. *Subcellular Distribution of ^{3}H-Metaraminol in the Brain of Control Rats and in Rats Pretreated with p-Chloroamphetamine (10 mg/kg i.p., Two Hours before the Experiments). 10 μCi ^{3}H-Metaraminol Was Administered Intraventricularly One Hour before the Experiments Started*

Fractions	Protein %	Relative specific concentration ^{3}H-metaraminol: Control	Relative specific concentration ^{3}H-metaraminol: after p-chloroamphetamine
Nuclear	7.0	0.40	0.36
Mitochondrial	56.6	0.20	0.21
Microsomal	12.3	0.69	0.69
Supernatant	24.1	3.22	3.02
Submitochondrial Fractions			
M_1	58.0	0.52	0.45
M_2	17.4	0.49	0.56
M_3	24.6	2.53	3.03

Table 6. *The Effect of Amphetamine and its Derivatives on Pentetrazol and ES Thresholds, and on Motility in Mice Pretreated with Reserpine. Reserpine or the Amphetamine and Its Derivatives, Respectively, Were Administered Two Hours, Respectively Half an Hour, before the Experiments*

Pretreatment	Treatment	ml/10 g Pentetrazol Mean ± S.E.	ES number of mice with convulsion / number of experimental mice	Motimeter Value in 30 min Mean ± S.E.
0.9% NaCl	0.9% NaCl	0.183 ± 0.012	15/16 (10.5 mA)	228 ± 26
Reserpine 2.5 mg/kg i.p.	0.9% NaCl	0.092 ± 0.003	16/16 (8.5 mA)	8 ± 1.5
Reserpine 2.5 mg/kg i.p.	Amphetamine 50 μM/kg s.c.	0.126 ± 0.012	14/16 (8.5 mA)	800 ± 137
Reserpine 2.5 mg/kg i.p.	Methamphetamine 50 μM/kg s.c.	0.161 ± 0.012	10/16 (8.5 mA)	168 ± 41
Reserpine mg/kg i.p.	p-Cl-Amphetamine 50 μM/kg s.c.	0.183 ± 0.015	8/16 (8.5 mA)	145 ± 36
Reserpine mg/kg i.p.	Trifluormethyl-Amphetamine 75 μM/kg s.c.	0.160 ± 0.02	12/16 (8.5 mA)	20 ± 6

relative specific concentration of ^{3}H-MA. Relative specific concentration means the percentage of drug concentration divided by the percentage of protein concentration in each fraction. The amount of MA is considerably higher in the soluble fractions, both in the primary and in the submitochondrial fractions, than in the particulate one. P-Cl-amphetamine did not change the subcellular distribution of MA, thus the inhibition of CA depletion cannot be explained in this way.

As we have shown before, p-chloro-amphetamine suspended the convulsion facilitating effect of reserpine without having any influence upon amine depletion. We supposed that the molecule is able to replace the CA-s functionally. We investigated the effect of amphetamine and some amphetamine derivatives on the convulsion facilitating effect and the spontaneous motility decreasing effect of reserpine. This problem is of some interest because the mechanism of amphetamine in the CNS has not been settled yet. The most accepted theory is that the action of amphetamine is mediated through CA and does not depend on its storage, but on the amount of newly synthesized CA (*Sulser* and *Sanders-Bush,* 1971). Some of our data suggest that several actions of amphetamine can be considered as direct ones.

Table 6 shows that, in mice, amphetamine, methamphetamine, p-Cl-amphetamine and trifluoromethyl amphetamine diminish the convulsion facilitating effect of reserpine, using either the intravenous pentetrazol test or electrochock. At the same time when amphetamine hypermotility occurs, the excitatory effect of metamphetamine and p-chloroamphetamine is inhibited, the motility of the experimental

Table 7. *The Effect of Amphetamine and Its Derivatives on the ES Threshold of Control Rats and in Rats Pretreated with Thyroxine (1 mg/kg/day s.c., 3 times). The Dosages of the Amphetamines Were the same as Shown in Table 6*

Treatment	ES number of rats with convulsion / number of experimental rats	
	0.9% NaCl 16 mA	Thyroxine 15 mA
0.9% NaCl	5/10	7/10
Amphetamine	0/10	2/10
Methylamphetamine	1/10	0/10
p-Chloroamphetamine	1/10	0/10
Trifluormethylamphetamine	1/10	0/10

mice being slightly less than that of the control animals. In the case of trifluoromethyl-amphetamine (which does not influence the spontaneous motility if given alone)—the tranquillizing action of reserpine manifests itself completely. These experiments show that the excitation and the ability to decrease susceptibility to seizures caused by amphetamine and its derivatives are independent effects. These results are confirmed by our experiments on rats using thyroxine. According to *Issekutz* and *Dirner* (1937), thyroxine increases the susceptibility to pentetrazol seizures in guinea-pigs. According to our previous experiments this also holds for ES in rats (*Pfeifer et al.*, 1960).

Table 7 shows that all amphetamines investigated inhibit the convulsion facilitating effect of thyroxine and elevate the convulsive threshold in control animals. It is worth mentioning that in rats pretreated with thyroxine all compounds investigated caused a greater excitation than in the control animals. Even trifluoromethyl-amphetamine became an excitant.

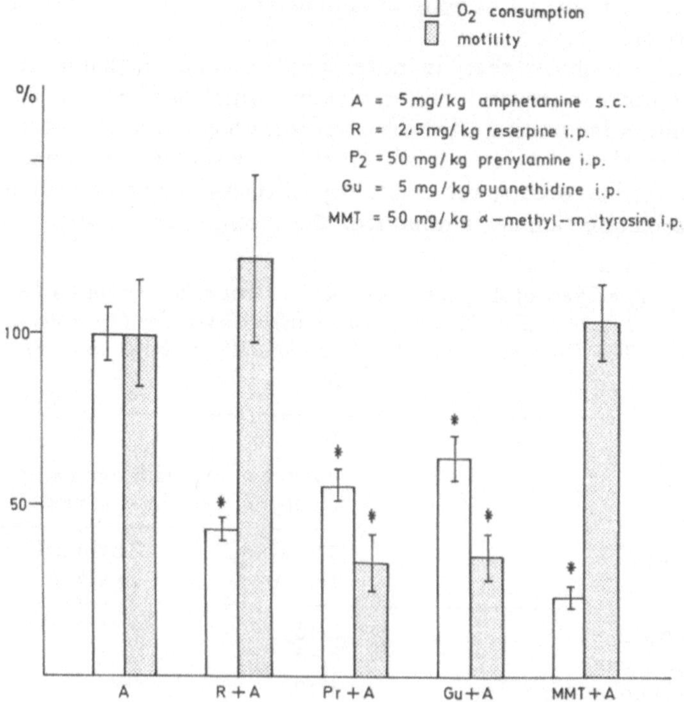

Fig. 3. The effect of amine depletors on increased metabolism and hypermotility produced by DL-amphetamine in mice

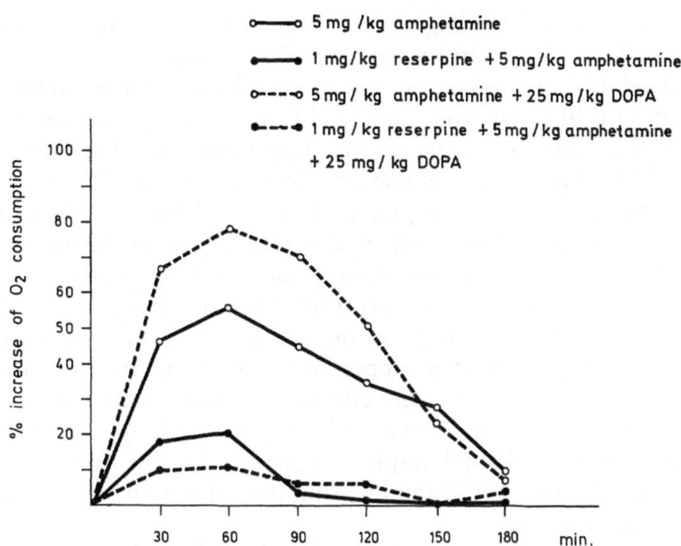

Fig. 4. The effect of reserpine and DOPA on the calorigenic effect of DL-amphetamine in rats

The effect of amphetamine and of amphetamine derivatives on the amine metabolism in brain is different. According to *Sulser* and *Sanders-Bush* (1971), amphetamine releases the newly synthesized CA and increases the turnover of 5-HT, while p-Cl-amphetamine depletes brain 5-HT, inhibits tryptophan hydroxylase and releases stored CA according to *Strada et al.* (1970). Following our experiments methamphetamine and trifluoromethylamphetamine do not influence the monoamine level of the brain. In spite of the different action of amphetamine and its derivatives on the amine metabolism, these compounds have the same effect on the susceptibility to seizures: they antagonize the convulsion-facilitating action of reserpine. Our theory is that they substitute functionally the missing CA-s.

In order to support the idea that some action of amphetamine may be direct in the CNS we compared the hypermotility and increased metabolism produced by amphetamine. The calorigenic effect of amphetamine is a central action as was proved by *Issekutz* and *Gyermek* (1949).

Hypermotility and the O_2 consumption increasing effect of amphetamine are not equally influenced by CA depletors in mice (Fig. 3). While reserpine and MMT diminish the calorigenic effect, they do not influence hypermotility; prenylamine and guanethidine both significantly decrease these effects. Previously we have shown that guanethidin decreases the NA level in the brain. This was confirmed by *Cass* and *Callingham* (1964). Our results are at variance with the idea that amphetamine action depends on the newly synthesized NA. Prenylamine and guanethidine do not inhibit CA synthesis just like reserpine and MMT; however, prenylamine and guanethidine inhibit amphetamine hypermotility. Therefore, we assume that amphetamine exerts its effect on several different brain areas. Perhaps the CA depletors have a different effect, both quantitatively and qualitatively, in different regions of the brain.

Reserpine also inhibits the calorigenic effect of amphetamine in rats, but DOPA does not restore this. On the other hand, DOPA potentiates the O_2 consumption-increasing effect of amphetamine significantly (Fig. 4). This is supported by the fact that not only the amine depletors influence the effects of amphetamine distinctly. Phenoxybenzamine, an alpha adrenergic blocking agent, and tremorine, a central cholinergic stimulant, do not influence the calorigenic effect of amphetamine, but they diminish hypermotility considerably (Fig. 5).

Thus the mechanism of the central action of amphetamine is still not quite clear. This question is of major interest because some of the amphetamine effects mimick the effects of the transmitter substances

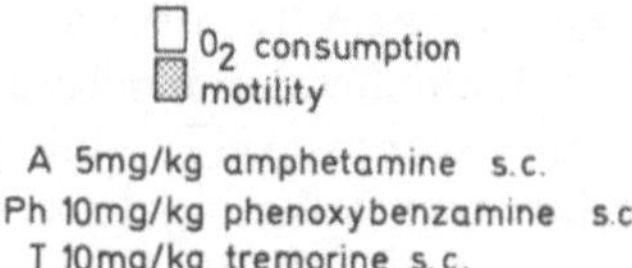

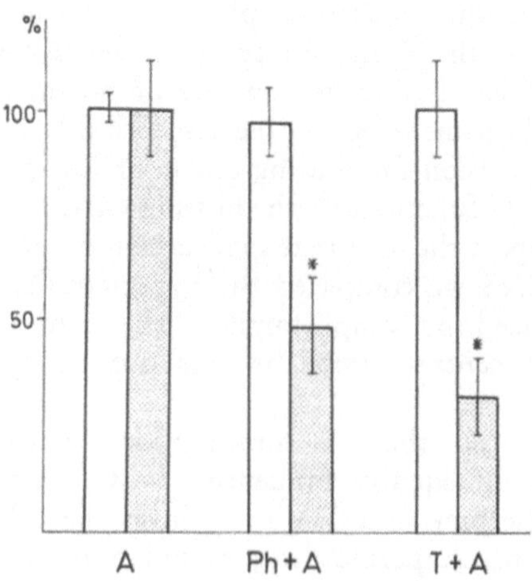

Fig. 5. The effect of tremorine and phenoxybenzamine on increased metabolism and hypermotility produced by DL-amphetamine in mice

Table 8. *The Subcellular Distribution of ^{3}H-D-Amphetamine and p-Chloroamphetamine in the Brain of Control Rats and in Rats Pretreated with Reserpine (2.5 mg/kg i.p., 4 Hours before). 5 mg/kg D-Amphetamine + 10 μCi ^{3}H-D-Amphetamine i.p., and 15 mg/kg p-Chloroamphetamine i.p. Were Injected One Hour before Decapitation*

Fractions	Protein %	Relative specific concentration					
		Amphetamine		p-Chloroamphetamine			
		Control	after reserpine	Control	after reserpine	NA*	DA*
Nuclear	34.8	0.41	0.48	0.58	0.52	0.72	0.90
Mitochondrial	35.0	0.45	0.39	1.04	0.85	1.04	0.91
Microsomal	14.6	0.50	0.40	1.50	2.17	1.42	0.93
Supernatant	15.6	4.22	4.09	1.40	1.34	0.88	1.25
Submitochondrial Fractions							
M_1	79.6	0.51	0.50	0.79	0.68	0.40	0.49
M_2	12.6	0.85	0.60	1.83	1.65	2.56	2.46
M_3	7.7	6.31	6.81	1.91	3.27	1.93	1.72

* *E. De Robertis*, Pharm. Rev. *18*, 413 (1966).

NA and DA. Does amphetamine bind to the catecholaminergic receptors or does it exert its effect trough the mediators? On the basis of our knowledge both theories are possible so far. We have tried to get some more information by comparing the subcellular distribution of amphetamine and p-Cl-amphetamine in rats and investigating whether reserpine has any influence in this respect. The subcellular distribution of the two compounds is different (Table 8).

P-chloroamphetamine shows a much greater affinity to the particulate fractions than amphetamine does. Amphetamine can be found mostly in the soluble fractions. Reserpine has not any effect on the subcellular distribution of amphetamine, but this compound increases the relative specific concentration of p-chloroamphetamine in the microsomal fraction and in the axoplasm. Table 8 also shows the relative specific concentration of NA and DA according to *De Robertis* (1966). The distribution of p-chloroamphetamine has some similarity to that of NA.

I would like to emphasize that, despite of the low relative specific concentration of amphetamine in the M_2 fraction (mostly synaptic vesicles, which are the main storage site of CA) the absolute amount of amphetamine is 0.35 μg/g while that of NA is only 0.034 μg/g according to *De Robertis* (1966). The amount of p-chloroamphetamine in the synaptic vesicles fraction is even higher, *i.e.* 1.87 μg/g. This means that the catecholamine storage sites take up and bind the amphetamines, suggesting that they may bind to the receptor site just like the physiological mediators do.

Are amphetamines in the soluble fraction present in either a bound or a free form? Soluble fractions were prepared from brains of ^{3}H-amphetamine-treated rats and filtered on a Sephadex-G-25 column. There was no difference between the two soluble fractions

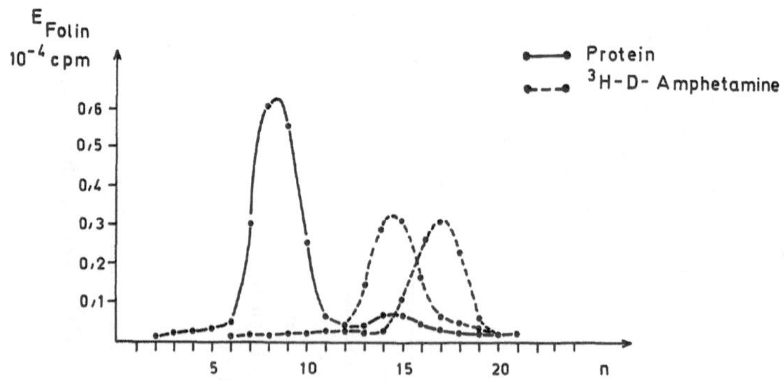

Fig. 6. The elution curves of protein and ^{3}H-D-amphetamine prepared from rat brain axoplasm

(Fig. 6). When the axoplasm was treated with TCA or sodium hydroxyde the amphetamine was eluated later. It can be assumed that amphetamine is present in the soluble fractions of the brain in a bound form. It cannot yet be decided whether this result is of any pharmacological or physiological significance. In any case it is worth mentioning that, according to our preliminary experiments, ^{3}H-NA is also partly present in a bound form in the axoplasm.

To sum up, it appears that the discharge of convulsions depends on the monoamine level in the brain. We assume that amphetamine and some of its derivatives are able to substitute cerebral CA functionally. Accordingly, it can be supposed that some of the central actions of amphetamine are direct ones.

References

Andén, N.-E.: On the mechanism of noradrenaline depletion by α-methylmetatyrosine and metaraminol. Acta pharmac. Toxicol. *21*, 260—271 (1964).

Carlsson, A., and *Margit Lindqvist*: In vivo decarboxylation of α-methyl-DOPA and α-methyl-m-Tyrosine. Acta physiol. scand. *54*, 87—94 1962).

Cass, R., and *B. A. Callingham*: Some effects of drugs which influence sympathetic transmission on tissue catecholamine level in the rat. Biochem. Pharmac. *13*, 1619—1625 (1964).

De Robertis, E.: Adrenergic endings and vesicles isolated from brain. Pharmac. Rev. *18*, 413—424 (1966).

Fuller, R. W., *C. W. Hines*, and *J. Mills*: Lowering of brain serotonin level by chloramphetamine. Biochem. Pharmac. *14*, 483—488 (1965).

Issekutz, B., and *Z. Dirner*: Wirkungsort des Thyroxins. Arch. exp. Path. Pharmak. *185*, 685—706 (1937).

Issekutz, B., and *L. Gyermek*: Die Wirkung von Dihydroergotamin und Dihydroergocornin auf den Gaswechsel. Arch. int. Pharmac. *178*, 174—196 (1949).

Miller, F. P., *R. H. Cox, Jr.*, *W. R. Snodgrass*, and *R. P. Maickel*: Comparative effects of p-Chlorophenylamine, p-Chloroamphetamine and p-Chloro-N-methylamphetamine on rat brain norepinephrine, serotonin and 5-hydroxy-indole-3-acetic acid. Biochem. Pharmac. *19*, 435—442 (1970).

Pfeifer, A. K., and *E. Galambos*: The effect of (±)-p-chloroamphetamine on the susceptibility to seizures and on the monoamine level in brain and heart of mice and rats. J. Pharm. Pharmac. *19*, 400—402 (1967).

Pfeifer, A. K., *I. Pataky*, *É. Sátory*, and *P. Vértes*: Observations on the early pharmacological effects of thyroxine. Arch. int. Pharmac. Ther. *127*, 44—57 (1960).

Pletscher, A., G. Bartholini, H. Brudere, W. P. Burkard, and *K. F. Gey*: Chlorinated arylalkylamines affecting the cerebral metabolism of 5-hydroxy-tryptamine. J. Pharmac. exp. Ther. *145*, 344—350 (1964).
Strada, S. J., E. Sanders-Bush, and *F. Sulser*: p-Chloroamphetamine. Temporal relationship between psychomotor stimulation and metabolism of brain norepinephrine. Biochem. Pharmac. *19*, 2621—2629 (1970).
Sulser, F., and *E. Sanders-Bush*: Effect of drugs on amines in the CNS. An. Rev. Pharmac. *11*, 209—230 (1971).

Reprint requests should be sent to: Dr. *A. Schaefer*, Institute for Experimental Medicine, Hungarian Academy of Sciences, Budapest, Hungary.

Discussion

Szentágothai: Do any fluorescence microscopic studies exist which show some histochemical equivalents of the pharmacological effects demonstrated?

Pfeifer: We did not perform any fluorescence microscopic studies. However, we demonstrated with electronmicroscopic investigations that the MAO inhibitor Nialamid increased the number of granulated synaptic vesicles in the ventromedial nucleus of the rat hypothalamus. The time couse of the increase in the number of granulated synaptic vesicles ran parallel with the spectrofluorometrically measured increase of noradrenaline.

Taxi: I am interested in the chemical characteristics of the substance to which noradrenaline is bound.

Pfeifer: We established that amphetamine is present in a bound form in the soluble fractions of rat brain. We have only preliminary data showing that NA is also bound. The substance to which amphetamine is bound gives the Folin reaction and its molecular weight is less than 5000.

Journal of Neural Transmission, Suppl. XI, 181—185 (1974)

An Attempt to Localize Cyclic Phosphodiesterase at Adrenergic Nerve Endings

M. A. Gerebtzoff and **J. Ziegels**

Institute of Anatomy, University of Liège, Belgium

With 3 Figures

Summary

Derived from the Gomori lead method for phosphatases, the Shanta et al. technique for cyclic phosphodiesterase presents a good specificity but does not allow an appreciation of enzyme activity levels nor a good localization. It has been tried by the authors on different tissues and particularly on cerebellar cortex.

In the concept of a double messenger system in hormonal and neurohumoral control, proposed by Sutherland and collaborators (*Sutherland et al.*, 1965; *Robison et al.*, 1971), cyclic 3′, 5′-adenosine-monophosphate (cyclic AMP) constitutes an essential link. Stimulation of an endocrine gland or a nerve fibre produces the release of a hormone or of a neurotransmitter, the first messengers, which in turn activate adenyl cyclase synthetizing from ATP cyclic AMP, the second messenger, at membrane levels of target cells. Cyclic AMP is then metabolized by a cylic phosphodiesterase to 5′-AMP (Fig. 1).

A method for the detection of this last enzyme was published by *Shanta et al.* in 1966. They incubate fresh tissue cryostat-cut slices with cyclic AMP and snake venom containing 5′-nucleotidase, and capture the phosphate thus freed with lead, the lead phosphate being visualized as lead sulphide, following the principle of the Gomori techniques.

The specifity of this method can be controlled by inhibitors of phosphodiesterase, such as methylxanthines. We have introduced a further control: inhibition of 5′-nucleotidase by a nickel salt. Preparations incubated without snake venom reveal the secondary activity of tissue 5′-nucleotidase; those incubated with 5′-AMP show the full activity of this enzyme in presence of an adequate substrate.

While the principle of the technique is sound and the above mentioned series of controls leave no doubt about its specificity, there is no real possibility to appreciate different levels of enzyme activity: the results are irregular even in the same preparation. This could be a minor drawback if localization was precise. But that is not the case.

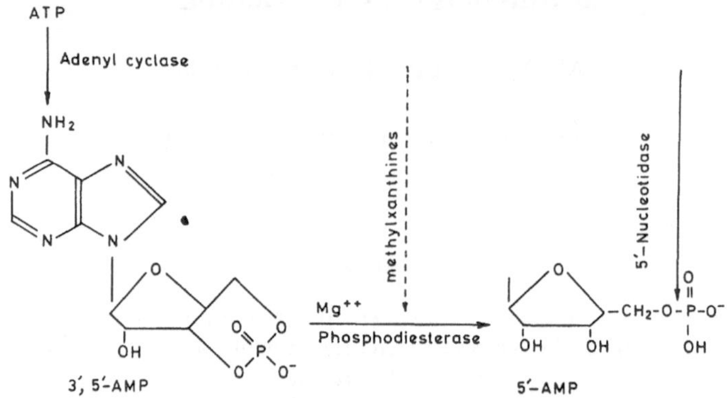

Fig. 1

In some materials investigated, such as the superior cervical ganglion, seminal vesicle, heart and pancreas, the lead sulphide precipitate was found on nuclei and nucleoli as well as on cell membranes, vessel walls, and lipid granules. The reaction was abolished by aminophylline, so that it was due to phosphodiesterase, but the precipitate appeared to scatter at random.

We investigated particularly the cerebellar cortex, since *Siggins et al.* (1969, 1971) have demonstrated that noradrenaline has an inhibitory effect on Purkinje cells. Here, the Shanta technique gives clearly defined results: cyclic phosphodiesterase activity appears to be more or less limited to the axons and terminals of basket cells (Figs. 2 and 3); but nucleoli, and sometimes nuclei, of neurones and neuroglia are positive, as in most techniques derived from the Gomori lead or cobalt methods. Some rare fibres of undetermined nature, present in the granular layer, also give a positive reaction.

Localization of cyclic phosphodiesterase in basket cells, known to inhibit the activity of Purkinje cells (*Eccles et al.*, 1967), is in good agreement with the physiological observations of the Siggins group. But, while *De Robertis et al.* (1967) showed by ultracentrifugation that this enzyme is mostly concentrated in synaptic membranes, we

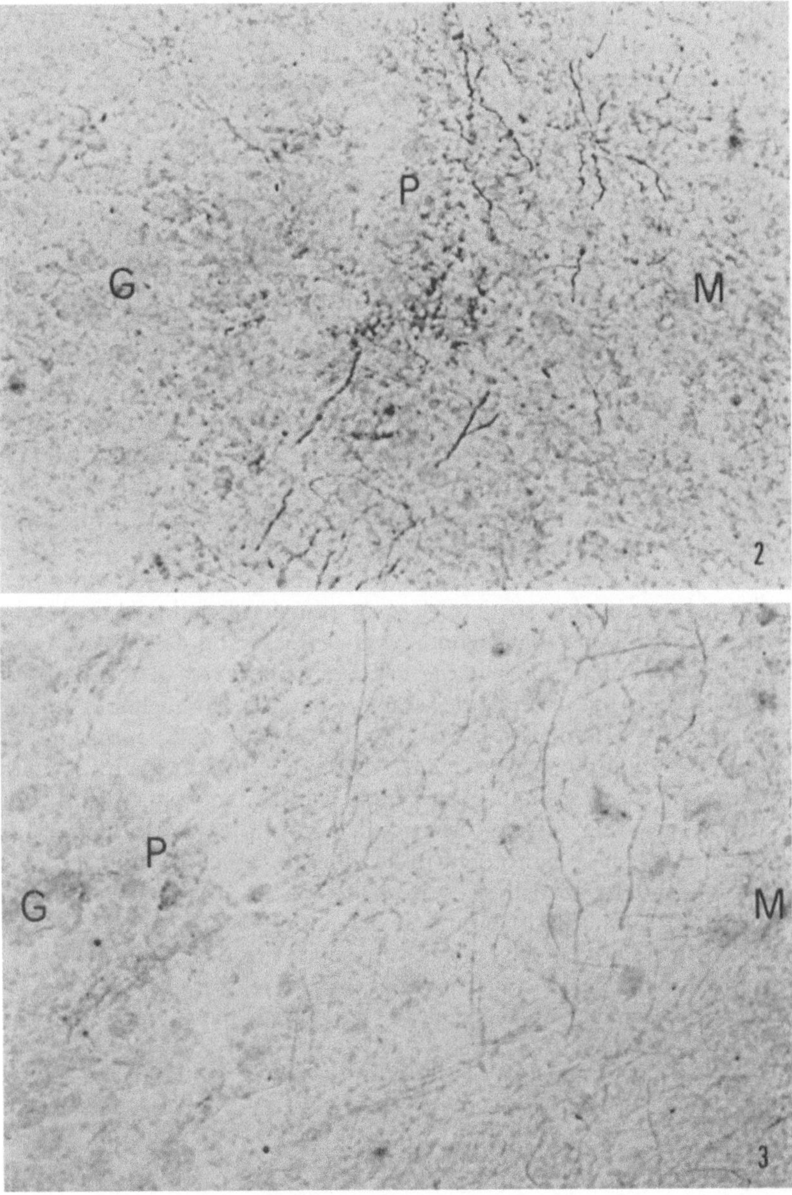

Figs. 2 and 3. Activity of cyclic phosphodiesterase in rat cerebellum. Technique of *Shanta et al.* (1966), ×400

Fig. 2. Positive reaction in axons of basket cells

Fig. 3. Positive axons branching round Purkinje cells. G, granular layer; P, Purkinje cells layer; M, molecular layer

find it also along axons. It would appear that the lead precipitate acts as an impregnation medium at the site and *near the site* of phosphate liberation. This is fatal for precise localization and might inspire false interpretations.

For 5′-nucleotidase localization, we obtain results similar, as a rule, to those of *Tewari* and *Bourne* (1963) and of *Scott* (1967), but quite different from the distribution recently described by *Davidoff* and *Galabov* (1971). This enzyme is active not only along basket cell axons, but also in afferent fibres, in cerebellar glomeruli and in T-axons of granule cells. The same reserves are valid for this localization as for that of cyclic phosphodiesterase.

Nevertheless, it seems that the first catabolic step for cyclic AMP is confined, in the cerebellum, to basket cells, while the second step, liberation of phosphate by 5′-nucleotidase, has a diffuse localization in the molecular and granular layers.

It is evident that the Shanta technique can furnish interesting results, but is, on the whole, unreliable. That is why we tried other approaches. The rupture of the oxygen bond between the 3′ carbon and the phosphate ion of cyclic AMP is accomplished by hydratation and not by reduction so that electron transfer techniques cannot be used. On the other hand, both cyclic AMP and 5′-AMP possess hydroxyl functions. During the two steps involved in cyclic AMP metabolisation by phosphodiesterase and, later, 5′-nucleotidase, we pass from one 2′ hydroxyl to two hydroxyls: in 2′ and 3′ positions. If a substrate containing either cyclic AMP or 5′-AMP is oxydized with liberation of electrons and transfer to a tetrazolium salt, by means of a dehydrogenating system, this would give a precise localization of the dehydrogenase concerned. But, in the absence of a specific dehydrogenase for the ribose hydroxyl functions, we might localize an enzyme that, in normal conditions, has nothing to do with the cyclic AMP system. Only a convergence of results with the Shanta technique and with biochemical investigations might reassure us on the localization obtained. Nevertheless, we are still working in that direction.

In conclusion, we feel that histochemists are still far from proficient in the study of the cyclic AMP system.

References

Davidoff, M., and *G. Galabov*: Die saure 5′-Nucleotidase im Zentralnervensystem der weißen Ratte. Histochemie *27*, 320—330 (1971).

De Robertis, E., G. Rodriguez de Lores Arnaiz, and *M. Alberici*: Subcellular distribution of adenyl cyclase and cyclic phosphodiesterase in rat brain cortex. J. biol. Chem. *242*, 3487—3493 (1967).

Eccles, J. C., M. Itô, and *J. Szentágothai*: The cerebellum as a neuronal machine. Berlin: Springer-Verlag. 1967. 335 p.

Robison, G. A., R. W. Butcher, and *E. W. Sutherland*: Cyclic AMP. New York: Academic Press. 1971. 531 p.

Scott, T. G.: The distribution of 5'-nucleotidase in the brain of the mouse. J. Comp. Neur. *129*, 97—114 (1967).

Shanta, T. R., W. D. Woods, M. B. Waitzman, and *G. H. Bourne*: Histochemical method for localization of cyclic 3', 5'-nucleotide phosphodiesterase. Histochemie *7*, 177—190 (1966).

Siggins, G. R., B. J. Hoffer, and *F. E. Bloom*: Cyclic adenosine monophosphate: possible mediator for norepinephrine effects on cerebellar Purkinje cells. Science *165*, 1018—1020 (1969).

Siggins, G. R., A. P. Oliver, B. J. Hoffer, and *F. E. Bloom*: Cyclic adenosine monophosphate and norepinephrine: effects on transmembrane properties of cerebellar Purkinje cells. Science *171*, 192—194 (1971).

Sutherland, E. W., I. Øye, and *R. W. Butcher*: The action of epinephrine and the role of the adenyl cyclase system in hormone action. Recent Progr. Hormone Res. *21*, 623—646 (1965).

Tewari, H. B., and *G. H. Bourne*: Histochemical studies on the distribution of alkaline and acid phosphatases and 5'-nucleotidase in the cerebellum of rat. J. Anat. *97*, 65—72 (1963).

Author's address: Prof. Dr. *M. A. Gerebtzoff*, Institute of Anatomy, University of Liège, Belgium.

Discussion

Wollemann: I would propose to try the nitro blue tetrazolium method using myokinase, hexokinase and glucose-6-phosphate-dehydrogenase in excess, and to include specific phosphodiesterase inhibitors. Since phosphodiesterase is localized mainly in the cytoplasm, only a diffuse intracellular localization can be expected.

Gerebtzoff: This suggestion might be quite interesting, in spite of the complexity of the proposed technique and the probable interference of intrinsic glucose-6-phosphate dehydrogenase.

Csillik: I want to point out the importance of such critical studies in enzyme histochemistry, like this one. The original Shanta technique is performed on non-fixed cryostat sections. What is the result after fixation?

Gerebtzoff: In my experience, formaldehyde fixation does not yield more favorable results, and inhibition of enzyme activity begins very soon. A short glutaraldehyde fixation appears to be promising. Postfixation is quite disastrous. Concerning the possible damage or displacement of the enzyme and/or the reaction product by freezing, I admit such a possibility. So far, I do not have experience with techniques omitting freezing.

Journal of Neural Transmission, Suppl. XI, 187—193 (1974)

Effects of Papaverine Derivatives on Cyclic 3', 5'-AMP Phosphodiesterase and Relaxation of Rabbit Ileum

S. F. Berndt and **H.-U. Schulz**

Department of Neurology, University of Würzburg, and Department of Pharmacology, Medical School of Hannover, Federal Republic of Germany

With 6 Figures

Summary

The effects of papaverine derivatives on cyclic 3′, 5′-AMP phosphodiesterase (PDE) and on the spontaneous contractions of isolated rabbit ileum were investigated. There was a strong correlation between the relaxing activity and the inhibition of PDE for eupaverine, ethylpapaverine and papaverine. Eupaverine was the most effective spasmolytic agent and the most potent inhibitor of PDE activity, followed by ethylpapaverine and papaverine. In contrast we found a very strong relaxing effect but only slight inhibition of PDE activity by tetrahydropapaveroline (THP). Decreased contractions could be mimicked by exogenously given cyclic AMP and cyclic dibutyryl AMP. Our results support the assumption that smooth muscle relaxation in rabbit ileum is mediated by cyclic AMP.

It is well established that papaverine has a strong spasmolytic effect on smooth muscles. Evidence was presented from *Kukovetz* and *Pöch* (1970) that papaverine is a very potent inhibitor of coronary artery phosphodiesterase (PDE). Their results on circular strips of coronary arteries represent inhibition of phosphodiesterase in a dose-dependent manner, in a good agreement with the relaxant effect. These authors suggest that spasmolytic action of papaverine is due to the decreased activity of phosphodiesterase and mediated by consecutive accumulation of cyclic AMP.

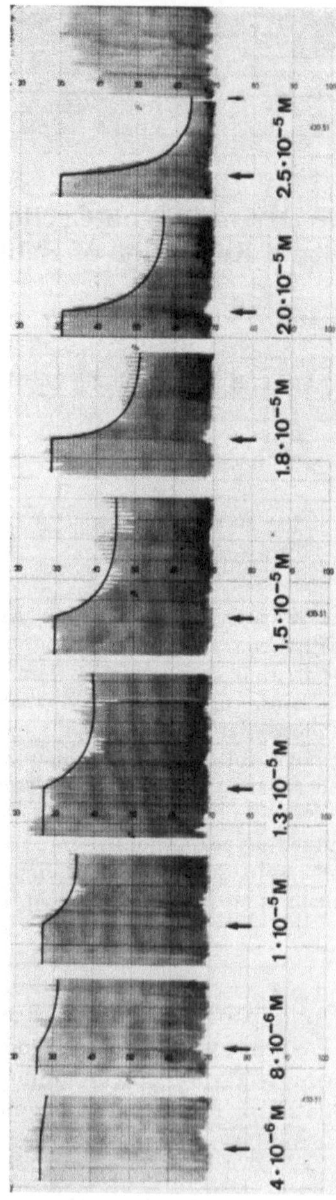

Fig. 1. Effects of papaverine on spontaneous contractions of isolated rabbit ileum in tyrode solution (See text)

Accepting this hypothesis it was of interest to study the inhibitory effect of papaverine derivatives on partially purified PDE *in vitro* in comparison to their relaxing activity on isolated rabbit ileum. Fig. 1 shows a typical experiment with isolated rabbit ileum in an organ bath in which the inhibition of spontaneous contractions by varying dosis of papaverine is demonstrated. At the end of each test, papaverine has been washed out twice with tyrode-solution. In this special experiment, the 50 % inhibition of spontaneous contractions is reached at about 17.5 μM. Even at the end of this run the washing out effect is completely satisfying.

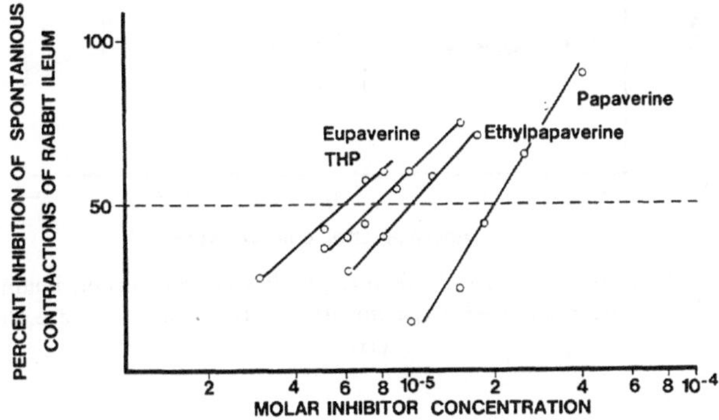

Fig. 2. Dose response curves of papaverine and papaverine derivatives. Each point is the mean of 3 experiments

Similar experiments have been carried out with several papaverine derivatives. The results of these investigations are summarized in Fig. 2. Tetrahydropapaveroline (THP) is shown the most effective spasmolytic agent, followed by eupaverine, ethylpapaverine and papaverine. Correspondent concentrations at 50 % inhibition were 6, 7.5, 10, 20 μM showing that papaverine was the weakest spasmolytic agent.

In the following investigation the inhibition of rabbit ileum PDE by papaverine derivatives was studied, using a partially purified PDE. Three adult male rabbits, fed conventionally, were stunned and exsanguinated. Pieces of about 3 cm of the distal ileum were excised from each animal, rinsed with cold saline and the fat was carefully removed. Then the pieces were pooled and pulverized in a mortar with liquid nitrogen. 5 g of the powder was immediately homogenized in 3 volumes of ice-cold 0.154 M KCl, filtered and centrifuged at 100,000$\times$g for 30 min at 3° C. The supernatant was adjusted to

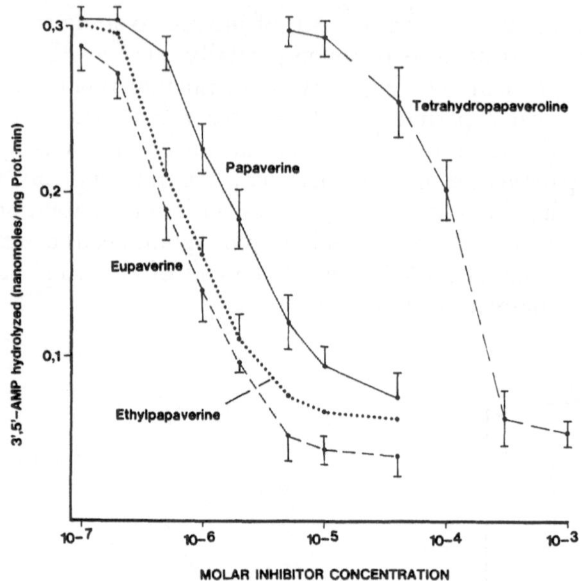

Fig. 3. Activity of 3′, 5′-AMP phosphodiesterase of rabbit ileum. Each point is the mean of 4—6 experiments. Vertical bars indicate the standard error

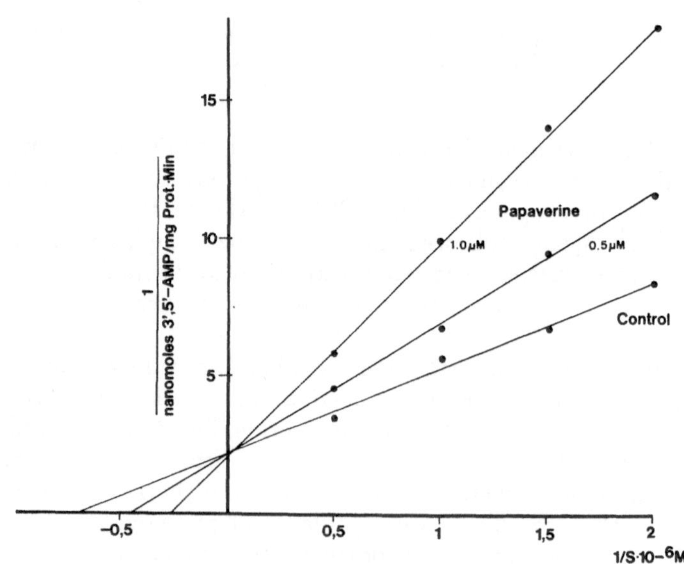

Fig. 4. Lineweaver-Burk plot of rabbit ileum phosphodiesterase. Each point represents the mean of 3 experiments

30 % saturation with ammonium sulfate maintaining the pH at 7.5 and allowed to stand at 3° C overnight. The precipitate was collected by centrifugation at 3000×g for 30 min, dissolved in 0.01 M Tris-HCL-buffer pH 7.5 and dialysed overnight at 3° C against 100 volumes of the same buffer. This purified phosphodiesterase could be stored in frozen state at —25° C until used without loss of activity.

Activity of cyclic AMP phosphodiesterase is demonstrated in Fig. 3. The values obtained using the method of *Thompson* and *Appleman* (1971) with several modifications due to nonspecific binding of nucleosides to the anion exchange resin are plotted as hydrolyzed nmoles cyclic AMP/mg protein against a wide concentration range of inhibitor substances. The data obtained with the modified method were in good agreement with corresponding values obtained by using ^{14}C-cyclic AMP as substrate for PDE and subsequent separation of cyclic AMP by thin layer chromatography. All compounds demonstrated in this graph represent typical dose-response-curves.

Eupaverine proved to be the strongest inhibitor of PDE whereas THP was the weakest one. The curve line of ethylpapaverine is very close to that of eupaverine, but there is a rather distinct distance to papaverine. 100-fold more concentrations of THP are necessary to get equal degrees of PDE inhibition, as seen in Fig. 3. In order to study the kinetic behaviour of the isolated rabbit ileum phosphodiesterase, *Lineweaver-Burk* plots were applied (Fig. 4). The linear regression curves are computer-calculated, resulting in Michaelis-Menten constant of 1.5 μM. The intercept of the control with the papaverine inhibition curve is located directly on the ordinate, which means a competitive type of inhibition. The same type of inhibition was seen for papaverine derivatives, suggesting a high specifity for inhibition of rabbit ileum phosphodiesterase.

Table 1. *Comparison of Inhibition Constants and Spasmolytic Action (See Text)*

	INHIBITION CONSTANT K_i (μM)	SPASMOLYTIC ACTION I_{50} (μM)
EUPAVERINE	0,6	7,5
ETHYLPAPAVERINE	0,8	10
PAPAVERINE	2	20
TETRAHYDRO-PAPAVEROLINE	350	6

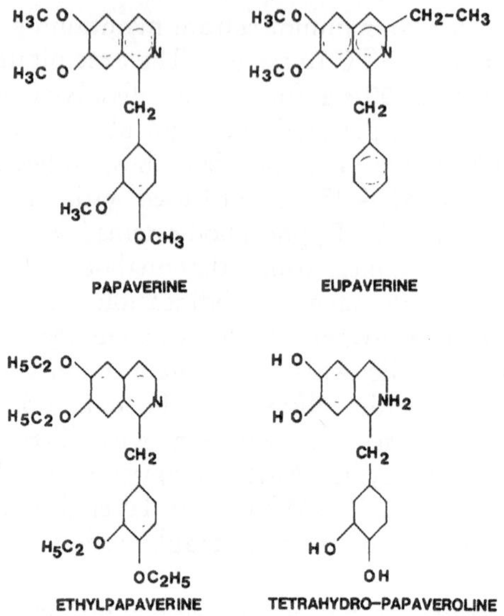

Fig. 5. Chemical formulas of papaverine derivatives

The inhibition constants K_i calculated by Dixon-plots and data for 50 % inhibition of spontaneous contractions (I_{50} in μM) are illustrated in Table 1. Regarding these two parameters there is a very good correlation between the first three drugs. In this case the decrease in sensitivity corresponds to the spasmolytic action for each substance with the only exception of THP. While having a very potent spasmolytic action, the lowest I_{50} 6 μM there is only weak inhibition of phosphodiesterase. The K_i-value is 350 μM.

An explanation for these findings is possibly given by the molecular conformation of this papaverine derivative (Fig. 5). In contrast to papaverine, eupaverine and ethylpapaverine the molecule of THP contains rather unchanged catechol structure. As catechols are known to stimulate adenylcyclase directly, leading to consecutive accumulation of cyclic AMP, it seems to be likely that THP has the same mechanism of action.

In further experiments cyclic AMP or dibutyryl cyclic AMP (DBA) was given exogenously to the organ bath (Fig. 6). As already shown by other authors, both substances are able to relax isolated rabbit ileum. Surprisingly, the effect of cyclic AMP was even more marked than that of DBA. The 50 % inhibition of cyclic AMP is at about 45 μM, whereas DBA is 10 times less potent.

We obtained a good correlation between spasmolytic action of papaverine, eupaverine and ethylpapaverine and their inhibitory effect on PDE of rabbit ileum in organ bath. The assumption that cyclic AMP is the mediator of smooth muscle relaxation is further supported by the potent spasmolytic action of THP, probably due to the direct stimulation of adenylcyclase.

Direct estimations of cyclic AMP levels in rabbit ileum in an organ bath are in progress in our laboratory now. The preliminary results are in good agreement with the data presented in this paper.

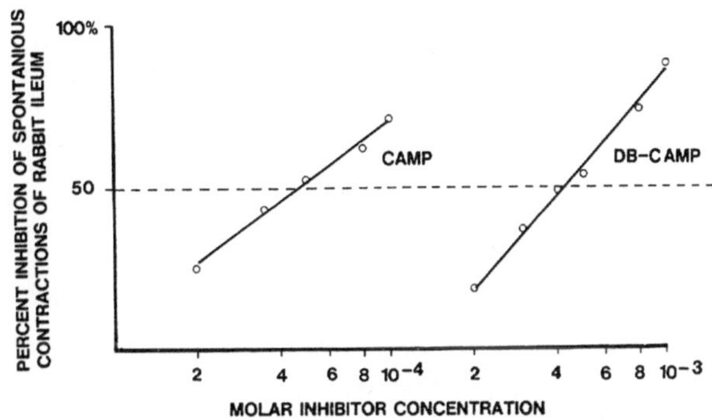

Fig. 6. Dose response curves of cycl. AMP and dibutyryl cycl. AMP. Each point represents the mean of 3 experiments

References

Kukovetz, W. R., and *G. Pöch*: Inhibition of cyclic 3′, 5′-nucleotide-phosphodiesterase as a possible mode of action of papaverine and similarly acting drugs. Naunyn-Schniedeberg's Arch. Pharmak. *267*, 189—194 (1970).

Thompson, W. J., and *M. M. Appleman*: Multiple cyclic nucleotide phosphodiesterase activities from rat brain. Biochem. *10*, 311—316 (1971).

Author's address: Dr. *S. F. Berndt*, Department of Neurology, University of Würzburg, Josef-Schneider-Straße 2, D-8700 Würzburg, Federal Republic of Germany.

Journal of Neural Transmission, Suppl. XI, 195—212 (1974)

Spinal Transmission of Autonomic Processes

M. Réthelyi

First Department of Anatomy, Semmelweis University Medical School, Budapest, Hungary

With 10 Figures

Summary

Some of the knowledge now available concerning the spinal autonomic processes in the term of sympathetic and parasympathetic reflex paths have been summarized in this chapter.

The scanty information about the spinal parasympathetic reflex path refers exclusively but not unequivocally to the localization of preganglionic parasympathetic neurons in the sacral spinal cord.—The sympathetic reflex path has been separated into three components: 1. primary visceroafferent fibers, 2. interneurons and 3. efferent, preganglionic sympathetic neurons. 1. The primary visceroafferent fibers appear to terminate in the Vth lamina of the spinal grey matter, ventrally of the main termination area of the cutaneous primary afferent fibers. Histological data suggest the possible termination of visceroafferent fibers in the Clarke's column in the mid-thoracic segments. No indication can be found about a direct (monosynaptic) connection between primary sensory fibers and efferent neurons. 2. As the sympathetic reflex activity is travelling along spinal and supraspinal routes, both spinal and supraspinal interneurons have to be considered. According to indirect electrophysiological data the spinal interneurons would be localized in the Vth lamina and in more ventral regions of the spinal grey matter. Thus they may serve as convergence points of impulses arriving from viscera as well as from the skin and from muscles. Neurophysiological observations indicate that the supraspinal interneurons are localized in the medulla. 3. Light microscopic and ultrastructural analysises of the preganglionic sympathetic neurons in the intermedio-lateral nucleus have shown that I. the dendritic tree of the neurons is oriented in cranio-caudal direction; II. the axon of the neurons after having left the nucleus turns ventrally and courses along the lateral margin of the ventral horn;

III. the presynaptic fibers approach the nucleus at the lateral circumference and the fibers establish repeated climbing-type contacts with the dendrites and perikarya; IV. three types of axon terminals can be found containing different types of vesicles.—Finally a tentative scheme of the neuronal organization of the sympathetic reflex path is given based upon the results of some preliminary degeneration experiments of the author.

The spinal transmission of autonomic processes progresses along the autonomic reflex pathways. As the efferent visceral outflow from the spinal cord is made up of sympathetic and parasympathetic divisions, the visceral stimuli may pass by two different chains of neurones.

The parasympathetic reflex path has not been given much attention sofar. The scanty information available concerns mainly the efferent sacral parasympathetic nucleus. Earlier data, based almost entirely on normal material (*Jacobson*, 1908; *Laruelle*, 1936; *Rexed*, 1954) have been complemented by recent results. The localization of the sacral parasympathetic neurones was shown by antidromically evoked potentials (*Schnitzlein et al.*, 1963) and by retrograd changes following section of the pelvic nerve (*Oliver et al.*, 1969). These studies indicate that the sacral parasympathetic neurones occupy a well circumscribed area in the lateral part of the intermediate region in the sacral segments. The intermedio-lateral nucleus is involved in this area, but parasympathetic neurones were also found medial to this nucleus. No direct histological or neurophysiological data could be obtained on the afferent fibres and interneurones participating in the parasympathetic reflex path.

All components of the *sympathetic reflex path* are relatively well known due to the extensive neurophysiological and neurohistological studies. As this review deals with the spinal transmission mechanisms of autonomic processes, the peripheral part of the sympathetic reflex arc will be omitted. The central course of the afferent neurones, the interneurones and the efferent (preganglionic) neurones will be outlined separately. Finally, the neuronal organization of the sympathetic reflex path will be described. It is generally agreed that the sympathetic reflex activity in preganglionic sympathetic neurones is mediated via spinal as well as through supraspinal pathways (*Sell et al.*, 1958; *Coote* and *Downman*, 1966; *Koizumi et al.*, 1968). In the following description both components will be considered.

Central Course and Termination of First Order Visceroafferent Neurones

The main difficulty in tracing, by anatomical means, the central course of first order visceroafferent fibres is the location of their perikarya in the spinal ganglia. Here they are intermingled with the cells of origin of somatic afferents and the two cell types do not differ histologically. Thus we have no direct means to separate the intraspinal course of the two kinds of neurones by degeneration methods using dorsal radicotomy. In order to overcome this difficulty, *Szentágothai* (1966) tried to compare the degeneration findings after dorsal radicotomy in segments having numerous visceroafferents with those in segments having less. As only segments with relatively similar structural arrangement of the grey matter can be compared, the midthoracic segments (Th_{7-9}) were taken as having a considerable number of visceroafferents while the upper lumbar segments (L_{3-4}) were chosen as segments having a similar overall arrangement but being obviously poor in visceroafferents. *Szentágothai* (1966) concluded that no difference could be observed in the dorsal horn. The massive degeneration in the intermedio-medial nucleus, pointed out by him, cannot really be decisive because it is equally present in the two representative segmental groups. The only obvious difference present was in the distribution of the dorsal root fibres in Clarke's column. In the upper lumbar segments the dorsal root fibres enter Clarke's column only 2—3 segments above their entrance into the spinal cord, while in the lower thoracic segments they reach it at the level of their entrance. This difference gave *Szentágothai* (1966) the idea that Clarke's column, or at least some of its neurones in the midthoracic segments where they have no direct connections with the proprioceptive afferents from the limb—might relay impulses from visceroafferents.

Microelectrode recordings revealed important details concerning the distribution and termination of visceroafferent fibres and about the interaction of viscero- and somatoafferents in the spinal cord. The visceroafferent fibres terminate in lamina V of *Rexed* (1954), just below the main termination area of the cutaneous primary afferent fibres (*Pomeranz et al.*, 1968; *Selzer* and *Spencer*, 1969). On the other hand, a group of visceroafferents, mainly the large diameter myelinated fibres (A-beta fibres) fail to generate local segmental reflexes (*Widen*, 1955; *Downman*, 1955; *Franz et al.*, 1966; *Pomeranz et al.*, 1968). They probably bypass dorsal horn grey matter which might mean that primary visceroafferent fibres reach directly supraspinal regions via the dorsal column system. Microelectrode recordings and degeneration studies both point to the probability that no direct connection occurs between the primary sensory fibres and the preganglionic sympathetic neurones (*Schimert*, 1939; *Szentágothai*, 1948; *Beacham* and *Perl*, 1964).

It should be emphasized that primary afferent fibres from the skin (A-alpha, beta, gamma and C-fibres) as well as from muscles (Group II, III and IV afferents) may represent the afferent neuron in the sympathetic reflex arc (*Sato* and *Schmidt*, 1966; *Sato et al.*, 1969; *Schmidt* and *Schönfuss*, 1970; *Coote* and *Perez-Gonzalez*, 1970; *Kirchner et al.*, 1970).

Interneurones Involved in the Segmental (A) and Supraspinal (B) Sympathetic Reflex Arcs

A. Spinal interneurones can be characterized functionally by their input and histologically by their localization. Interneurones receiving visceral input can be excited and inhibited by cutaneous and to a lesser extent by muscle primary afferent fibres (*Pomeranz et al.*, 1968; *Selzer* and *Spencer*, 1969). This arrangement yields a good morphological basis for referred pain (Rouch's convergence projection theory, quoted and explained by *Selzer* and *Spencer*, 1969). No indication can be found how these neurones are connected with descending fibres.

Interneurones with visceral input are located (taking benefit from the termination of the primary visceroafferents) in the Vth lamina and perhaps in more ventral layers, *i.e.*, in the base of the dorsal horn and in the more dorsal part of the intermediate zone. Using dendroarchitectonic criteria, the spinal interneuron pool can be divided into two groups. Interneurones in the four dorsal laminae (the head of the dorsal horn) are provided with longitudinally oriented dendrites. Interneurones in the intermediate zone have radially oriented dendrites confined to the transverse plane (*Scheibel* and *Scheibel*, 1968; *Szentágothai* and *Réthelyi*, 1973). The Vth lamina and thus the interneurones of the spinal component of the sympathetic reflex arc belong to the latter group. The primary afferent fibres show a similar distribution: fibres in the head of the dorsal horn run longitudinally, whereas fibres terminating in the intermediate zone are confined to the transverse plane (*Sterling* and *Kuypers*, 1967; *Scheibel* and *Scheibel*, 1969). It is premature to speculate how the spatial distribution of the interneurones and that of the presynaptic fibres determines the connection and function of the interneurones. One point seems to be worth mentioning. Due to the radial orientation of dendrites in the intermediate zone and to the similar course of presynaptic axons, these neurones are able to collect the activity of a great deal of fibres. In other words, a high degree of convergence of stimuli can be supposed to occur in these neurones. This can be applied well to interneurones of the autonomic reflex path, because they receive fibres from the viscera as well as from the skin and from muscles (*Pomeranz et al.*, 1968).

Splanchnic impulses reach higher brain centres including the cerebral cortex via the spinothalamic tract (*Aidar et al.*, 1952; *Korn*, 1969). Histologically this indicates that axons of the interneurones, or their collaterals, join the spinothalamic tract.

B. The perikarya of interneurones establishing the supraspinal component of the sympathetic reflex path are localized in the medulla, more precisely in the medial reticular formation at the obex level (*Alexander*, 1946; *Sell et al.*, 1958; *Weidinger et al.*, 1961). Central excitatory and inhibitory networks are supposed to exist here that can be affected independently (*Koizumi et al.*, 1968). The axons of these interneurones descend in the lateral funiculus: excitatory fibres in the dorsolateral part of the lateral funiculus (*Kerr* and *Alexander*, 1964) while the opinion on the exact location of the inhibitory tracts diverge (*Downman* and *Houssian*, 1958; *Prout et al.*, 1964; *Illert* and *Seller*, 1969).

Recently, attention became focused on the descending interneurones of the supraspinal component since the fluorescence microscopical technique became available. *Carlsson et al.* (1964) and later *Dahlström* and *Fuxe* (1965) found monoaminergic nerve fibres in the lateral funiculus and a dense plexus of monoaminergic (NA and 5-HT containing) terminals in the intermedio-lateral nucleus. After a total transection of the spinal cord, practically all monoaminergic terminal disappear below the lesion. Conversely, there is a marked accumulation of monoaminergic fibres just above the lesion. Perikarya showing fluorescence activity could not be encountered in the spinal cord, but they are quite numerous in the medulla, in the region where the supraspinal sympathetic centres are assumed to be localized. The findings listed above led to the conclusion that a descending monoaminergic fibre tract connects monosynaptically the bulbar autonomic nuclei with the spinal intermedio-lateral nucleus (*Dahlström* and *Fuxe*, 1965).

The Neuroarchitecture of the Intermedio-Lateral Nucleus (Preganglionic Sympathetic Neurones)

Due to recent neurophysiological observations (*Hongo* and *Ryall*, 1966; *Polosa*, 1967) it is generally accepted that the cell bodies of

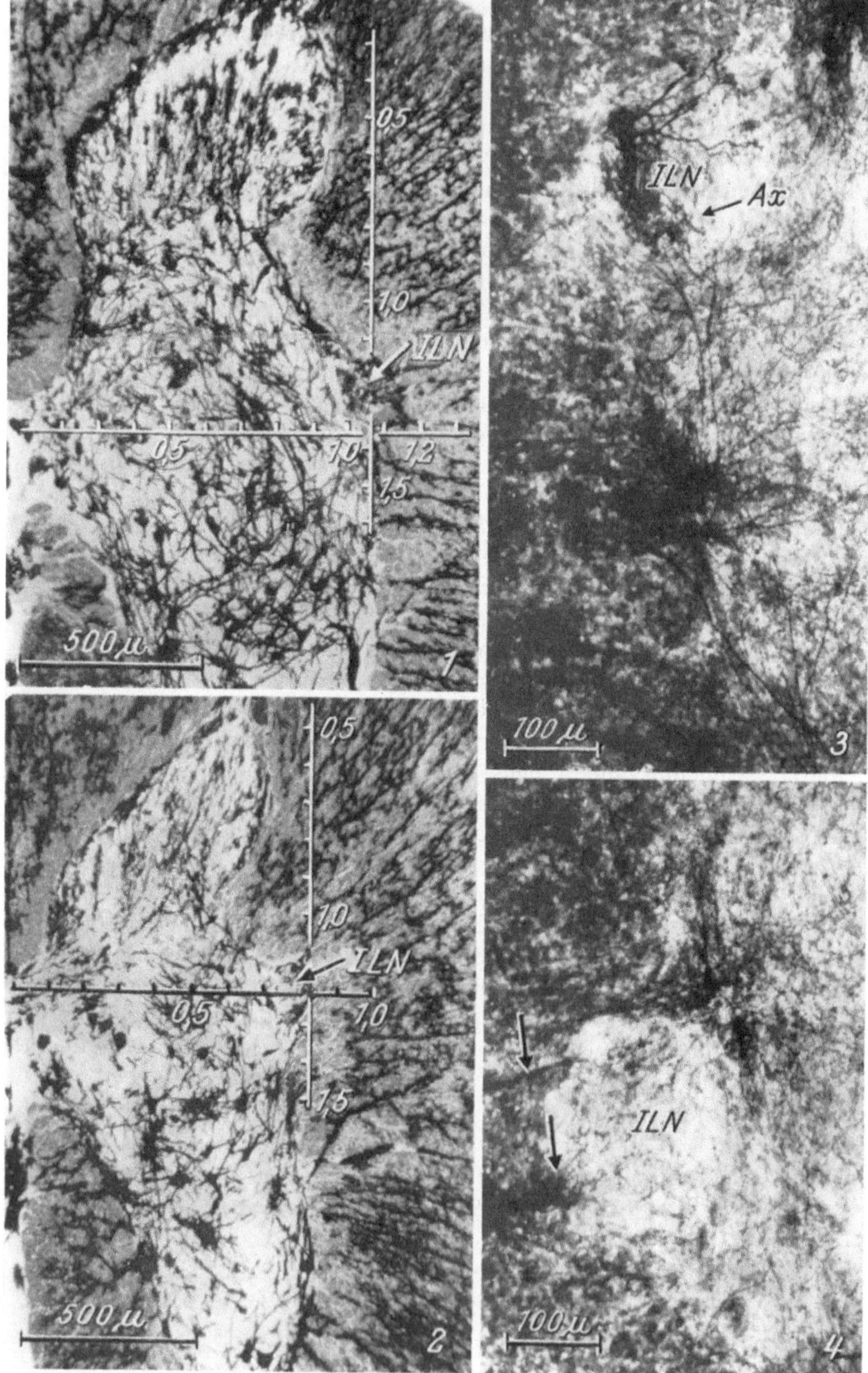
0,5
1,0
ILN
0,5
1,0
1,2
1,5
500 μ
1
ILN
Ax
100 μ
3
0,5
1,0
ILN
0,5
1,0
1,5
500 μ
2
ILN
100 μ
4

preganglionic sympathetic neurones are restricted to the intermedio-lateral nucleus in the spinal cord. Classical as well as recent neuro-histological analysis (*Poljak*, 1924; *Bok*, 1928; *Gagel*, 1928; *Laruelle*, 1936; *Henry* and *Calaresu*, 1972) furnished a large body of information about this nucleus including the numerical segmental distribution of the preganglionic sympathetic neurones. A detailed Golgi and electron microscope analysis of this nucleus will appear in the near future (*Réthelyi*, 1972). Some of the results of this work are described here.

The intermedio-lateral nucleus lies about 1.1 mm from the midline and at 1.2—1.3 mm from the dorsal surface of the spinal cord in the Th_2 segment (Fig. 1). Both coordinates are somewhat shorter in the Th_4 segment (0.8 and 1.1—1.2 mm, respectively) (Fig. 2). These figures coincide with the results of *Polosa* (1967) who could detect the greatest antidromic activity of the preganglionic sympathetic neurones at depths of 1 and 2 mm from the dorsolateral surface the average being 1.4 mm. The axons of the preganglionic sympathetic neurones emerge from the soma, but more frequently from one of the dendrites. They turn ventralwards and can be traced between the ventral grey matter and the lateral funiculus (Fig. 3). No initial collateral branches are given off along their intraspinal course.

The dendrites of the preganglionic sympathetic neurones spread mainly in a longitudinal direction, although transverse dendrites do indeed occur, more frequently in kittens. Longitudinal sections reveal elongated perikarya and the longitudinally oriented dendrites (Fig. 5, inset). This orientation contrasts remarkably with the dendritic pattern of the neighbouring grey matter. The dendritic skeleton of the intermedio-lateral nucleus can be found all along the thoracic segments in the lateral margin of the grey matter at the appropriate horizontal level. The longitudinal orientation of the dendrites also appears in the electron microscope pictures (Fig. 5) this being one of

---

Fig. 1. Localization of the intermedio-lateral nucleus (ILN) in cross section. The coordinates indicate the distance from the midline and from the dorsal surface of the spinal cord in mm. Adult cat, Th_2, Golgi-Kopsch procedure

Fig. 2. Localization of the intermedio-lateral nucleus (ILN) in cross section. The coordinates indicate the distance from the midline and from the dorsal surface of the spinal cord in mm. Adult cat, Th_4, Golgi-Kopsch procedure

Fig. 3. Dendritic arborization, the origin and course of the axon (Ax) of a preganglionic sympathetic neuron. ILN, intermedio-lateral nucleus. Kitten, upper thoracic region, rapid Golgi procedure

Fig. 4. Presynaptic fibre bundles (arrows) enter the intermedio-lateral nucleus (ILN) at its lateral border. Kitten, upper thoracic region, rapid Golgi procedure

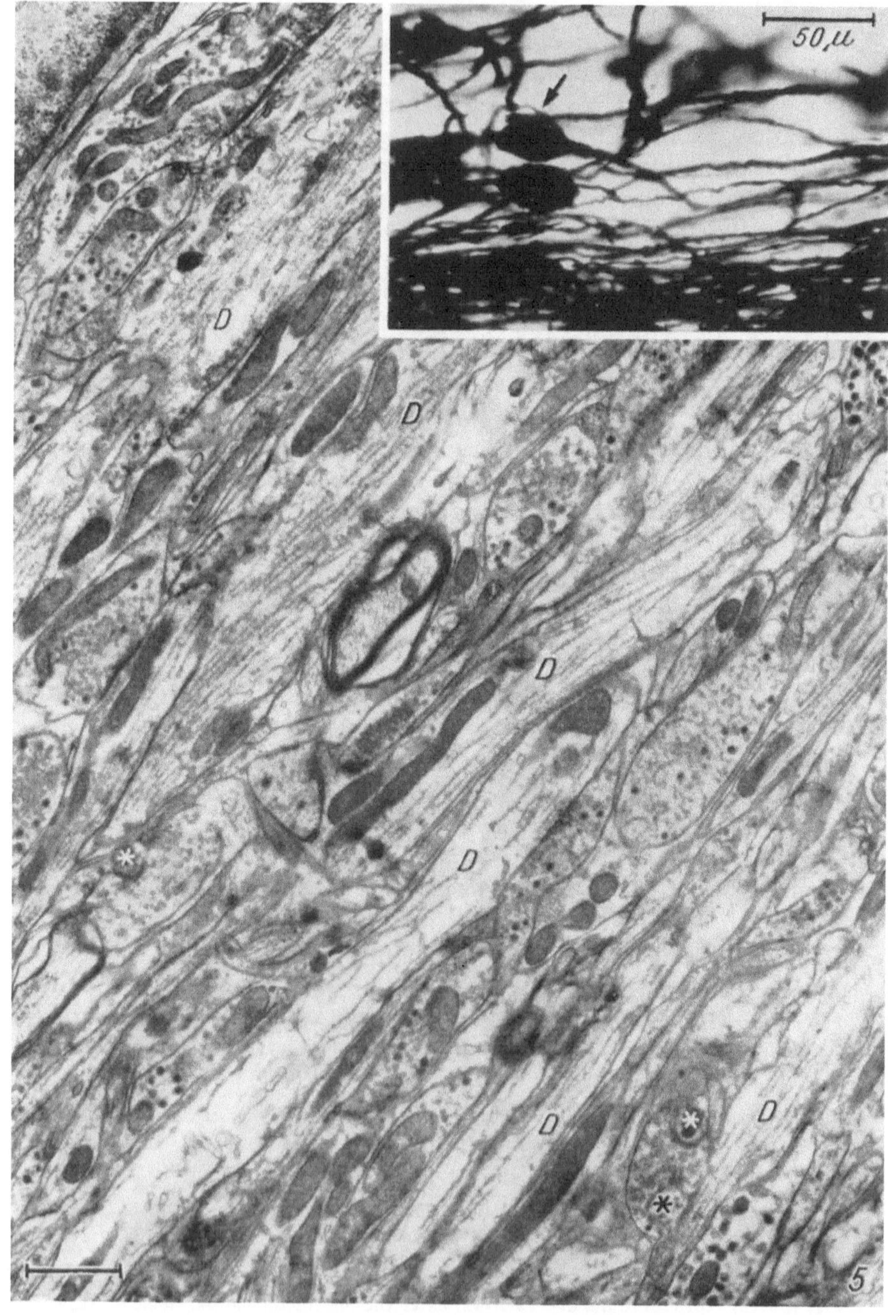
50 μ
D
D
D
D
D
D
5

the ultrastructural characteristics which helps us to identify the nucleus in ultrasections.

The presynaptic fibres of the neuropil approach the intermedio-lateral nucleus in small bundles at its lateral side (Fig. 4). On entering the nucleus the fibre bundles break up and the fibres turn in a longitudinal direction, so that their further course can be better examined in longitudinal sections. Here the axons run parallel with the dendrites and bifurcate. Small side branches with terminal thickenings can be rarely observed. The axons are beaded, thin and thick sections of the axons follow each other. Few if any fibres can be seen that would enter the intermedio-lateral nucleus at its medial border arriving from the intermediate zone of the spinal grey matter.

The electronmicroscopic analysis of the neuropil of the intermedio-lateral nucleus shows three characteristics features:

1. Myelinated fibres are practically missing in the nucleus (Fig. 5). The few occurring might be the axons of the preganglionic sympathetic neurones. The grey matter in the spinal cord generally contains a great deal of myelinated fibres, thus the difference between the intermedio-lateral nucleus and the adjacent grey matter is obvious even in semi-thin sections.

2. The axons follow the orientation of the dendrites and establish repeated, climbing type connections both with dendrites and perikarya. Characteristic long axon profiles can frequently be seen: two thick portions of an axon are connected by a thin bridge (Fig. 6); an axon winds around a dendrite (Fig. 7); the axon passes along a perikaryon and its widened part establishes a synaptic contact with this perikaryon (Fig. 8). All these profiles favour the assumption that repeated contacts are established.

3. The axon terminals contain different types of vesicles and this yields a good basis for their classification into three groups. Terminals may contain either spheric synaptic vesicles exclusively, as present in the bouton-like terminals which generally contact small dendritic appendages (Fig. 5), or spheric synaptic vesicles mixed with dense-core vesicles of 900—1200 Å in diameter (Fig. 6, and many axons in Fig. 5), or flattened synaptic vesicles exclusively (Figs. 7 and 8). This heterogeneity of the contents of the terminals might point to the different origin of the axons to which they belong.

Fig. 5. Low-power electron micrograph of the intermedio-lateral nucleus in longitudinal section. Dendrites (D) run parallel, they are surrounded by axon profiles containing synaptic and dense-core vesicles. Small dendritic appendages, indicated by asterisks, are generally contacted by bouton-like terminals. Scale: 1 micron. Inset upper right: Golgi-Kopsch section demonstrating the longitudinal orientation of the dendrites in the intermedio-lateral nucleus. Arrow points to a perikaryon

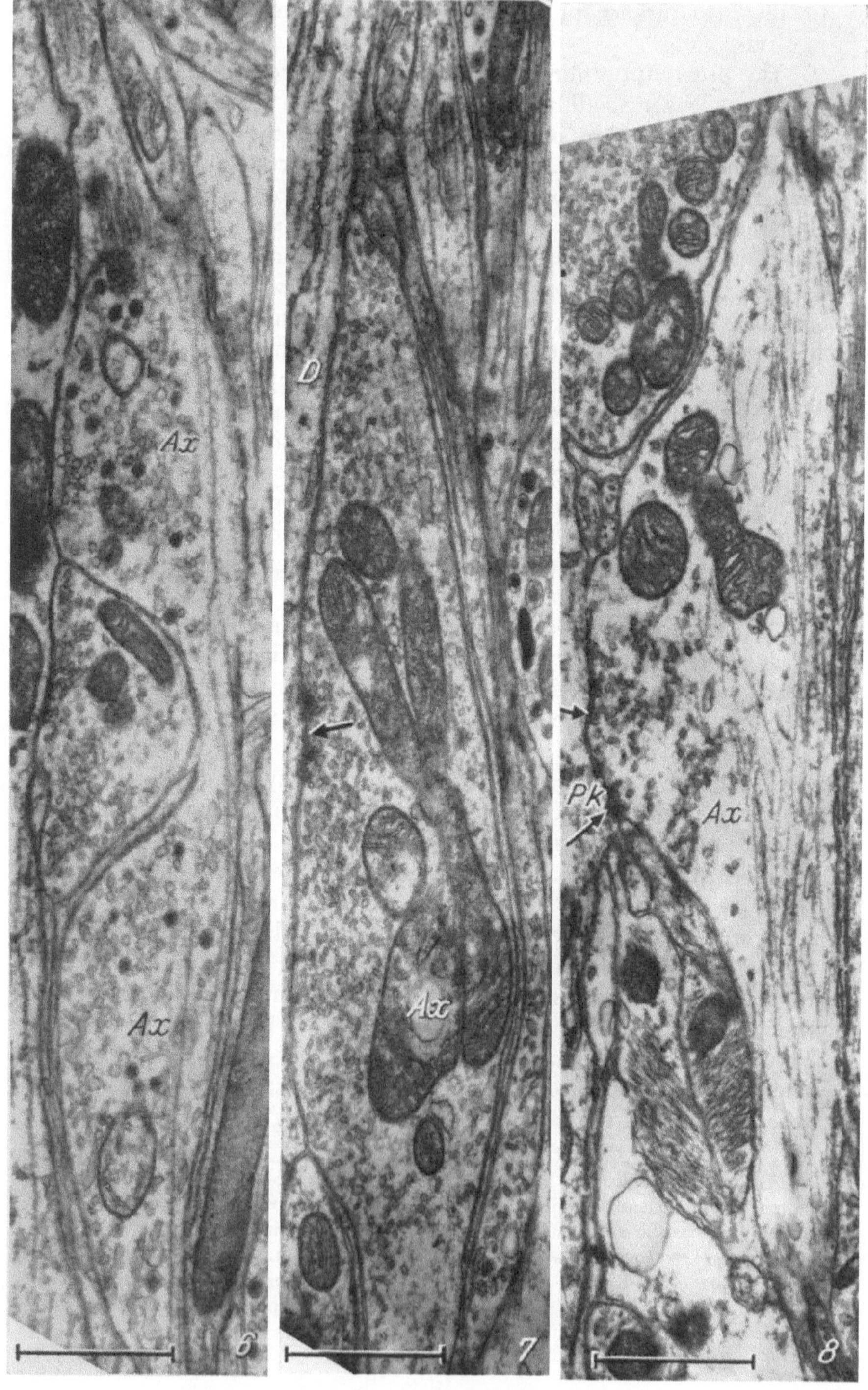

Ax
Ax
6
D
Ax
7
Pk
Ax
8

Connections between Neurones of the Sympathetic Reflex Arc

Figure 10 summarizes two different views on the local transmission arrangement as well as the basic uncertainty prevailing. The controversial question is whether the descending fibres from supraspinal centres (axons of the medullary interneurones, MIN) reach the preganglionic sympathetic neurones directly (solid line) or via interneurones (dashed line). In the first arrangement, which is based on the results of fluorescence analyses (*Carlsson et al.*, 1964; *Dahlström* and *Fuxe,* 1965), the two components, spinal and supraspinal, would be completely independent. Conversely, neurophysiological observations would suggest that the components are not independent, the supraspinal centres controlling the activity of the spinal component (*Khajutin* and *Lukoshkova,* 1970; *Malliani et al.*, 1971). For such an assumption the second arrangement would be more appropriate in which the spinal and supraspinal component converge on a common group of interneurones (SIN_2) impinging upon the preganglionic sympathetic neurones. Obviously this is a crude oversimplification of the neuronal connectivity really present, because at least three types of axon terminals can be found in the intermedio-lateral nucleus which excludes the possibility of a single type of interneuron terminating here. In spite of this important discrepancy, degeneration experiments favour the second arrangement.

Nyberg-Hansen (1965) could not find degeneration in the intermedio-lateral nucleus after the destruction of the reticular formation in the medulla. I have tried to approach the same problem performing two types of lesions in the spinal cord.

1. The whole lateral funiculus was destroyed at the level of C_2. No degeneration was found two days later in the intermedio-lateral nucleus at the levels of Th_2 and Th_3. This failure can be explained in two ways. Either no descending fibres reach the intermedio-lateral nucleus directly, or the survival time was not chosen correctly. The two days might seem to be extremely short, but, in the central nervous system, degeneration of non-myelinated fibres happens very rapid (*Heimer* and *Wall,* 1968; *Réthelyi* and *Szentágothai,* 1969; *Réthelyi* and *Halász,* 1970; *Raisman,* 1972). As a control, the substantia

Fig. 6. Two widened parts of an axon (Ax) are connected by a narrow bridge. The axon contains spheric synaptic vesicles and dense-core vesicles. Scale: 1 micron

Fig. 7. Axon (Ax) containing flattened synaptic vesicles climbs up around a dendrite (D) and establishes a synaptic contact (arrow). Scale: 1 micron

Fig. 8. Synapse de passage (arrows) between an axon (Ax) and a perikaryon (Pk). The axon contains flattened synaptic vesicles. Scale: 1 micron

gelatinosa was observed which is made up mainly of non-myelinated fibres. A large number of fine and medium size degenerated fibres could be found in the substantia gelatinosa adjacent to the transection indicating the rapid involvement of non-myelinated axons in the degeneration process.

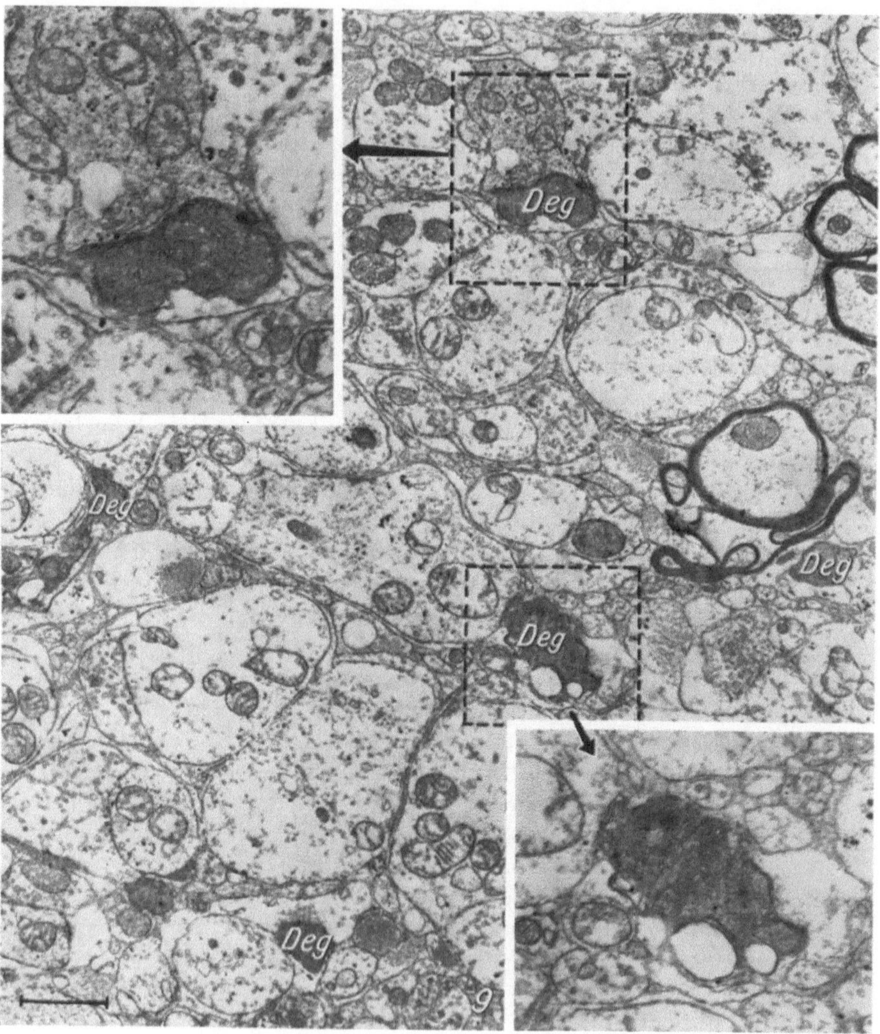

Fig. 9. Degenerated axon fragments (Deg) in the intermedio-lateral nucleus 2 mm in cranial direction from the level of the transection which destroyed the dorsolateral funiculus and the base of the dorsal horn at the level of Th_5. Survival time is 24 hours. Scale: 1 micron

2. In the second experiment, which confirmed the result of the first yielding also new details, the dorsal part of the lateral funiculus was destroyed at the level of the Th_5 segment and the lesion was also extended into the dorsal horn. The next day the animal was killed and the spinal cord was cut into 2 mm thick discs. The intermedio-lateral nucleus was examined using electron microscopy in three consecutive discs above and in three consecutive discs below the lesion. Degenerated axons were found only in discs adjacent to the lesion (Fig. 9). The existence of long descending fibres is not proved by this

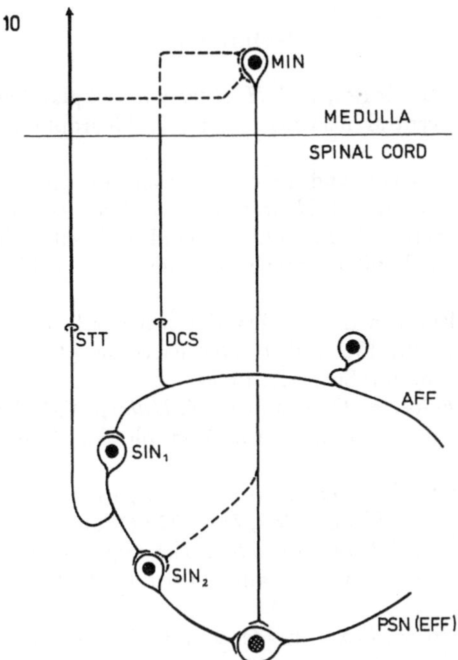

Fig. 10. Diagram indicating the central part of the sympathetic reflex path. The afferent neurones (AFF) terminate in the spinal cord on interneurones (SIN_1) and/or ascend in the dorsal column system (DCS) and reach the medullary interneurones (MIN) via an unknown neuron link (dashed line). Spinal interneurones (SIN_1) contact other interneurones (SIN_2) and/or their axons ascend in the spino-thalamic tract (STT) to reach the medullary interneurones (MIN) presumably through interneurones (dashed line). The efferent elements, *i.e.*, the preganglionic sympathetic neurones [PSN(EFF)], receive impulses from the spinal interneurones (SIN_2) and possibly directly from the medullary interneurones (MIN) which constitute the medullary sympathetic centres (I. arrangement). As an alternative, the axon of the medullary interneurones terminates on spinal interneurones (dashed line to SIN_2) and the descending impulses reach the preganglionic sympathetic neurones indirectly, via interneurones (II. arrangement)

experiment either. Instead, it raises the possibility of the presence of short interneurones the axons of which enter the intermedio-lateral nucleus without leaving the boundaries of the segment. Furthermore, this observation proves that very short survival times (even 24 h) give good results.

Our understanding of the spinal autonomic mechanism is still quite marginal. However, at least the identity of the preganglionic sympathetic neurones is now clear. The characteristic ultrastructural arrangement of their synapses offer certain cues for further studies of their segmental and supraspinal connections.

References

Aidar, O., W. A. Geohogan, and *H. Ungewitter*: Splanchnic afferent pathways in the central nervous system. J. Neurophysiol. *15*, 131—138 (1952).

Alexander, R. S.: Tonic and reflex functions of medullary sympathetic cardiovascular centers. J. Neurophysiol. *9*, 205—217 (1946).

Beacham, W. S., and *E. R. Perl*: Background and reflex discharge of sympathetic preganglionic neurons in the spinal cat. J. Physiol. *172*, 400 to 416 (1964).

Bok, S. T.: Das Rückenmark. In: Handbuch der mikroskopischen Anatomie des Menschen. Bd. IV, Teil I. Nervensystem (*Möllendorff, W.*, ed.), pp. 478—578. Berlin: Springer. 1928.

Carlsson, A., B. Falck, K. Fuxe, and *N. A. Hillarp*: Cellular localization of monoamines in the spinal cord. Acta physiol. Scand. *60*, 112—119 (1964).

Coote, J. H., and *C. B. B. Downman*: Central pathways of some autonomic reflex discharges. J. Physiol. *183*, 714—729 (1966).

Coote, J. H., and *J. F. Perez-Gonzalez*: The response of some sympathetic neurones to volleys in various afferent nerves. J. Physiol. *208*, 261—278 (1970).

Dalström, A., and *K. Fuxe*: Evidence for the existence of monoamine neurons in the central nervous system. II. Experimentally induced changes in the intraneuronal amine levels in bulbospinal neuron systems. Acta physiol. Scand. *64*, Suppl. 247, 7—85 (1965).

Downman, C. B. B.: Skeletal muscle reflexes of splanchnic and intercostal nerve origin. J. Neurophysiol. *18*, 218—235 (1955).

Downman, C. B. B., and *A. Houssian*: Spinal tracts and supraspinal centres influencing visceromotor and allied reflexes in cats. J. Physiol. (London) *141*, 484—499 (1958).

Franz, D. N., M. H. Evans, and *E. R. Perl*: Characteristics of viscerosympathetic reflex in the spinal cat. Am. J. Physiol. *211*, 1292—1298 (1966).

Gagel, O.: Zur Histologie und Topographie der vegetativen Zentren im Rückenmark. Zschr. Anat. Entwgesch. *85*, 213—250 (1928).

Heimer, L., and *P. D. Wall*: The dorsal root distribution to the substantia gelatinosa of the rat with a note on the distribution in the cat. Exp. Brain Res. *6*, 89—99 (1968).

Henry, J. L., and *F. R. Calaresu*: Topography and numerical distribution of neurons of the thoraco-lumbar intermediolateral nucleus in the cat. J. Comp. Neurol. *144*, 205—214 (1972).

Hongo, T., and *R. W. Ryall*: Electrophysiological and microelectrophoretic studies on sympathetic preganglionic neurones in the spinal cord. Acta physiol. Scand. *68*, 96—104 (1966).

Illert, M., and *M. Seller*: A descending sympathoinhibitory tract in the ventrolateral column of the cat. Pfügers Arch. *313*, 343—360 (1969).

Jacobsohn, L.: Über die Kerne des menschlichen Rückenmarkes. Anhang zu den Abhandlungen der königlichen preußischen Akademie der Wissenschaften. Berlin: Reimer. 1908.

Kerr, F. W. L., and *S. Alexander*: Descending autonomic pathways in the spinal cord. Arch. Neurol. (Chic.) *10*, 249—261 (1964).

Khayutin, V. M., and *E. V. Lukoshkova*: Spinal mediation of vasomotor reflexes in animals with intact brain studied by electrophysiological methods. Pfügers Arch. *321*, 197—222 (1970).

Kirchner, F., *A. Sato*, and *H. Weidinger*: Central pathways of reflex discharges in the cervical trunk. Pfügers Arch. *319*, 1—11 (1970).

Koizumi, K., *A. Sato*, *A. Kaufman*, and *Mc. C. Brooks*: Studies of sympathetic neuron discharges modified by central and peripheral excitation. Brain Res. *11*, 212—224 (1968).

Korn, H.: Splanchnic projection to the orbital cortex of the cat. Brain Res. *16*, 25—38 (1969).

Laruelle, L.: Contribution a l'étude du nevraxe végétatif. C.R. Assoc. Anat. (Paris) *31*, 210—229 (1936).

Malliani, A., *M. Pagani*, *G. Recordati*, and *P. Schwartz*: Spinal sympathetic reflexes elicited by increases in arterial blood pressure. Am. J. Physiol. *220*, 128—134 (1971).

Nyberg-Hansen, R.: Sites and mode of termination of reticulo-spinal fibers in the cat. An experimental study with silver impregnation methods. J. Comp. Neurol. *124*, 71—100 (1965).

Oliver, J. E., *Jr.*, *W. E. Bradley*, and *T. F. Fletcher*: Identification of preganglionic parasympathetic neurons in the sacral spinal cord of the cat. J. Comp. Neurol. *137*, 321—325 (1969).

Poljak, S.: Die Struktureigentümlichkeiten des Rückenmarkes bei den Chiropteren. Zschr. Anat. Entwgesch. *74*, 507—576 (1924).

Polosa, C.: The silent period of sympathetic preganglionic neurons. Canad. J. Physiol. Pharmacol. *45*, 1033—1045 (1967).

Pomeranz, B., *P. D. Wall*, and *W. V. Weber*: Cord cells responding to fine myelinated afferents from viscera, muscle and skin. J. Physiol. (London) *199*, 511—532 (1968).

Prout, B. J., *J. H. Coote*, and *C. B. B. Downman*: Supraspinal inhibition of a cutaneous vascular reflex in the cat. Amer. J. Physiol. *207*, 303—307 (1964).

Raisman, G.: A second look at the parvicellular neurosecretory system. In: Brain-Endocrine Interaction. Median Eminence: Structure and Function. Int. Symp. Munich 1971, pp. 109—118. 1972.

Réthelyi, M.: Cell and neuropil architecture of the intermedio-lateral (sympathetic) nucleus of cat spinal cord. Brain Res. *46*, 203—213 (1972).

Réthelyi, M., and *B. Halász*: Origin of the nerve endings in the surface zone of the median eminence of the rat hypothalamus. Exp. Brain Res. *11*, 145—158 (1970).

Réthelyi, M., and *J. Szentágothai*: The large synaptic complexes of the substantia gelatinosa. Exp. Brain Res. *7*, 258—274 (1969).

Rexed, B.: A cytoarchitectonic atlas of the spinal cord in the cat. J. Comp. Neurol. *100*, 297—379 (1954).

Sato, A., and *R. F. Schmidt*: Muscle and cutaneous afferents evoking sympathetic reflexes. Brain Res. *2*, 399—401 (1966).

Sato, A., A. Kaufman, K. Koizumi, and *Ch. Mc C. Brooks*: Afferent nerve groups and sympathetic reflex pathways. Brain Res. *14*, 575—587 (1969).

Scheibel, M. E., and *A. B. Scheibel*: Terminal axonal patterns in cat spinal cord. II. The dorsal horn. Brain Res. *9*, 32—58 (1968).

Scheibel, M. E., and *A. B. Scheibel*: Terminal patterns in cat spinal cord. III. Primary afferent collaterals. Brain Res. *13*, 417—443 (1969).

Schimert, J.: Das Verhalten der Hinterwurzelkollateralen im Rückenmark. Zschr. Anat. Entwickl. Gesch. *109*, 665—687 (1939).

Schmidt, R. F., and *K. Schönfuss*: An analysis of the reflex activity in the cervical sympathetic trunk induced by myelinated somatic afferents. Pflügers Arch. *314*, 175—198 (1970).

Schnitzlein, H. N., H. H. Hoffman, D. M. Hamlett, and *E. M. Howell*: A study of the sacral parasympathetic nucleus. J. Comp. Neurol. *120*, 477—485 (1963).

Sell, R., A. Erdélyi, and *H. Schaefer*: Untersuchungen über den Einfluß peripherer Nervenreizung auf die sympathische Aktivität. Pflügers Arch. *267*, 566—581 (1958).

Selzer, M., and *W. A. Spencer*: Convergence of visceral and cutaneous afferent pathways in the lumbar spinal cord. Brain Res. *14*, 331—348 (1969).

Sterling, P., and *H. G. J. M. Kuypers*: Anatomical organization of the brachial spinal cord of the cat. I. The distribution of dorsal root fibers. Brain Res. *4*, 1—15 (1967).

Szentágothai, J.: Anatomical considerations of monosynaptic reflex arcs. J. Neurophysiol. *11*, 445—454 (1948).

Szentágothai, J.: Pathways and subcortical relay mechanisms of visceral afferents. Acta neuroveg. (Wien) *28*, 103—120 (1966).

Szentágothai, J., and *M. Réthelyi*: Cyto- and neuropil architecture of the spinal cord. In: New Developments in Electromyography and Clinical Neurophysiology (*Desmedt, J. E.*, ed.), Vol. 3, pp. 20—37. Basel: Karger. 1973.

Weidinger, W. H., L. Fedina, H. Kehrel, and *H. Schaefer*: Über die Lokalisation des „bulbären sympathischen Zentrums" und seine Beeinflussung durch Atmung und Blutdruck. Zschr. Kreisl. Forsch. *50,* 229—241 (1961).

Widen, L.: Cerebellar representation of high threshold afferents from splanchnic nerve. Acta physiol. scand. Suppl. *117,* 1—69 (1955).

Author's address: Dr. *M. Réthelyi,* Second Department of Anatomy, Semmelweis University Medical School, Budapest, Hungary.

Discussion

Taxi: With respect to the three types of presynaptic endings in the intermedio-lateral nucleus mentioned, I should like to ask whether there is any correlation between them and their functional meaning. May they contain, for instance, different neurotransmitters?

Réthelyi: It is rather difficult to correlate the vesicle content of the axons and their function, although a working hypothesis has been put forward in my original paper (*Réthelyi,* 1972). These difficulties are: 1. in the spinal cord the significance of spheroid versus flattened vesicles is not quite clarified; 2. using potassium permanganate fixation it has until now been impossible to find the small granulated vesicles in the axon terminals, thus we do not even know which type of axons contain catecholamines as a transmitter. A great deal of effort is still needed to approach this problem.

Malinsky: Is there any difference in the occurrence of axons terminals containing large dense-core vesicles in various regions of the spinal grey matter?

Réthelyi: Everywhere in the spinal grey matter dense-core vesicles do occur. We observed such vesicles in some of the axon terminals in Clarke's column and in the substantia gelatinosa, but in none of these two areas can as many dense-core vesicles been found as in the intermedio-lateral nucleus. The only region in the CNS where dense-core vesicles occur in the same quantity as in this nucleus, is the median eminence of the hypothalamus, containing the nerve terminals surrounding the portal vessels.

Csillik: I want to point to the peculiar discrepancy between the fact that no degeneration was found in the intermedio-lateral nucleus several segments below the transection of the lateral tract, whereas, according to the Fuxe group, all monoamine fluorescence disappears below the transection. Is it possible that these monoaminergic fibres descend in the substantia gelatinosa, known to contain a fair number of fluorescent fibres, or may some unknown interneurones be present in this descending pathway?

Réthelyi: Though none of the proposed explanations can be excluded, it might be that the time dependency of the degeneration process is responsible for the lack of histological signs of degeneration.

Szentágothai: I should like to add that, in the spinal cord, we are confronted with the same discrepancy between results of conventional degeneration experiments and the disappearance of fluorescence as experienced elsewhere. This discrepancy is only temporal at certain sites, for example in the peripheral branches of vegetative ganglia. In other cases the same interference (for example in the hypothalamus) causes complete disappearance of fluorescence in terminals, but no conventional degeneration. The convenient explanation—that this may be due to transneuronal changes in the catecholamine producing neurones—may not be the correct one. These cases should be investigated more carefully with a multi-methodic approach.

Csillik: What are the present views about parasympathetic dorsal root fibres, the existence of which has been postulated by Ken Kuré and other authors in the late '30-es.

Réthelyi: I do not have any personal experience with regard to this question.

Journal of Neural Transmission, Suppl. XI, 213—225 (1974)

Role of Short-Range Transmitters in Hypothalamic Activities

E. Endröczi

Department of Experimental and Clinical Laboratory Investigations, Postgraduate Medical School, Budapest, Hungary

With 9 Figures

Summary

Ovariectomy produced a decrease of the hypothalamic catecholamine content which could be restored by estradiol. Local implantation of estradiol resulted in a significant rise of the catecholamine concentration in the basal and medial hypothalamus. Testosterone administration was ineffective. During lactation the catecholamine content of the hypothalamus was higher than in nonlactating females.

Bilateral lesions in the preoptic area produced a decrease of the catecholamine concentration in the median eminence and the estradiol administration did not induce increase in ovariectomized and preoptic lesioned rats. It is assumed that rostral afferentation to the median eminence is involved in the function of catecholaminergic neurons.

Humoral factors involved in the regulation of hypothalamic activities can be divided into categories as follows: (1) chemical transmitters in a classical sense (acetylcholine, noradrenaline, 5-hydroxytryptamine, etc.), (2) the humoral agents which exert specific influences on the target cells located in medium range (*e.g.*, prostaglandins), and (3) the long-range acting hormones (*e.g.*, pituitary hormones, steroids, etc.). Interactions between chemical transmitters of different origins and their mutual participation in the organization of specific nervous activities at the hypothalamic level have been the subjects of numerous studies in the recent decades. The present paper gives some contributions to the role of short-range transmitters in controlling the hypothalamo-pituitary activities, especially to the changes of catecholamines in response to castration and substitution with sex steroids, on the one hand, and to prolactin administration as well as during lactation.

Catecholamines and Pituitary-Ovary Function

Quantitative changes in hypothalamic catecholamines during the estrous cycle have been established by several authors. By the use of different methods they found a decrease of the noradrenaline and dopamine concentration in proestrous and a peak of these monoamines in diestrous (*Fuxe* and *Hökfelt,* 1970; *Stefano* and *Donoso,* 1967), although these findings could not be replicated. Similarly, opposite data have been reported on the effects of sex steroids upon hypothalamic catecholamines; thus, estradiol administration resulted in a decrease of noradrenaline concentration (*Donoso* and *Stefano,* 1967) which is opposed to the observations of *Kurachi et al.* (1968) and *Tonge* and *Greengrass* (1970) who found a restoration of the hypothalamic catecholamine content by estrogen treatment in castrated rats.

Studying the noradrenaline and dopamine concentration of the hypothalamus with fluorimetric methods (*Chang,* 1964; *Udenfried* and *Zaltzman-Nierenberg,* 1963), we found that 3 to 4 weeks after ovariectomy the catecholamine content is lower than in intact female rats. The administration of estradiol, which was given in 10 per cent ethanol/physiological saline subcutaneously for 7 days, led to the restoration of the decreased catecholamine concentration in the ovariectomized rats (Fig. 1).

In contrast to observations on ovariectomized animals, it was found that estradiol administration to intact females produced a biphasic change of the hypothalamic catecholamine level. Thus, estradiol treatment led to a significant drop of the noradrenaline concentration within 24 hours although there was no significant change in the dopamine level. On the other hand, daily administration of 2 μg/100 g estradiol for 7 days produced a moderate increase of both catecholamines in comparison to the intact groups, treated with physiological saline (Fig. 2).

Intrahypothalamic implantation of 5 to 10 μg estradiol-17β into the median eminence region of castrated females produced a marked rise of the noradrenaline and dopamine levels 7 to 10 days after local hormone application. The animals were castrated one month prior to

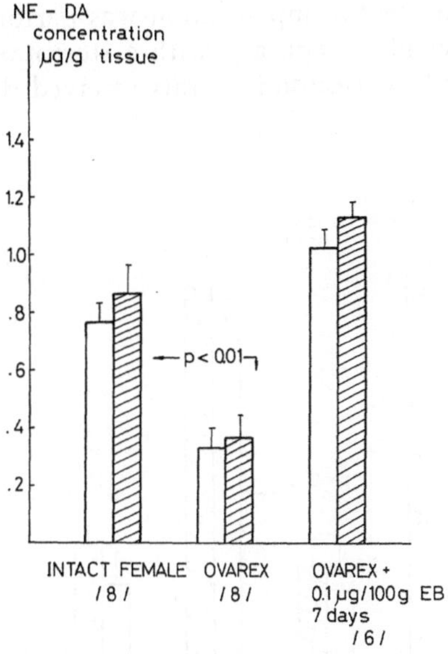

Fig. 1. Effect of estradiol administration on the noradrenaline and dopamine content of the hypothalamus and its restoration by estradiol administration. Empty bars, noradrenaline (NE); hatched bars, dopamine (DA). Mean and standard errors are presented

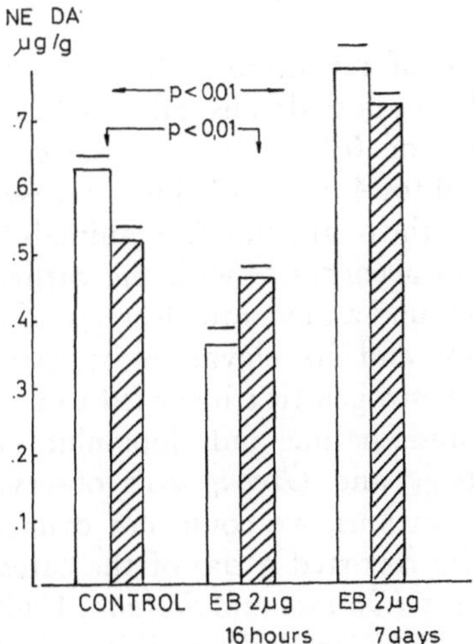

Fig. 2. Effect of estradiol administration on the hypothalamic noradrenaline (empty bars) and dopamine (hatched bars) concentration in female rats

brain surgery. The steroid implantation was performed on the tip of a 0.2 mm diameter glass capillary with a stereotaxic apparatus. The sham-operated and ovariectomized rats received cholesterol implants (Fig. 3).

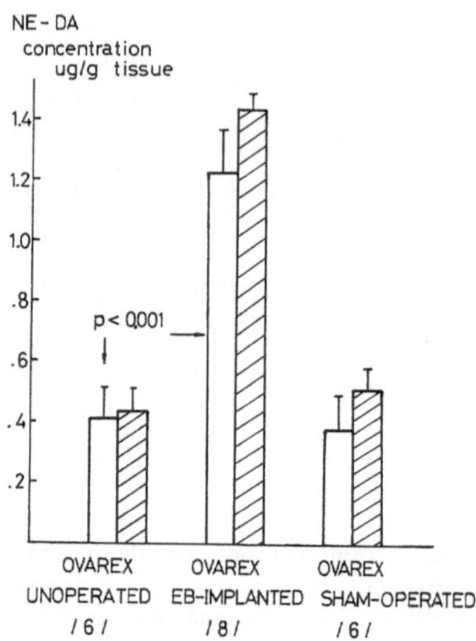

Fig. 3. Effect of intrahypothalamic implantation of estradiol on hypothalamic catecholamines in ovariectomized rats. Sham-operated animals were given cholesterol implants

The administration of testosterone did not restore the catecholamine content of the hypothalamus after ovariectomy. Thus, the subcutaneous injection of 0.1 mg/100 g testosterone propionate in 0.1 ml propylene glycol for 4 days failed to increase the noradrenaline and dopamine levels of the ovariectomized animals (Fig. 4).

Our findings are in accordance with the earlier data of *Kurachi et al.* (1968) who also found a moderate decrease of the catecholamines following ovariectomy and its elevation by estradiol substitution therapy. Similarly, the estrogen treatment led to a marked increase of the hypothalamic noradrenaline and dopamine content in spayed rats according to *Tonge* and *Greengrass*'s observations (1970). In contrast to these observations, we could not confirm the findings of other investigators who reported a rise of the catecholamine concentration after ovariectomy (*Donoso* and *Stefano*, 1967).

The biphasic action of estradiol administration in intact female rats may be explained by the multiple influence of this steroid hormone on the hypothalamo-pituitary axis. A direct influence of the estrogens is very likely but we must be aware that other factors may be also involved in these events. Thus, the follicle stimulating hormone (FSH) exerts a direct facilitatory influence on the turnover rate of noradrenaline (*Anton-Tay et al.*, 1967) in hypophysectomized and gonadectomized rats.

The other factor which may be involved in the estrogen-induced changes in hypothalamic catecholamines is prolactin (LTH). *Hökfelt* and *Fuxe* (1972) reported the selective facilitatory influence of prolactin on the dopamine disappearance rate from the median eminence region in both gonadectomized and hypophysectomized animals. It is known that both, follicle stimulating hormone and prolactin, are undergoing changes following estrogen administration, and these alterations can be involved in the changes occurring in the catecholamine concentration of the hypothalamus.

An early decrease of the hypothalamic catecholamine level after estradiol administration seems to be in accordance with the observations that the lowest level of hypothalamic monoamines is present in the proestrous phase (*Donoso* and *Stefano*, 1967; *Kurachi et al.*, 1968). A secondary rise of the hypothalamic catecholamines in the response to estrogen treatment may be due to a direct effect of the estrogens or other factors as a result of the changes in the hypothalamo-pituitary axis.

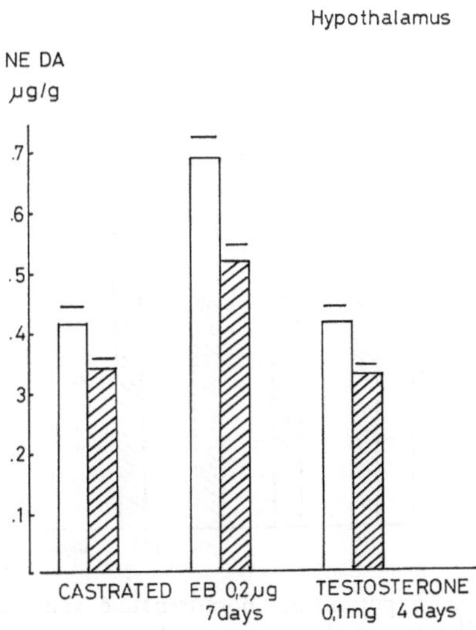

Fig. 4. Influence of estradiol and testosterone administration on hypothalamic catecholamines in ovariectomized rats

Hypothalamic Catecholamines during Lactation and after Prolactin Administration

It is generally known that depletion of the hypothalamic catecholamines results in an increase of prolactin secretion and it was suggested that catecholamines control the prolactin inhibitory factor of the hypothalamus. After reserpine-induced depletion of the brain monoamines, the prolactin inhibitory activity of the hypothalamus was found to be lower than in intact animals. But the other possibility should also be considered, namely, that removal of brain catecholamines produces an increase of the prolactin stimulating factor.

Hypothalamic noradrenaline and dopamine concentrations show characteristic changes during lactation in rats. In comparison to intact females, the hypothalamic content of both noradrenaline and dopamine is higher in the first three weeks during lactation, roughly until the weaning period (Fig. 5).

Studying the possible role of prolactin on hypothalamic catecholamines, 0.1 mg prolactin (NIH-LTH bovine) was implanted into the median eminence of castrated female rats. The prolactin was suspended in agar-agar and the gel brought into the hypothalamus in an amount of several microliter volumes with the aid of a glass capillary. Sham-operated castrated animals received only agar-agar. The animals were sacrificed on the 4th postoperative day and the

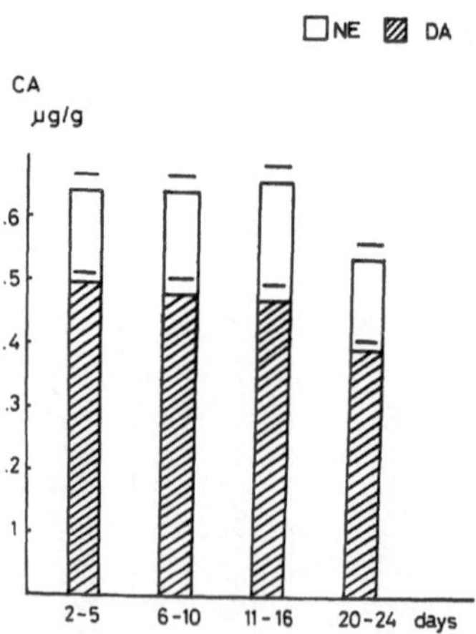

Fig. 5. Changes in the hypothalamic noradrenaline and dopamine concentration during lactation until weaning. Four groups of lactating rats (5—6 animals) were sacrificed in different periods of the lactation. The values obtained on the 20th to 24th days correspond to the normal level measured in non-lactating females

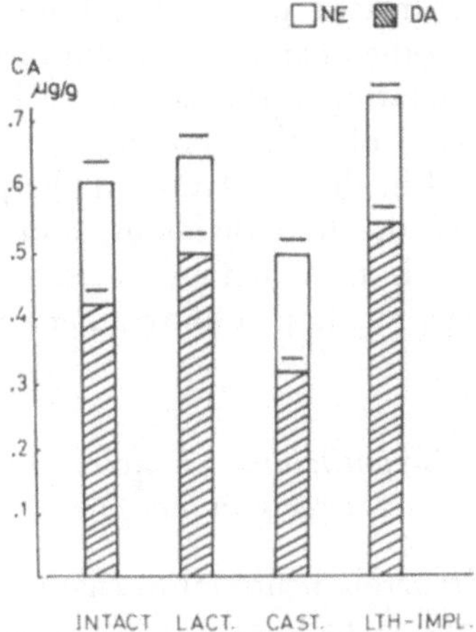

Fig. 6. Effect of intrahypothalamic prolactin implantation in ovariectomized rats. In comparison, both intact and lactating values are also shown in the figure

hypothalamic catecholamines measured according to the methods described in the earlier part of this paper. Prolactin implantation produced a significant rise of the hypothalamic catecholamines (Fig. 6).

A direct influence of prolactin on the dopaminergic neurones of the median eminence had already been observed by *Hökfelt* and *Fuxe* (1972). The present investigations confirm the suggestions of these authors that the prolactin has a direct effect on the hypothalamic dopaminergic system. However, we have difficulties in interpreting the possible role of an increased catecholamine pool during lactation and in the regulation of prolactin. Intrahypothalamic implantation of reserpine, producing depletion of monoamines at the median eminence

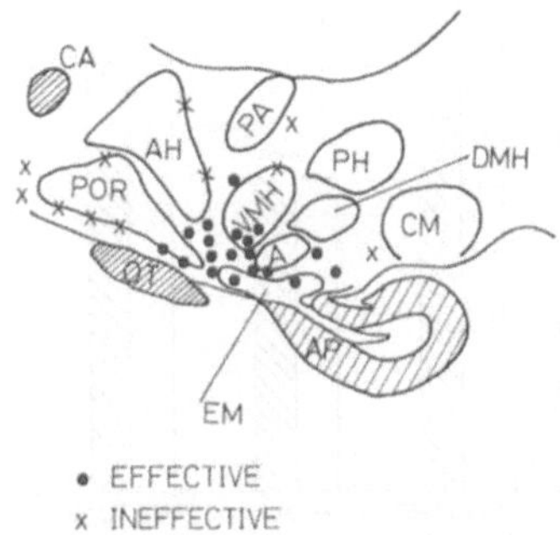

Fig. 7. Schematic illustration of the localization of prolactin implants with and without effect on the noradrenaline and dopamine content of the basal and medial hypothalamus

level, resulted in an increased prolactin release (*van Maanen* and *Smelik*, 1968). On the other hand, the hypothalamic catecholamine content is higher in lactating rats than in cycling females. If we consider that an increase of either noradrenaline or dopaminergic activity can be involved in the regulation of the prolactin inhibiting or releasing activity of the hypothalamus, and that the prolactin autofeedback regulation is mediated by catecholaminergic neuronal mechanisms, the findings mentioned above merit some consideration (Fig. 7).

Influence of Anterior Hypothalamic Lesions on the Hypothalamic Catecholamines

Involvement of adrenergic neuronal mechanisms in the regulation of pituitary-ovary function have already been suggested more than two decades ago (*Markee et al.*, 1948; *Sawyer et al.*, 1949; *Everett*, 1964; etc.). Roughly, two monoaminergic systems of different origins can be distinguished in the hypothalamus: the main ascending noradrenergic and dopaminergic pathways are reaching the median eminence from brainstem structures while short monoaminergic fibres have their origin and connections within the hypophysiotropic area. In recent years, the question of their participation in the control of

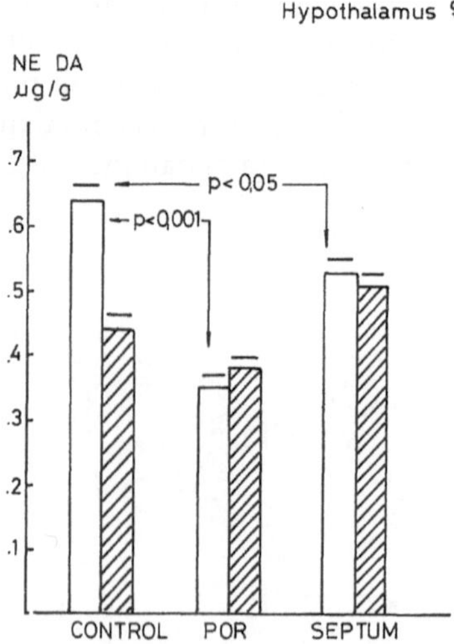

Fig. 8. Effect of bilateral midline preoptic and rostral septal lesions on the noradrenaline (empty bars) and dopamine (hatched bars) concentration of the hypothalamus

different pituitary activities has been raised by many authors (see reviews by *Fuxe* and *Hökfelt*, 1970; *McCann*, 1970; *Sawyer*, 1963; etc.). There is no evidence that monoaminergic neurones from the rostral and basal preoptic area project to the external layer of the median eminence or to the nuclei of the hypophysiotropic area. A bilateral electrolytic lesion of the midline preoptic region results in a

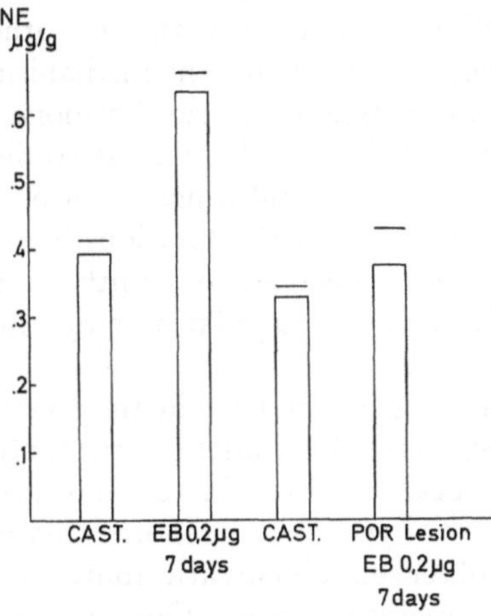

Fig. 9. Alterations in the hypothalamic noradrenaline content of ovariectomized rats due to estradiol administration in sham-operated and preoptic lesioned conditions

change of the pituitary-ovary function which may be characterized by constant estrous, while the electrical stimulation of this part of the brain leads to LH discharge in the proestrous phase (see review by *Everett*, 1964). In recent investigations we found that the midline preoptic lesion produced a significant decrease of the catecholamine concentration of the basal and medial hypothalamus. The electrolytic damage was made by anodic polarization with 3 mA for 10 sec. It destroyed the middle part of the preoptic area above the chiasma opticium but did not extend into the rostral median eminence region (Fig. 8).

The preoptic lesion inducing a decrease of hypothalamic catecholamines may be important in the light of studies showing that an estradiol-induced increase of monoamines is lacking in ovariectomized rats. The animals were castrated one month prior to lesioning the preoptic area and 0.2 μg/100 g estradiol was given subcutaneously for 7 days. In contrast to the non-lesioned ovariectomized rats, estradiol treatment did not increase the hypothalamic catecholamine level following the preoptic lesion (Fig. 9).

General Conclusion

Numerous observations indicate that hypothalamic monoaminergic neurones play a role in the control of pituitary activities. The noradrenaline and dopamine metabolism of the presynaptic vesicles is a complex process; release and reaccumulation versus synthesis may be independently influenced which is followed by a change of the catecholamine pool with or without alterations in quantity of the monoamine content. The mechanism of catecholamine changes as a result of ovariectomy, sex hormone administration or brain lesions remained unclear in the present studies. Thus, if release is changed in response to each influence mentioned the same result might be observed as in case of the alterations in reaccumulation rate. Moreover, an increase or decrease of synthesis in the absence of changes of either release or reaccumulation may also contribute to the complexity of the picture.

Contradictory findings of changes in the hypothalamic catecholamines during the estrous cycle or after ovariectomy and sex hormone administration have been reported in recent years. A moderate but significant decrease of hypothalamic noradrenaline and dopamine concentrations was observed after ovariectomy becoming manifest in the 3rd and 4th postoperative weeks. Our findings are in agreement with recent data of *Kurachi et al.* (1968), although these authors used different physico-chemical methods for the estimation of brain monoamines. Moreover, an increase of brain catecholamines after estrogen treatment was found by *Tonge* and *Greengrass* (1970) in castrated rats. This seems to be in accordance with our data, but the authors measured the rise of catecholamine levels much earlier than in our cases.

The hypothalamus is regarded as a complex of nuclei having a considerable autonomous function. Both, neural afferentation and humoral factors of different origins play a basic role in the integration of hypothalamic activities. Receptor sites and specific hormone-binding cells could be observed in both the hypothalamus and the rostral limbic structures. Recent data clearly demonstrated that the hippocampus can accumulate more corticosterone than any other brain area, and specific receptors for estradiol and thyroid hormones have been also found at the archicortical level (see review by *Endröczi,* 1972). The hippocampo-hypothalamic and septal-hypothalamic connections play a role in the maintenance of the circadian rhythm of hypothalamic activities, and proved to be important in the regulation of hormone-adjusted behavioural reactions. By the use of agents blocking serotonin synthesis, it appears that hormonally sensitive limbic activities involve at least serotoninergic transmission at the septal-hippocampal level and both, catecholaminergic and serotoninergic neuronal networks, at the hypothalamic relays. If we consider the sequence of neuroendocrine processes in a hierarchic order, we may come to the conclusion that a great number of long-, medium- and short-range transmitters are involved in each process. The hypothalamus with its diversified and highly specific humoral transmissions plays a key role in controlling the homeostasis of the living organism.

References

Anton-Tay, F., R. W. Pehlam, and *R. J. Wurtman*: Increased turnover of H-norepinephrine in rat brain following castration or treatment with ovine follicle-stimulating hormone. Endocrinol. *84*, 1489—1492 (1969).

Anton-Tay, F., and *R. J. Wurtman*: Norepinephrine turnover in rat brain after gonadectomy. Science *159*, 1245 (1968).

Chang, C. C.: A sensitive method for spectrophotofluorometric assay of catecholamines. Int. J. Neuropharmacol. *3*, 643—649 (1964).

Donoso, A. O., and *F. J. E. Stefano*: Sex hormones and concentration of noradrenaline and dopamine in the anterior hypothylamus of castrated rats. Experientia (Basel) *23*, 665—666 (1967).

Dupont, A., E. Bastarach, E. Endrőczi, and *C. Fortier*: Hippocampal uptake of estradiol and triiodothyronine and thyroxine in the rat. (In preparation for publication.)

Endrőczi, E.: Limbic system, learning and pituitary-adrenal function, p. 160. Budapest: Akadémia kiadó. 1972.

Everett, J. W.: Central neural control of reproductive functions of the the adenohypophysis. Physiol. Rev. *44*, 373—431 (1964).

Fuxe, K.: Cellular localization of monoamines in the median eminence and infundibular stem of some mammals. Z. Zellforsch. *61*, 710—724 (1964).

Fuxe, K.: Evidence for the existence of monoamine neurons in the central nervous system. III. The monoamine nerve terminal. Z. Zellforsch. *65*, 573—595 (1965).

Fuxe, K., and *T. Hökfelt*: Central monoaminergic system and hypothalamic function. In: The Hypothalamus (*Martini, L., F. Motta,* and *F. Fraschini*, eds.), pp. 123—138. New York-London: Academic Press. 1970.

Hökfelt, T., and *K. Fuxe*: Effects of prolactin and ergot alkaloids on the tubero-infundibular dopamine (DA) neurons. Neuroendocrinol. *9*, 100 to 122 (1972).

Kurachi, K., R. Iwata, and *K. Hirota*: Experimental studies on the metabolism of catecholamine in rat brain and sexual function. Integr. Mechanism of Neuroendocrine Systems, Hokkaido Univ. Med. Library, Series, Vol. 1, pp. 151—162 (1968).

Maanen, J. V. van, and *P. G. Smelik*: Induction of pseudopregnancy in rats following local depletion of monoamines in the median eminence of the hypothalamus. Neuroendocrinol. *3*, 177—186 (1968).

Markee, J. E., C. H. Sawyer, and *W. H. Hollinshead*: Adrenergic control of the release of luteinizing hormone from the hypophysis of the rabbit. Rec. Prog. Hormone Res. *2*, 117—131 (1948).

Meites, J.: Pharmacological control of prolactin secretion and lactation. In: Proc. of the 1st Pharmacol. Meeting, Stockholm (*Guillemin, R.*, ed.), Vol. 1, pp. 151—181. Pergamon Press. 1962.

McCann, S. M.: Neurohumoral correlates of ovulation. Fed. Proc. *29*, 1888—1894 (1970).

McEwen, B. S., J. M. Weiss, and *L. S. Schwartz*: Retention of corticosterone by cell nuclei from brain regions of adrenalectomized rats. Brain Res. *17*, 471—482 (1970).

Sawyer, C. H.: Mechanisms by which drugs and hormones activate and block release of pituitary gonadotropins. In: Proc. 1st Int. Pharmacol. Meeting, Stockholm (*Guillemin, R.*, ed.), Vol. 1, pp. 27—45. New York-London: Pergamon Press. 1963.

Sawyer, C. H., J. E. Markee, and *B. F. Townsend*: Cholinergic and adrenergic components in the neurohumoral control of the release of LH in the rabbit. Endocrinol. *44*, 18—37 (1949).

Stefano, F. J. E., and *A. O. Donoso*: Norepinephrine levels in the rat hypothalamus during the estrous cycle. Endocrinol. *81*, 1405—1406 (1967).

Tonge, S. R., and *P. M. Greengrass*: The acute effects of estrogen and progresterone on the monoamine levels of rat brain of ovariectomized rats. Psychopharmacol. (Berlin) *21*, 374—381 (1971).

Udenfriend, S., and *P. Zaltzman-Nirenberg*: Norepinephrine and 3, 4-dihydroxyphenylethylamine turnover in guinea pig brain in vivo. Science *142*, 394—396 (1963).

Authors' address: Dr. *E. Endrőczi,* Department of Experimental and Clinical Laboratory Investigations, Postgraduate Medical School, Budapest, Hungary.

Discussion

Ariëns Kappers: I should like to refer to the sagittal section of the rat brain in which some axons of the dopaminergic cells are seen to terminate on the primary plexus of the hypothalamo-hypophyseal portal system or on dendrites of CRF neurons. In the literature there are two opinions regarding the sites of these terminals: they would either end on the primary plexus or on terminals of CRF neurons, influencing the release of CRF by this axo-axonic contact. According to both opinions, the terminals of the dopaminergic axons are in a close topographical relationship with the primary plexus in the median eminence, but their function would be different. Is there any reason or personal preference for either one of the above opinions concerning the site of the dopaminergic axon terminals? Would it not be better to indicate in this figure that the endings of the dopaminergic axons are, according to several authors, in close contact with the axon terminals of the CRF neurons, rather than with their perikarya or dendrites?

Endrőczi: From a physiological point of view, the modifying influence of dopaminergic terminals upon axon terminals of CRF neurones is very plausible; however, I have no direct evidence to suggest a predominance of an axo-axonic or an axo-dendritic connection. The figure showed only possible connections without specification of the character of the terminals.

Gerebtzoff: I would suggest to include the pineal gland in the general scheme of interrelations resulting from the present investigations.

Endrőczi: I agree that the pineal gland exerts an influence on the hypothalamus, although our studies did not involve such effects in this phase of the investigations.

Csillik: Are the alterations observed in the hypothalamic catecholamine content accompanied by (or as a consequence of) metabolic alterations of brain stem catecholamine neurones, being the parent cells of most of the hypothalamic catecholamine terminals?

Endrőczi: I think your question is very basic in judging the effect of hormones on hypothalamic catecholamines. In studying the changes of catecholamine concentration in both the brain stem and the hypothalamus, we have found that prolonged administration of estradiol produces alterations in both levels and in the same direction. A single injection of estradiol to ovariectomized rats resulted in a change of the hypothalamic catecholamines without any significant influence at the brain stem level. A direct effect of estrogens on the intrinsic hypothalamic monoaminergic system was also indicated by the studies using local application of estradiol into the median eminence area.

Akert: Was 6-hydroxy-dopamine also applied in the preparations, and is it conceivable to assume that similar specific binding sites exist also in higher forebrain centers for peptides?

Endrőczi: Both electrophysiological (*Steiner*, 1970) and behavioural studies (see *Sawyer* and *Gorski*, 1971—UCLA Symposium) have demonstrated that the rostral limbic areas are sensitive to peptide hormones. In contrast to the accumulation of steroids by the nervous tissue, the study of specific binding sites for peptides is very difficult and I think this to be a very interesting aspect of neuroendocrine research.

Stach: I should like to know whether there are literature data concerning the effect of insulin on the amine metabolism of the hypothalamus.

Endrőczi: Sorry, we did not study the effect of insulin on brain monoamines and I do not recall anyone who reported data about the direct influence of insulin in this respect.

Journal of Neural Transmission, Suppl. XI, 227—253 (1974)

GABA as a Transmitter in the Central Nervous System of Vertebrates*

J. Storm-Mathisen

Norwegian Defence Research Establishment, Division for Toxicology, Kjeller, Norway

With 6 Figures

Summary

The topographical and subcellular distribution of GAD has been studied in Deiters' nucleus, the cerebellum, hippocampus and substantia nigra, and the effect of lesions of nerve pathways has been examined. The results showed that GAD is localized in Purkinje cells and probably also in neurones confined to the cerebellar cortex. In the hippocampus GAD is localized in intrinsic neurones, some of which are likely to be the basket cells. In the substantia nigra GAD is localized in nerve terminals the density of which may be highest in the pars compacta. The axons to which these terminals belong probably originate in the neostriatum or globus pallidus.

All the available data put together go far towards proving that GABA is the transmitter of the Purkinje cells. They indicate that the same is true for the cerebellar stellate, basket and Golgi cells, the basket cells of the hippocampus, local inhibitory neurones in the cerebral cortex and the striato-nigral fibres. These may be but a small selection of all GABA-neurones.

Introduction

For a substance to be regarded as the transmitter at a specific type of synapse it has to meet several criteria which may be divided into three groups:

---

* *Abbreviations used:* AChE, acetylcholinesterase; ChAc, choline acetyltransferase; DOPA, dihydroxyphenylalanine; GABA, γ-aminobutyric acid; GAD, glutamate decarboxylase; IPSP, inhibitory postsynaptic potential.

1. *Action:* When exogenously applied to the synapse it should mimic in great detail the action of the actual transmitter. This involves the characteristics of any conductance change and the effect of antagonists.
2. *Presence:* The substance should be available in the presynaptic terminal at a rate sufficient to keep up with its release. This seems often to be accomplished by concentration of the substance and the apparatus for its synthesis in the presynaptic element. A mechanism for its rapid removal from the receptor site should also be localized at the synapse. This may involve fast re-uptake by membrane transport or (and) rapid enzymatic degradation.
3. *Release:* The transmitter candidate should be released from the presynaptic element during the transmission process.

Primarily due to the work of Kravitz and his coworkers (*Otsuka et al.*, 1966) we know that these three groups of criteria are met to a considerable extent by GABA in the case of certain inhibitory neurones in Crustacea. This work established the role of GABA as a neurotransmitter substance. Lately, substantial evidence has been brought up that GABA may be a transmitter substance also in the brain of higher animals.

In the following presentation I will deal in some detail with the experiments that our laboratory has contributed in this field[1], and briefly review the most relevant literature.

Our own work deals with subjects under group II in the above scheme. We have measured the capacity for synthesis of GABA, *i.e.*, the activity of GAD, rather than GABA itself. This is based on the assumption that the synthesis rate is more significant than the actual amount of GABA, and on the observation that GAD is concentrated in synaptosomes (*Fonnum*, 1968). Further, GAD is more stable, and is technically much easier to measure than GABA in such small tissue samples as are necessary to this type of work. Finally, there is a remarkably close relationship between the GABA level and the GAD activity in different locations (*Baxter*, 1970). GAD activity is measured by a version (*Fonnum et al.*, 1970) of the method of *Albers* and *Brady* (1959) which depends on the evolution of (^{14}C) CO_2 during decarboxylation of L-(1—^{14}C) glutamate (Fig. 1). The enzyme GAD II (*Haber et al.*, 1970 a, b), which, unlike GAD, is bound to mitochondria and occurs in glia and non-nervous tissues, shows only a very low activity under the assay conditions used (*Storm-Mathisen* and *Fonnum*, 1971). The microdissected tissue samples are obtained

[1] Most of the work has been done in collaboration with Dr. *F. Fonnum.*

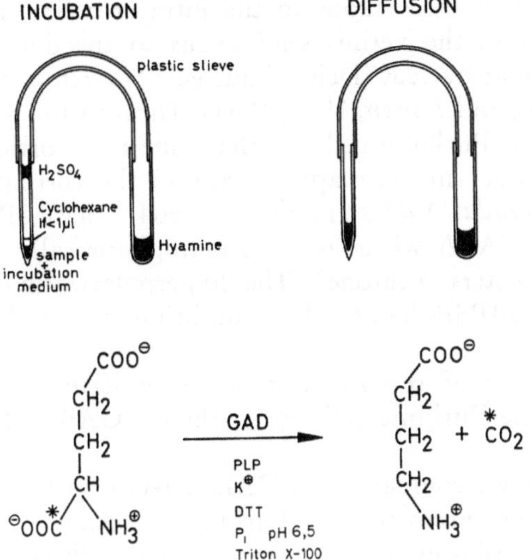

Fig. 1. Assay procedure for GAD after *Albers* and *Brady* (1959); (*Fonnum et al.*, 1970)

according to the procedure of *Lowry* (1953). Thin sections are cut from fresh frozen tissue, freeze-dried and dissected by hand with splinters of razor blades. The samples (0.2—2 μg) are then weighed on a quartz fibre microbalance.

How then to attack *localization problems* in the CNS where different kinds of cell processes are so intimately intermingled? We have been using three different types of approach:

A. *Microdissection* and microchemical analysis of GAD in selected brain regions where the different tissue elements are partially separated, such as the hippocampus and cerebellum;
B. *subcellular* fractionation of discrete regions to obtain an indication of what proportion of GAD is localized in the nerve terminals;
C. *denervation* experiments, observing whether any loss of GAD is induced in a region by anterograde degeneration of afferent nerve pathways after their transection.

The Purkinje Cells

The Purkinje cell of the cerebellum is at present the only reasonably well-established GABA-ergic neurone in mammalian brain. The work leading up to this conclusion rests on the knowledge of the Oslo group of neuroanatomists (Fig. 2). Thus, the Purkinje cells in the hemispheres send their axons to the intracerebellar nuclei, whereas Purkinje cells in the vermis send axons to the dorsal part of the lateral vestibular nucleus (Deiters' nucleus) (*Jansen* and *Brodal*, 1940, 1958; *Walberg* and *Jansen*, 1961, 1964). Through the work of Ito and coworkers, the Purkinje cells of the respective cortical areas were found to produce monosynaptic IPSPs in the corresponding nuclei (*Ito* and *Yoshida*, 1966; *Ito et al.*, 1964). Such IPSPs could be mimicked by GABA when this was iontophoretically applied in the vicinity of Deiters' neurones. The hyperpolarization produced by GABA and the IPSPs had similar equilibrium potentials (*Obata et al.*, 1967).

On the basis of the above data we undertook a study of the capacity of the Purkinje cells to synthesize GABA (*Fonnum et al.*, 1970).

In the cat we compared the GAD activity in the dorsal part of Deiters' nucleus, which receives Purkinje cell axons, with that in the anatomical distribution of the axons of these cells established by the ventral part, which does not receive such axons. The dorsal part

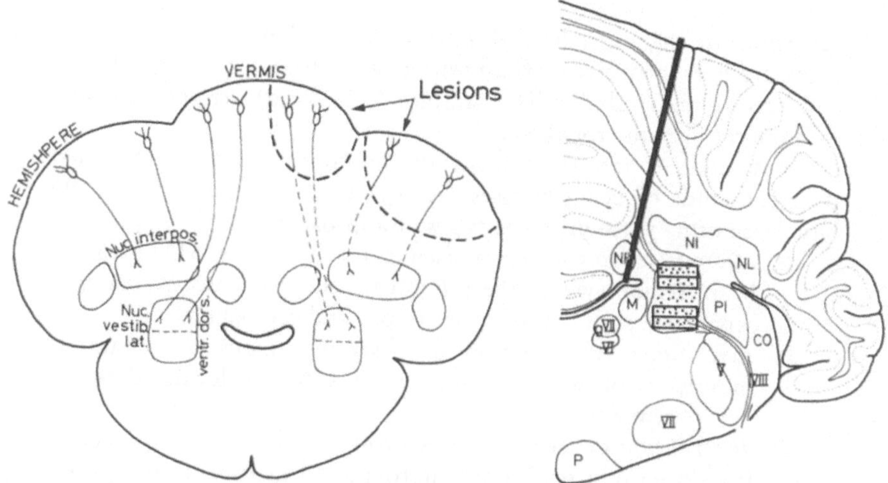

Fig. 2. Left part, schematic drawing showing projection of Purkinje axons. Right part, camera lucida drawing of a section through the cerebellum and brain stem of a rat to show site of transection (bar) of Purkinje axons from vermis to the lateral vestibular (Deiters') nucleus, and dissection of the latter (rectangles) in two animals (6 and 7). Symbols: CO, cochlear nuclei; dots, giant cells of Deiters; M, nucleus vestibularis medialis; NF, NI and NL, nuclei fastigii, interpositus and lateralis; P, pyramidal tract; PI, inferior cerebellar peduncle; V, spinal tract and spinal tract nucleus of trigeminal nerve; VI and VII, nuclei of trochlear and facial nerves; g VII, knee of facial nerve fibres; VIII, fibres of statoacoustic nerve

Table 1. *GAD Activity in the Dorsal and Ventral Parts of Lateral Vestibular Nucleus in Normal and Operated Cats*

Cat	Side	Activity		Dorsal/Ventral
		Dorsal	Ventral	
C1 (normal)	R and L	15 ± 3.8 (10)	7.1 ± 2.0 (9)	2.1
C6 (normal)	R and L	18 ± 3.8 (9)	6.0 ± 0.9 (9)	3.0
C7 (normal)	R and L	15 ± 2.2 (9)	6.2 ± 1.7 (9)	2.5
C2 (operated)	L	9.6 ± 2.0 (4)	7.7 ± 1.4 (6)	1.3
C2 (operated)	R	6.7 ± 1.6 (6)	6.1 ± 0.9 (7)	1.1
C3 (operated)	L	9.8 ± 3.5 (5)	7.0 ± 1.2 (5)	1.4
C3 (operated)	R	6.5 ± 2.9 (5)	7.4 ± 1.2 (6)	0.9
C4 (operated)	L	11.1 ± 2.7 (3)	6.4 ± 0.6 (4)	1.7
C4 (operated)	R	5.6 ± 1.4 (4)	6.2 ± 0.6 (4)	0.9

The results are given as μmoles CO_2/h/g dry wt. Mean values ± S.D. No. of samples in parentheses. The survival times were 7 days for C2 and 12 days for C3 and C4. From *Fonnum et al.* (1970).

contained 2—3 times the activity in the ventral part (Table 1). Further, the activity in the intracerebellar nuclei was similar to that in the dorsal part of Deiters' nucleus (Table 2).

Table 2. *GAD and ChAc Activity in Nucleus Interpositus in Normal Cats and in Cats with Lesions in the Left Hemisphere*

Cat	Enzyme	Activity	
		Right side	Left side
C7 (normal)	GAD	18 ± 5.0	19 ± 3.9
C5 (operated)	GAD	15 ± 1.8	7.3 ± 1.4
C6 (operated)	GAD	14 ± 3.0	4.4 ± 2.1
C5 (operated)	ChAc	1.1 ± 0.2	1.3 ± 0.2

The results are expressed as μmoles/h/g dry wt. Mean ± S.D. From *Fonnum et al.* (1970).

After lesions in the cerebellar vermis destroying the axons to the dorsal part of Deiters' nucleus, the activity dropped in the dorsal part by up to 70 % and to an extent depending on the size of the lesion, whereas no change occurred in the ventral part (Table 1). Lesions in the cerebellar hemispheres produced similar loss of activity in the intracerebellar nuclei (Table 2). These results have been confirmed in the rat (Table 3) (*Storm-Mathisen*, 1972). On the other hand, under-

Table 3. *GAD in Lateral Vestibular Nucleus after Interruption of Fibres from Cerebellar Vermis*

Part of nucleus (see Fig. 2)	Animal (days survival)	Activity on operated side (% of control)	Activity on control side (μmoles/h/g dry wt.)
Dorsal 1	7 (3)	29 ± 5 (4,4)	69.7 ± 5.1
	6 (7)	24 ± 4 (3,3)	70.9 ± 7.7
Dorsal 2	7 (3)	27 ± 4 (4,4)	59.1 ± 6.3
	6 (7)	30 ± 5 (3,3)	67.5 ± 3.3
Ventral 1	7 (3)	89 ± 10 (3,3)	18.0 ± 1.8
	6 (7)	72 ± 7 (2,2)	23.3 ± 2.3
Ventral 2	7 (3)	91 ± 12 (3,3)	19.3 ± 0.7
	6 (7)	97 ± 14 (2,3)	17.3 ± 2.5

Values are presented as mean ± S.E.M. Number of samples on operated and control side respectively are given in parentheses. (From *Storm-Mathisen*, 1972.)

cutting of the cerebellar peduncles did not alter the activity of GAD in the intracerebellar nuclei, whereas the activity of ChAc was reduced by 80 % (Table 4) (*Fonnum* and *Storm-Mathisen,* see *Fonnum,* 1972 a).

After destruction of Purkinje cells there was no change in choline acetylase activity or in dry weight per tissue volume, nor in AChE, lactate dehydrogenase, succinate dehydrogenase or cell structure as judged by light microscopy of suitably stained sections. Thus, there were no signs of unspecific tissue damage. The possibility that the decline of activity observed could be due to transneuronal effects cannot be strictly excluded, but results from other sites in the CNS *(vide infra)* indicate that GAD is not prone to such effects. We can,

Table 4. *GAD and ChAc in Rat Cerebellum after Transection of the Peduncles*

			After lesion	
Location	Enzyme	Normal	4 days	7 days
Cortex	GAD	47 ± 3.6	54 ± 2.9	46 ± 2.2
	ChAc	1.8 ± 0.17	0.93 ± 0.01	0.15 ± 0.05
Nucleus	GAD	150 ± 20	130 ± 13	
interpositus	ChAc	15 ± 4.0	3.3 ± 0.3	

The results are expressed in μmoles/h/g dry wt. Mean values ± S.D. Six samples from each animal. (*Fonnum* and *Storm-Mathisen;* from *Fonnum,* 1972 a.)

therefore, conclude with reasonable certainty that at lesat 70 % of the GABA-synthesizing capacity in the nuclei receiving Purkinje axons is situated within these axons and their terminals.

Subcellular fractionation indicated that in Deiters' nucleus about 60 % of the GAD activity was present in nerve terminals while in the nucleus interpositus a somewhat lesser proportion was found (Table 5).

Electron microscopy showed that nerve terminals made up roughly 4—5 % of the tissue volume in the dorsal part of Deiters' nucleus and axons 30—35 %. The GAD activity in white matter immediately dorsal to nucleus interpositus which is pierced by the Purkinje axons, was only 12 % of that in the nucleus. The cerebellar cortex which contains several types of alleged GABA neurones, had only about 30 % of the GAD activity of the nuclei. These data indicate that GAD is heavily concentrated in the nerve terminals relative to the axons and bodies of the Purkinje cells.

Table 5. *Subcellular Distribution of GAD and Lactate Dehydrogenase in Lateral Vestibular Nucleus and Nucleus Interpositus*

Enzyme	Lateral vestibular nucleus	Nucleus interpositus
GAD	59 ± 12 (3)	42 ± 8 (5)
Lactate dehydrogenase	24 ± 3 (3)	24 ± 5 (5)

The results are expressed as percentage of enzyme recovered in particulate form. The results are expressed as mean value ± S.D. (number of experiments). The recoveries were between 85—105 %. (From *Fonnum et al.*, 1970.)

By electron microscopy of the dorsal part of Deiters' nucleus up to 40 % of the nerve terminals were found to degenerate after lesions in the vermis of the anterior cerebellar lobe (*Fonnum* and *Walberg*, 1973). Although this figure may be somewhat underestimated, it indicates that as little as about 2 % of the tissue volume contains most of the GAD activity of this nucleus. Similar results were obtained for the cerebellar nuclei.

Simultaneous with our observations, *Otsuka et al.* (1971) observed by ultramicroassay methods that GABA is more concentrated in giant Deiters' cells dissected from the dorsal part as compared to ones dissected from the ventral part of the lateral vestibular nucleus. After lesions in the vermis, the GABA concentration decreased in the dorsal cells, but not in the ventral. These authors, therefore, concluded that GABA was concentrated in the terminals of Purkinje fibres originating from the vermis, which are known to be localized on the bodies and proximal dendrites of Deiters' cells in the dorsal part of the lateral vestibular nucleus (*Mugnaini* and *Walberg*, 1967).

Furthermore, GABA is released into the fourth ventricle during massive stimulation of the cerebellar Purkinje cells if GABA breakdown is inhibited by amino-oxyacetic acid. By such stimulation the release of GABA was increased up to seven times, whereas the total content and chromotographic pattern of amino acids in the perfusate stayed unchanged (*Obata* and *Takeda*, 1969).

By microiontophoretic techniques GABA has been found to be more effective than glycine in inhibiting dorsal Deiters' cells (*Bruggencate* and *Engberg*, 1969 a). Like the potentials produced by GABA, the IPSPs evoked in these cells by stimulation of the cerebellar vermis are unaffected by strychnine (*Bruggencate* and *Engberg*, 1969 b), but blocked by picrotoxine (*Obata et al.*, 1970) and bicuculline (*Curtis et al.*, 1970 a). These features are thought to be specific characteristics of GABA synapses in general.

The hypothesis has been put forward that at GABA-synapses, like at monoaminergic synapses, the rapid inactivation of the transmitter after it has interacted with the postsynaptic receptor, is obtained by its uptake into the presynaptic terminals (*Iversen* and *Neal*, 1968). The Purkinje fibre system would be ideal for testing this hypothesis, *i.e.*, whether such an uptake occurs specifically into the GABA-ergic nerve terminals.

Cerebellar Cortex

As we have seen, the output of the cerebellar cortex is inhibitory, mediated by the probably GABA-ergic Purkinje cells. In addition, the cortex contains at least three other types of inhibitory neurones. The basket and stellate cells are thought to inhibit Purkinje cells by axo-somatic and axo-dendritic synapses, respectively. The Golgi cells form inhibitory synapses on the dendrites of the granular cells. The basket and Golgi cells are in turn inhibited by recurrent collaterals of the Purkinje cells. A detailed review of these subjects has been published by *Eccles et al.* (1967).

Pharmacological studies indicate that GABA may be the transmitter of the basket, stellate and Golgi cells (*Curtis et al.*, 1971 b; *Woodward et al.*, 1971; *Bisti et al.*, 1971). *Kuriyama et al.* (1966) studied the distribution of GAD and GABA in the different layers of the cerebral cortex. GAD activity was about 30 % higher in the layer of Purkinje cells than in the adjacent layers. The concentration of GABA and its increase after administration of amino-oxyacetic acid were well correlated to the levels of GAD, although the differences between the layers were somewhat more marked. These observations are compatible with the concentration of GAD and GABA in the basket cell terminals, situated near the bases of the Purkinje cells. There tends to be a slightly higher content of GABA and GAD in the molecular than in the granular layer (*Albers* and *Brady*, 1959; *Hirsch* and *Robins*, 1962) corresponding to the presence of a larger proportion of the inhibitory local neurones in the molecular layer.

In agreement with the notion that the cerebellum does not receive GABA-ergic afferents, we were unable to change the level of GAD in cortex or nuclei by undercutting the peduncles (Table 4) (*Fonnum* and *Storm-Mathisen*, see *Fonnum*, 1972 a). On the other hand, ChAc decreased markedly, confirming earlier reports that some of the cerebellar afferents (probably mossy fibres) may be cholinergic (*Phillis*, 1965).

Slices of cerebellar cortex incubated in a physiological salt solution accumulate (^{3}H) GABA by an active transport mechanism (*Hökfelt* and *Ljungdahl*, 1971; *Iversen* and *Johnston*, 1971). By stereotaxic injection of the isotope *in vivo* it was possible to obtain an excellent preservation of structure and to establish by electron microscopy that GABA is heavily concentrated in the basket cell terminals on the bodies and initial parts of the axons of the Purkinje cells (*Hökfelt* and *Ljungdahl*, 1972). In addition, heavy accumulation of label was seen over certain cell bodies identified as stellate and basket cells, and possibly Golgi cells. There was no GABA uptake in the Purkinje cell bodies. This may perhaps be due to the fact that they are surrounded by terminals vigorously taking up GABA, or to a possible difference between the membrane properties of the Purkinje cells and the other alleged GABA-ergic neurones in the cerebellar cortex.

Cerebral Cortex and Hippocampus

These structures are included under the same heading since the hippocampus in some respects may be regarded as a "simplified" type of cortex where certain problems are more easily studied than in the neocortex, although the two regions are basically similar.

In the hippocampus the bodies of the pyramidal cells, the only efferent cells, are arranged in one discrete layer. Different types of nerve terminals are arranged in zones parallel to this layer. This

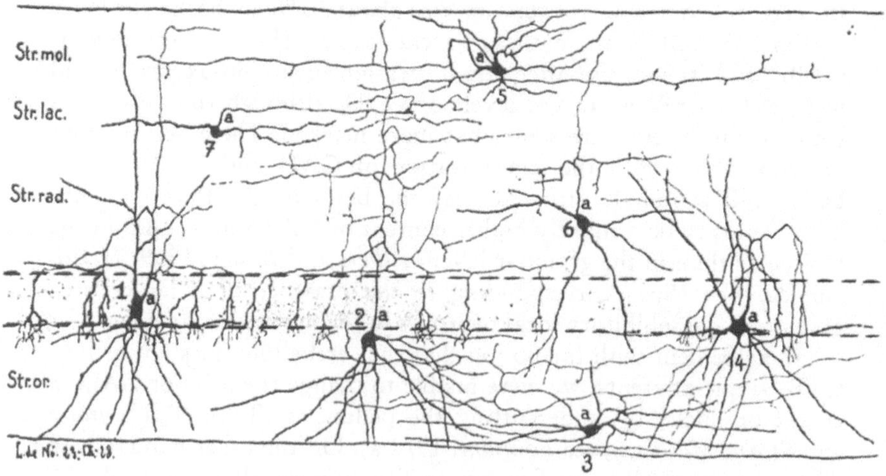

Fig. 3. Drawing of small neurones in the hippocampus regio superior of the mouse as seen in a Golgi section (Fig. 6 from *Lorente de Nó*, 1934). Cell 1, 2 and 4, typical basket cells sending their axons to stratum pyramidale. Cell 3, cell body on the border between alveus and stratum oriens. Cell 5, local neurone of stratum moleculare. a, axons. Stratum pyramidale is indicated by stipled lines

structural design facilitates electrophysiological and microchemical studies. The pyramidal cells are held to be inhibited by basket cells, which are intrinsic, short axoned neurones the terminals of which are localized on or close to the pyramidal cell bodies (Fig. 3) (*Andersen et al.*, 1964). An analogous situation holds for the adjacent area dentata (*Andersen et al.*, 1966).

The IPSPs produced by these synapses appear to be resistant to strychnine (*Andersen et al.*, 1963), and to be inhibited by bicuculline, as are the potentials elicited by GABA (*Curtis et al.*, 1970). Similar features are shown by inhibitory synapses on pyramidal cells in the neocortex (*Curtis et al.*, 1970 b), and the IPSPs here have been shown to be closely mimicked by GABA, but not by glycine, with respect to reversal potential and conductance change (*Dreifuss et al.*, 1969).

Jasper and *Koyama* (1969) found that in the spinal cat release of GABA from the cortical surface was augmented by lesions in the midbrain reticular formation, producing slow wave sleep. This release was, however, abolished by stimulation in the reticular formation, causing desynchronization of the EEG. Stimulation producing IPSPs in cortical cells causes release from neocortex of previously accumulated (^{3}H) GABA (*Mitchell* and *Srinivasan*, 1969) as well as of endogenous GABA (*Iversen et al.*, 1971). Cortical synaptosomes release GABA when electrically stimulated *in vitro* (*Bradford*, 1970).

Cortical tissue takes up GABA by an active membrane transport mechanism of high affinity and specificity (*Iversen* and *Neal*, 1968; *Kuriyama et al.*, 1969). The accumulated GABA appears to mix with the endogenous GABA pool and to be concentrated in nerve ending particles with the same density distribution as that containing GAD (*Neal* and *Iversen*, 1969; *Iversen* and *Johnston*, 1971). In neocortex about 80 % of the GAD activity is found in the nerve ending fraction (*Fonnum*, 1968) and a similar figure is obtained for the hippocampus (*Fonnum*, 1972 b). In comparison, about 70 % of the (^{3}H) GABA taken up by cortical slices and retained after processing for electron microscopy was present in nerve terminals and another 10 % in unmyelinated axons (*Bloom* and *Iversen*, 1971; *Iversen* and *Bloom*, 1972).

About 30 % of the nerve terminals in the neocortex seem to take up GABA with high affinity (*Iversen* and *Bloom*, 1972) and by gradient centrifugation these may be distinguished from nerve terminals taking up other transmitter candidates (*Wofsey et al.*, 1971). The GAD-containing terminals have a density distribution which is distinct from that of particles containing DOPA decarboxylase or ChAc, and in the hippocampus they are heavier than the latter ones (*Fonnum*, 1972 b). This is compatible with the concentration of

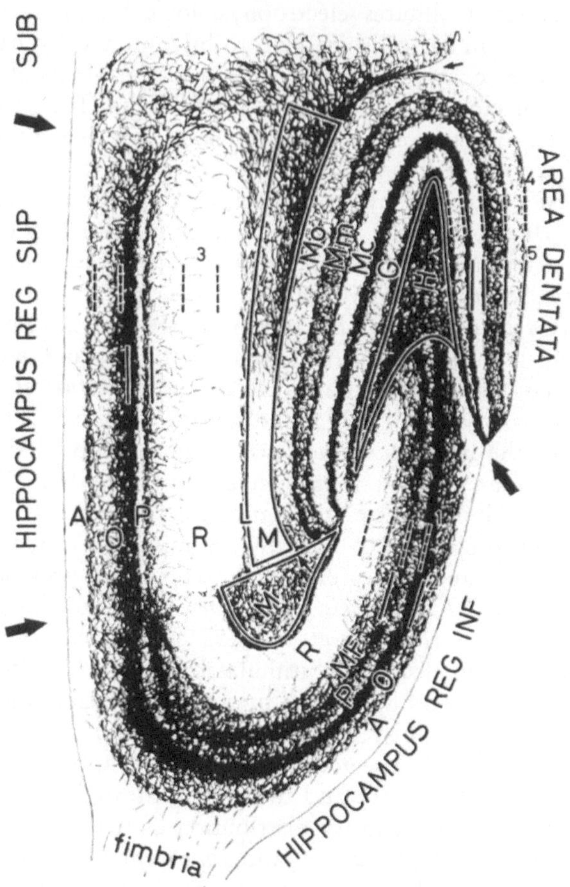

Fig. 4. Different zones in the rat archicortex illustrated by a camera lucida drawing of a section stained for AChE. The sites from where samples were dissected are indicated by solid or broken lines. *1*, dissection of hippocampus regio inferior (REG INF) in Table 8 "Mossy fibres". *2*, dissection of pyramidal zone of regio inferior elsewhere. *3*, dissection of regio superior (REG SUP) in Table 6. *4*, dissection of the zones of area dentata in Tables 6 and 7. *5*, dissection of area dentata in Table 8 (the zone Mc was discarded).

Symbols: A, alveus; O, stratum oriens; P, stratum pyramidale; R, stratum radiatum; L, stratum lacunosum; M, stratum moleculare of hippocampus and part of subiculum (SUB); MF, layer of mossy fibres in hippocampus regio inferior; Mo and Mm, outer and middle parts of stratum moleculare of area dentata; Mc, inner part of stratum moleculare containing the terminals of commissural afferents; G, stratum granulare; H, hilus. Coarse arrows indicate limits between cortical subfields, small arrows point to the bottom and orifice of the obliterated fissura hippocampi. (From *Storm-Mathisen*, 1972)

GAD in the axo-somatic terminals on the pyramidal cells, which contain a higher volume fraction of mitochondria than the axo-dendritic terminals (*Nafstad* and *Blackstad*, 1966) and, therefore, would be expected to sediment faster.

We have studied the laminar distribution of GAD in the hippocampal region by quantitative microchemical procedures (see Introduction) (*Storm-Mathisen* and *Fonnum*, 1971). The zones dissected are shown in figure 4. We found a peak of GAD activity in the zones containing the pyramidal and granular cell bodies, *i.e.*, where the inhibitory terminals are (Fig. 5). However, there was also another peak of activity, which was even somewhat higher, in the molecular layers of the hippocampus and the area dentata. In figure 5 the inner zone of the molecular layer was discarded, but later experiments have shown that this has indeed a lower activity than the adjacent zones

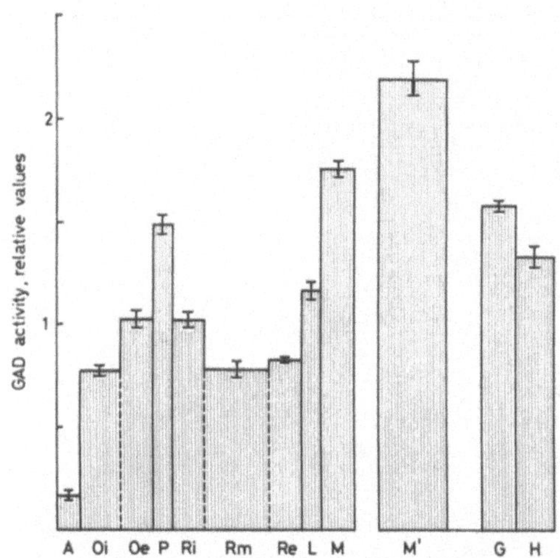

Fig. 5. Distribution of GAD in hippocampus regio superior and in area dentata. The activities are given relative to the average activity of samples cut at right angles through all layers of regio superior in each particular experiment. The mean activity of such samples from regio superior was in the series from hippocampus 44.5 ± 4.7 μmol/h/g dry wt. (S.D., 18 samples) and in that from area dentata 44.0 ± 4.0 μmol/h/g dry wt. (S.D., 8 samples). Heights of columns represent mean activities, top bars equal 2 × S.E.M. The data were from two animals. The numbers of samples for the different zones were for hippocampus: A 10, Oi 7, Oe 7, P 7, Ri 6, Rm 10, Re 4, L 7, and M 11; and for area dentata: M′ 8, G 9, and H 8 (symbols as in Fig. 4, except that the suffixes i, m and e denote internal, middle and external parts of layers, and that M′ corresponds to Mo + Mm of area dentata). The abscissa represents the approximate widths of zones, the total width of all layers in hippocampus regio superior being *ca.* 800 μm. (From *Storm-Mathisen* and *Fonnum*, 1971)

Table 6. *Distribution of GAD in Hippocampus and Area Dentata after Transection of Fimbria Compared to Sham Operated Animals*

Hippocampus regio superior				Area dentata			
Zone (see Fig. 4)	Animal*	Activity relative to zone G***	Activity as % of sham operated**	Zone (see Fig. 4)	Animal*	Activity relative to zone G***	Activity as % of sham operated**
M	17	1.21 ± .10 (4)	86 ± 6	Mo	17	1.24 ± .09 (5)	95 ± 3
	18	1.07 ± .06 (4)	90 ± 9		18	1.10 ± .04 (6)	86 ± 4
	10	1.13 ± .03 (4)			10	1.14 ± .04 (5)	
	15	.99 ± .09 (4)			15	1.12 ± .05 (5)	
R	17	.43 ± .03 (4)	102 ± 5	Mm	17	1.01 ± .07 (5)	112 ± 5
	18	.35 ± .01 (4)	80 ± 3		18	.87 ± .03 (6)	88 ± 5
	10	.40 ± .01 (4)			10	.84 ± .04 (5)	
	15	.39 ± .01 (4)			15	.86 ± .04 (6)	
O	17	.58 ± .05 (4)	100 ± 5	Mc	17	.69 ± .05 (5)	106 ± 11
	18	.54 ± .03 (4)	95 ± 10		18	.68 ± .02 (4)	92 ± 3
	10	.62 ± .03 (4)			10	.61 ± .06 (3)	
	15	.53 ± .04 (4)			15	.65 ± .02 (4)	
				G	17	1.00 ± .10 (6)	93 ± 7
					18	1.00 ± .02 (6)	88 ± 3
					10	1.00 ± .03 (6)	
					15	1.00 ± .04 (6)	
				H	17	.85 ± .07 (5)	90 ± 6
					18	.79 ± .01 (4)	83 ± 5
					10	.88 ± .05 (4)	
					15	.85 ± .06 (4)	

Mean values ± S.E.M. (numbers of samples) are given. (From *Storm-Mathisen*, 1972.)

* Lesioned animals: 17, 3 days survival and 18, 6 days survival. Sham operated animals: 10 and 15, undercutting of cerebellar peduncles, 7 days survival.

** Calculated for animals assayed simultaneously, *i.e.*, 17 based on 10 and 18 based on 15.

*** The mean activities of zone G in μmoles/h/g dry wt. ± S.E.M. were: Animal 17, 39.4 ± 2.7; 18, 47.1 ± .8; 10, 42.3 ± 1.1; 15, 53.6 ± 1.3.

(Table 6). Therefore, the bimodal distribution holds both in the hippocampus and the area dentata.

This distribution pattern is very similar to that found by autoradiography for (^{3}H) GABA taken up into hippocampal slices *in vitro* (*Hökfelt* and *Ljungdahl*, 1971). In the pyramidal layer the silver grains often form clusters around the pyramidal cells, while the latter are themselves free of activity. Scattered, densely labelled cell bodies are found mainly in the stratum oriens and the stratum lacunosum. Like for GAD, there is strong neuropil activity in the molecular layers of the hippocampus and the area dentata. In the parietal cortex the highest activity is present in the superficial layer, where the highest concentrations of endogenous GABA are also found (*Hirsch* and *Robins*, 1962). Again, scattered cell bodies, mainly localized in layer II and III, were densely labelled. In the electron microscope these were identified as stellate cells, which may be inhibitory interneurones in the cortex (*Hökfelt* and *Ljungdahl*, 1972).

In an effort to visualize GAD and GABA transaminase, *Csillik et al.* (1971) have studied the distribution of (^{14}C) thiosemicarbazide, an inhibitor of these enzymes, after intraperitoneal injection. This substance was shown to accumulate in the pyramidal and granular zones. Electronmicroscopically, the substance was found mainly in glial cells. However, negligible activity was found in the molecular layes, which we have shown to contain the highest activity of GAD (*vide supra*). It can therefore be concluded that thiosemicarbazide does not visualize the sites of GAD activity. *A priori* it would seem unlikely that it would do so, being a general reagent for aldehyde and keto groups and inhibiting GAD reversibly (*Killam et al.*, 1960) by interacting with pyridoxal phosphate, which is relatively loosely attached to this enzyme (*Roberts* and *Simonsen*, 1963). Furthermore, the convulsant effect of thiosemicarbazide may be partly unrelated to an interference with the GABA system (*Roberts et al.*, 1964) and it has been questioned whether the substance *in vivo* really reaches the sites of GAD activity (*Sze* and *Lovell*, 1970).

It would of course be a great advantage to be able to visualize GAD reliably in microscopical preparations. To be successful, however, one would have to find a ligand, binding very specifically and firmly to the enzyme.

After transection of the known afferent nervous pathways of the hippocampus there was only 10—20 % loss of GAD activity, even at extreme survival times (*Storm-Mathisen*, 1972). The pathways studied were the afferents from the contralateral hippocampus, the septum and those invading through the cingulum, fornix superior and postcommissural fornix (Tables 6 and 7), as well as that from the

Table 7. *GAD and ChAc in Hippocampus and Area Dentata Following Transection of Fimbria*

Region	Zone	Survival time (days)	GAD	ChAc
Regio Superior	P	4.5	90 ± 6 (10,8)	5 ± 1 (4,4)
		7	86 ± 7 (5,4)	11 ± 2 (7,5)
		22.5	88 ± 4 (4,4)	< 8 (4,4)
		625	84 ± 6 (4,4)	14 ± 4 (4,4)
	M	7	92 ± 16 (4,3)	
		22.5	99 ± 7 (3,3)	20 ± 3 (4,3)
		625	90 ± 7 (3,4)	
Regio Inferior	P	22.5	87 ± 3 (4,4)	12 ± 2 (4,4)
		625	89 ± 4 (4,4)	< 9 (4,4)
Area Dentata	Mo+Mm	7	96 ± 8 (3,3)	<13 (5,5)
	G	7	92 ± 8 (3,3)	10 ± 3 (5,5)
		22.5	84 ± 7 (4,4)	12 ± 2 (4,4)
		625	87 ± 5 (3,4)	< 8 (4,3)
	H	22.5	88 ± 7 (4,4)	16 ± 3 (4,4)
		625	90 ± 4 (3,4)	< 8 (4,3)

Activities on the operated side are expressed as a percentage of those on the control side ± S.E.M. (number of samples from operated and control sides given in parentheses). Symbols as in Fig. 4. (From *Storm-Mathisen*, 1972.)

medial and lateral parts of area entorhinalis, other fibres passing through the angular bundle, and the dentate-hippocampal mossy fibre system (Fig. 6, Table 8). While none of the lesions induced more than minor changes in GAD, there was a nearly total loss of ChAc after lesions of the afferents from the septum (Table 7) (*Lewis et al.*, 1967), but no decrease at all when these fibres were left intact. The loss was fully developed even at the shortest survival time tested (4.5 days).

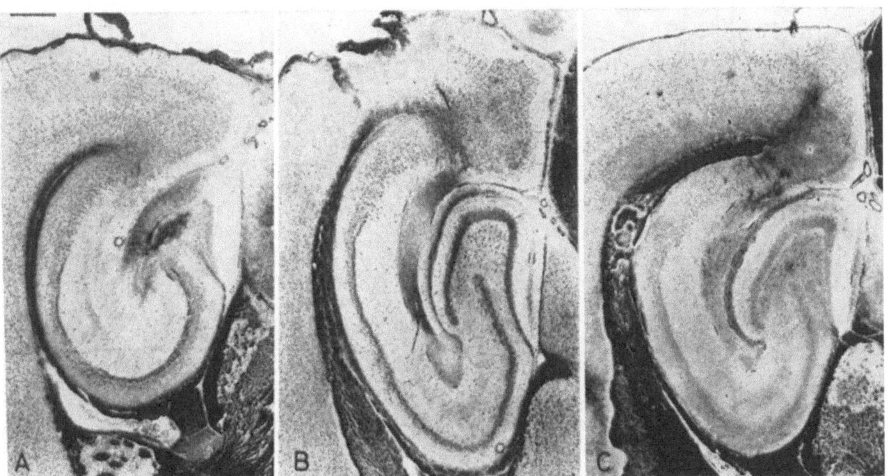

Fig. 6. Sections stained to show degenerating nerve terminals. Zones of degeneration are marked with small bars.
A: Massive degeneration in the layer of mossy fibres resulting from lesion of area dentata. B: Degeneration in the zones of termination of the perforant path in the molecular layers of area dentata and hippocampus after lesion in the medial part of area entorhinalis. C: Degeneration in the fields of termination of the fibres from the lateral part of area entorhinalis following a lesion in the angular bundle and deep parts of subiculum and hippocampus regio superior. Calibration bar for A, B and C 0.5 mm. (From *Storm-Mathisen*, 1972)

These experiments show fairly conclusively that in the hippocampal region nearly all, or possibly all, GAD activity must be located in intrinsic cells. The evidence listed above implies that these cells are GABA neurones and favours the basket cells as the source of the activity peak in the pyramidal and granular layers. The peak in the molecular layers has, at present, no obvious explanation, but it may be mentioned that also this layer contains local neurones being morphologically somewhat similar to the basket cells (*Ramón y Cajal*, 1968; *Lorento de Nó*, 1934).

Recently, *Kuhar et al.* (1972) found no reduction of GABA uptake in hippocampal synaptosomes after lesions in the medial septal area,

Table 8. *GAD in Hippocampus and Area Dentata Following Lesion of Nerve Pathways*

Region	Zone	Survival time (days)	Medial perforant path	Lateral perforant path	Mossy fibres
Regio Superior	M	5	96 ± 3 (8,8)	88 ± 4 (3,2)	—
		23	100 ± 4 (7,8)	101 ± 9 (4,4)	—
Regio Inferior	M	5	101 ± 5 (8,8)	—	—
		23	103 ± 3 (6,8)	118 ± 38 (2,2)	—
	R	6	—	—	107 ± 5 (3,3)
		23	—	—	94 ± 5 (3,4)
	MF	6	—	—	115 ± 15 (2,2)
		23	—	—	101 ± 18 (3,4)
	P	6	—	104 (1,1)	105 ± 11 (2,3)
		23	—	105 (1,1)	103 ± 18 (3,2)
Area Dentata	Mo	5	102 ± 5 (8,7)	110 ± 9 (4,4)	—
		23	127 ± 5 (8,7)	113 ± 4 (4,3)	—
	Mm	5	106 ± 9 (6,6)	97 ± 19 (2,2)	—
		23	110 ± 3 (7,8)	101 ± 8 (4,3)	—
	G	5	95 ± 6 (8,8)	87 ± 7 (2,2)	—
		23	95 ± 8 (6,6)	103 ± 9 (2,2)	—

Symbols and values as in Table 7. (From *Storm-Mathisen*, 1972.)

whereas there was a "large reduction" of choline uptake. GABA-neurones seem to be intrinsic cells also in the cerebral cortex, since chronically isolated cortical slabs have the same content as normal cortex of endogenous GABA and of (^{3}H) GABA, accumulated from the blood stream (*Gottesfeld et al.*, 1971).

In neocortex and hippocampus, 70—80 % of GAD is present in the nerve terminal fraction *(vide supra)*. Since the GABA-neurones are intrinsic neurones, their terminals must, therefore, either constitute a large proportion of their total volume or (and) the concentration of GAD must be much higher in the terminals than in the other parts (*Fonnum*, 1968, 1972 b).

Substantia Nigra

Among all regions of the brain, this nucleus contains the highest concentrations of GAD and GABA (*Albers* and *Brady*, 1959; *Chalmers et al.*, 1970; *Fahn* and *Côté*, 1968). The cell bodies of the nucleus are accumulated in the pars compacta, whereas most of the nerve terminals are found in the pars reticulata. The greater part of these seem to derive from the putamen and the nucleus caudatus (*Grofová* and *Rinvik*, 1970). We studied the distribution of GAD, DOPA decarboxylase and ChAc in the two parts of the nucleus in the cat (Table 9) (*Fonnum et al.*, unpublished). GAD showed the highest activity in the pars reticulata in contrast to DOPA decarboxylase and ChAc which were concentrated in the pars compacta. The heavy concentration of DOPA decarboxylase in the pars compacta corresponds to the localization of dopamine-containing cells in this part (*Dahlström* and *Fuxe*, 1964). The activity of ChAc is very low.

After homogenization of the substantia nigra (both parts) in isotonic sucrose, 85 % of GAD was obtained in a particulate form as compared to 20 % of lactate dehydrogenase. The nerve terminals are concentrated in the pars reticulata, but, here, GAD is only

Table 9. *Enzyme Activities in Substantia Nigra*

Location	Activity GAD	DOPA-dec	ChAc
Reticulata	240 ± 39	4.5 ± 0.9	1.8 ± 0.7
Compacta	170 ± 34	16 ± 3.9	3.5 ± 1.8

Activities expressed as μmoles/h/g dry wt. Mean values ± S.D. (Fonnum, Grofova, Rinvik, Storm-Mathisen and Walberg; from *Fonnum*, 1972 a.)

40 % higher than in the pars compacta. It, therefore, seems possible that GAD is confined to a population of nerve terminals preferentially situated near the cell bodies.

Extensive lesions of the putamen, and parts of the nucleus caudatus, but avoiding the globus pallidus, produced only small changes in GAD. However, when the globus pallidus was involved, GAD dropped substantially both in the pars reticulata and the pars compacta. In one animal with a large lesion in the globus pallidus the activity on the operated side was 17 ± 4 % of that on the control side. This is consistent with the results of *McGeer et al.* (1971). Similarly *Kim et al.* (1971) found a 50 % decrease of GABA in the substantia nigra 12 days after hemitransection at the subthalamic level. After lesions produced by suction and destroying most of the corpus striatum, but sparing "most of the pallidum", they found a 30 % decrease. These authors concluded that the substantia nigra receives GABA-containing afferents from the neostriatum. This is in agreement with the reports that monosynaptic inhibition, blocked by picrotoxine, can be elicited in nigral cells by stimulation of the head of the caudate nucleus (*Precht* and *Yoshida*, 1971; *Yoshida* and *Precht*, 1971).

So far we have not analysed cats with large lesions in the head of the caudate nucleus without concomitant lesions in the globus pallidus. This problem awaits further investigation[2]. In no case did we observe a decline in DOPA decarboxylase or ChAc.

Other Regions

GABA-neurones probably also occur in other regions of the central nervous system, although the evidence is less extensive. The listing below is not intended to be exhaustive. In rabbit retina, *Ehinger* and *Falck* (1971) found that (^{3}H) GABA is taken up into the neuropil in the inner plexiform layer, as well as into neurones probably representing amacrine cells and some neurones in the ganglion cell layer. *Lam* and *Steinmann* (1971) obtained similar results in dark adapted teleost retina, but in addition horizontal cells were labelled. Interestingly, during stimulation with light flashes there

[2] Since this manuscript was submitted for publication *Hattori, McGeer, Fibiger*, and *McGeer* (Brain Res. *54*, 103—114, 1973) have published a study in the rat corroborating their earlier conclusions that the GAD containing afferents to the substantia nigra originate in the globus pallidus. During the same time we have continued our studies of the origin of these fibres in the cat, arriving at the conclusion that they probably come from the neostriatum, and that the discrepancies in the literature may be explainable on the basis of the precise topographical organization of the neostriato-nigral projection (*Fonnum, Grofova, Rinvik, Storm-Mathisen*, and *Walberg*, to be published).

was a dramatic increase in the uptake into the horizontal cells, whereas the uptake in other locations stayed unchanged. Also the content of GABA and GAD are increased by light stimulation in the frog retina (*Graham et al.*, 1970). The laminar distribution of GAD and GABA is compatible with their association with the amacrine and horizontal cells, neurones which may have an inhibitory function in the retina (*Kuriyama et al.*, 1968).

There is an inhibitory, picrotoxine sensitive projection from the vestibular nuclei to the oculomotor nuclei (*Obata* and *Highstein*, 1970). This should be ideally suited for microchemical studies combined with lesions by virtu of its long course and well-defined terminal field. Also in the spinal cord GABA may act as an inhibitory transmitter besides glycine (*Curtis et al.*, 1971 a).

It is of special interest that GABA is likely to be a *presynaptic* inhibitory transmitter (*Davidson* and *Southwick*, 1971; *Davidoff*, 1972; *Huffman* and *McFadin*, 1972). In this case GABA causes a depolarisation rather than a hyperpolarisation of the presynaptic terminals. GABA, but not glycine, causes depolarisation also in the bodies of autonomic and sensory ganglion cells. This effect is blocked by picrotoxine, but not by strychnine. It has been suggested that the nature of the conductance change may be the same as in postsynaptic inhibition, but that the ionic composition in the cells is such that a depolarisation rather than a hyperpolarisation ensues (*De Groat*, 1972).

References

Albers, R. W., and *R. O. Brady*: The distribution of glutamate decarboxylase in the nervous system of the rhesus monkey. J. biol. Chem. *234*, 926 to 928 (1959).

Andersen, P., *J. C. Eccles*, and *Y. Løyning*: Pathway of postsynaptic inhibition in the hippocampus. J. Neurophysiol. *27*, 608—619 (1964).

Andersen, P., *J. C. Eccles*, *Y. Løyning*, and *P. E. Voorhoeve*: Strychnine resistant inhibition in the brain. Nature (London) *200*, 843 (1963).

Andersen, P., *B. Holmqvist*, and *P. E. Voorhoeve*: Entorhinal activation of dentate granule cells. Acta physiol. scand. *66*, 448—460 (1966).

Baxter, C. F.: The nature of γ-aminobutyric acid. In: Handbook of Neurochemistry (*Lajtha, A.*, ed.), Vol. III, pp. 289—353. New York: Plenum Press. 1970.

Bisti, S., *G. Iosif*, *B. F. Marchesi*, and *P. Strata*: Pharmacological properties of inhibitions in the cerebellar cortex. Exp. Brain Res. *14*, 24—37 (1971).

Bloom, F. E., and *L. L. Iversen*: Localizing ^{3}H-GABA in nerve terminals of rat cerebral cortex by electron microscopic autoradiography. Nature (London) *229*, 628—630 (1971).

Bradford, H. F.: Metabolic response of synaptosomes to electrical stimulation: Release of amino acids. Brain Res. *19*, 239—247 (1970).

Bruggencate, G. ten., and *I. Engberg*: Effects of GABA and related amino acids on neurons in Deiters' nucleus. Brain Res. *14*, 533—536 (1969a).

Bruggencate, G. ten., and *I. Engberg*: The effect of strychnine on inhibition in Deiters' nucleus induced by GABA and glycine. Brain Res. *14*, 536—539 (1969b).

Chalmers, A., E. G. McGeer, V. Wickson, and *P. L. McGeer*: Distribution of glutamic acid decarboxylase in the brains of various mammalian species. Comp. gen. Pharmacol. *1*, 385—390 (1970).

Csillik, B., A. M. Gerebtzoff, J. Kiss, and *E. Knyihar*: Zur Histochemie der limbischen Hemmung. Zytochemische und autoradiographische Untersuchungen über die Lokalisation der Enzyme des Gamma-amino-Buttersäure-Stoffwechsels im Hippocampus der Ratte. Histochemie *28*, 38—54 (1971).

Curtis, D. R., A. W. Duggan, and *D. Felix*: GABA and inhibition of Deiters' neurons. Brain Res. *23*, 117—120 (1970a).

Curtis, D. R., A. W. Duggan, D. Felix, and *G. A. R. Johnston*: GABA, bicuculline, and central inhibition. Nature (London) *226*, 1222—1224 (1970b).

Curtis, D. R., A. W. Duggan, D. Felix, and *G. A. R. Johnston*: Bicuculline, an antagonist of GABA and synaptic inhibition in the spinal cord of the cat. Brain Res. *32*, 69—96 (1971a).

Curtis, D. R., A. W. Duggan, D. Felix, G. A. R. Johnston, and *H. McLennan*: Antagonism between bicuculline and GABA in the cat brain. Brain Res. *33*, 57—73 (1971b).

Curtis, D. R., D. Felix, and *H. McLennan*: GABA and hippocampal inhibition. Brit. J. Pharmacol. *40*, 881—883 (1970).

Dahlström, A., and *K. Fuxe*: Evidence for the existence of monoamine-containing neurons in the central nervous system. I. Demonstration of monoamines in the cell bodies of brain stem neurons. Acta physiol. scand. 62, Suppl. 232, 1—55 (1964).

Davidoff, R. A.: Gamma-aminobutyric acid antagonism and presynaptic inhibition in the frog spinal cord. Science *175*, 331—333 (1972).

Davidson, N., and *C. A. P. Southwick*: Amino acids and presynaptic inhibition in the rat cuneate nucleus. J. Physiol. (London) *219*, 689—708 (1971).

De Groat, W. C.: GABA-depolarization of a sensory ganglion: Antagonism by picrotoxin and bicuculline. Brain Res. *38*, 429—432 (1972).

Dreifuss, J. I., J. S. Kelly, and *K. Krnjević*: Cortical inhibition and γ-aminobutyric acid. Exp. Brain Res. *9*, 137—154 (1969).

Eccles, J. C., M. Ito, and *J. Szentágothai*: The Cerebellum as a Neuronal Machine, 335 pp. Berlin-Heidelberg-New York: Springer. 1967.

Ehinger, B., and *B. Falck*: Autoradiography of some suspected neurotransmitter substances: GABA, glycine, glutamic acid, histamine, dopamine, and L-DOPA. Brain Res. *33*, 157—172 (1971).

Fahn, S., and *L. J. Côté*: Regional distribution of gamma-aminobutyric acid (GABA) in brain of the Rhesus monkey. J. Neurochem. *15*, 209 to 213 (1968).

Fonnum, F.: The distribution of glutamate decarboxylase and aspartate transaminase in subcellular fractions of rat and guinea-pig brain. Biochem. J. *106*, 291—298 (1968).

Fonnum, F.: Application of microchemical analysis and subcellular fractionation techniques to the study of neurotransmitters in discrete areas of mammalian brain. In: Studies in neurotransmitters at the synaptic level (*Costa, E.*, and *L. L. Iversen*, eds.). Adv. biochem. Psychopharmacol. *6*, pp. 75—88. New York: Raven Press. 1972a.

Fonnum, F.: Localization of cholinergic and γ-aminobutyric acid containing pathways in brain. In: Metabolic compartmentation in the brain (*Balázs, R.*, and *J. Cremer*, eds.), pp. 245—257. London: Macmillan. 1972b.

Fonnum, F., *J. Storm-Mathisen*, and *F. Walberg*: Glutamate decarboxylase in inhibitory neurones. A study of the enzyme in Purkinje cell axons and boutons in the cat. Brain Res. *20*, 259—275 (1970).

Fonnum, F., and *F. Walberg*: An estimation of the concentration of γ-aminobutyric acid and glutamate decarboxylase in the inhibitory Purkinje axon terminals in the cat. Brain Res. *54*, 115—127 (1973).

Gottesfeld, Z., *K. Krnjević*, and *R. J. Reiffenstein*: Penetration of circulating GABA into long-isolated cortical slabs. Canad. J. Physiol. Pharmacol. *49*, 70—78 (1971).

Graham, L. T., jr., *C. F. Baxter*, and *R. N. Lolley*: *In vivo* influence of light or darkness on the GABA system in the retina of the frog (Rana pipiens). Brain Res. *20*, 379—388 (1970).

Grofová, I., and *E. Rinvik*: An experimental electron microscopic study on the strionigral projection in the cat. Exp. Brain Res. *11*, 249—262 (1970).

Haber, B., *K. Kuriyama*, and *E. Roberts*: L-glutamic acid decarboxylase: A new type in glial cells and human brain gliomas. Science *168*, 589—599 (1970a).

Haber, B., *K. Kuriyama*, and *E. Roberts*: An anion stimulated L-glutamic acid decarboxylase in non-neural tissues: Occurrence and subcellular localization in mouse kidney and developing chick embryo brain. Biochem. Pharmacol. *19*, 1119—1136 (1970b).

Hirsch, H., and *E. Robins*: Distribution of γ-aminobutyric acid in the layers of the cerebral and cerebellar cortex. Implications for its physiological role. J. Neurochem. *9*, 63—70 (1962).

Hökfelt, T., and *Å. Ljungdahl*: Uptake of (^{3}H) noradrenaline and γ-(^{3}H) aminobutyric acid in isolated tissues of rat: An autoradiographic and fluorescence microscopic study. In: Histochemistry of nervous transmission (*Eränkö, O.*, ed.). Progr. Brain Res. *34*, 87—102 (1971).

Hökfelt, T., and *Å. Ljungdahl*: Autoradiographic identification of cerebellar and cerebral cortical neurons accumulating labelled gamma-aminobutyric acid (^{3}H-GABA). Exp. Brain Res. *14*, 354—362 (1972).

Huffman, R. D., and *L. S. McFadin*: Suppression of presynaptic inhibition and cerebellar disfacilitation by bicuculline. Life Sci. Part. *I, 11*, 113—121 (1972).

Ito, M., and *M. Yoshida*: The origin of cerebellar-induced inhibition of Deiters' neurons. I. Monosynaptic initiation of the inhibitory postsynaptic potentials. Exp. Brain Res. *2*, 330—349 (1966).

Ito, M., M. Yoshida, and *K. Obata*: Monosynaptic inhibition of the intracerebellar nuclei induced from the cerebellar cortex. Experientia (Basel) *20*, 575—576 (1964).

Iversen, L. L., and *F. E. Bloom*: Studies on the uptake of ^{3}H-GABA and (^{3}H) glycine in slices and homogenates of rat brain and spinal cord by electron microscopic autoradiography. Brain Res. *41*, 131—143 (1972).

Iversen, L. L., and *G. A. R. Johnston*: GABA uptake in rat central nervous system: Comparison of uptake in slices and homogenates and the effects of some inhibitors. J. Neurochem. *18*, 1939—1950 (1971).

Iversen, L. L., J. F. Mitchell, and *V. Srinivasan*: The release of γ-aminobutyric acid during inhibition in the cat visual cortex. J. Physiol. (London) *212*, 519—534 (1971).

Iversen, L. L., and *M. J. Neal*: The uptake of (^{3}H) GABA by slices of rat cerebral cortex. J. Neurochem. *15*, 1141—1149 (1968).

Jansen, J., and *A. Brodal*: Experimental studies on the intrinsic fibres of the cerebellum. II. Corticonuclear projection. J. comp. N. *73*, 267—321 (1940).

Jansen, J., and *A. Brodal*: Handbuch der mikroskopischen Anatomie des Menschen. IV. Nervensystem. Das Kleinhirn, 323 pp. Berlin-Göttingen-Heidelberg: Springer. 1958.

Jasper, H. H., and *I. Koyama*: Rate of release of amino acids from the cerebral cortex in the cat as affected by brainstem and thalamic stimulation. Canad. J. Physiol. Pharmacol. *47*, 889—905 (1969).

Killam, K. F., S. R. Dasgupta, and *E. K. Killam*: Studies of the action of convulsant hydrazides as vitamin B_6 antagonists in the central nervous system. In: Inhibition in the nervous system and gamma-aminobutyric acid (*Roberts, E.*, ed.), pp. 302—316. Oxford: Pergamon. 1960.

Kim, J. S., Y. Okada, R. Hassler, and *I. J. Bak*: The role of γ-aminobutyric acid (GABA) in extrapyramidal motor system. 2. Some evidence for the existence of a type of GABA-rich strionigral neurons. Exp. Brain Res. *14*, 95—104 (1971).

Kuhar, M. J., R. H. Roth, and *G. K. Aghajanian*: Choline uptake into synaptosomes from the hippocampus: Reduction after electrolytic destruction of the medial septal nucleus. Fed. Proc. *31*, 516 Abs. (1972).

Kuriyama, K., B. Haber, B. Sisken, and *E. Roberts*: The γ-aminobutyric acid system in rabbit cerebellum. Proc. nat. Acad. Sci. (Washington) *55*, 846—852 (1966).

Kuriyama, K., B. Sisken, B. Haber, and *E. Roberts*: The γ-aminobutyric acid system in rabbit retina. Brain Res. *9*, 165—168 (1968).

Kuriyama, K., H. Weinstein, and *E. Roberts*: Uptake of γ-aminobutyric acid by mitochondrial and synaptosomal fractions from mouse brain. Brain Res. *16*, 479—492 (1969).

Lam, D. M. K., and *L. Steinman*: The uptake of (γ-^{3}H) aminobutyric acid in the goldfish retina. Proc. nat. Acad. Sci. (Washington) *68*, 2777—2781 (1971).

Lewis, P. R., C. C. D. Shute, and *A. Silver*: Confirmation from choline acetylase analyses of a massive cholinergic innervation to the rat hippocampus. J. Physiol. (London) *191*, 215—224 (1967).

Lorento de Nó, R.: Studies on the structure of the cerebral cortex. II. Continuation of the study of the ammonic system. J. Psychol. Neurol. (Leipzig) *46*, 113—117 (1934).

Lowry, O. H.: The quantitative histochemistry of the brain. J. Histochem. Cytochem. *1*, 420—428 (1953).

McGeer, P. L., E. G. McGeer, J. A. Wada, and *E. Jung*: Effects of globus pallidus lesions and Parkinson's disease on brain glutamic acid decarboxylase. Brain Res. *32*, 425—431 (1971).

Mitchell, J. F., and *V. Srinivasan*: Release of ^{3}H-γ-aminobutyric acid from the brain during synaptic inhibition. Nature (London) *224*, 663—666 (1969).

Mugnaini, E., and *F. Walberg*: An experimental electronmicroscopical study on the mode of termination of cerebellar corticovestibular fibres in the cat lateral vestribular nucleus (Deiters' nucleus). Exp. Cell Res. *4*, 212—236 (1967).

Nafstad, P. H. J., and *T. W. Blackstad*: Distribution of mitochondria in pyramidal cells and boutons in hippocampal cortex. Z. Zellforsch. *73*, 234—245 (1966).

Neal, M. J., and *L. L. Iversen*: Subcellular distribution of endogenous and (^{3}H) γ-aminobutyric acid in rat cerebral cortex. J. Neurochem. *16*, 1245—1252 (1969).

Obata, K., and *S. M. Highstein*: Blocking by picrotoxin of both vestibular inhibition and GABA action on rabbit oculomotor neurons. Brain Res. *18*, 531—538 (1970).

Obata, K., M. Ito, R. Ochi, and *N. Sato*: Pharmacological properties of the postsynaptic inhibition by Purkinje cell axons and the action of γ-aminobutyric acid on Deiters' neurons. Exp. Brain Res. *4*, 43—57 (1967).

Obata, K., and *K. Takeda*: Release of γ-aminobutyric acid into the fourth ventricle induced by stimulation of the cat's cerebellum. J. Neurochem. *16*, 1043—1047 (1969).

Obata, K., K. Takeda, and *H. Shinozaki*: Further study on pharmacological properties of the cerebellar-induced inhibition of Deiters' neurones. Exp. Brain Res. *11*, 327—342 (1970).

Otsuka, M., L. L. Iversen, Z. W. Hall, and *E. A. Kravitz*: Release of gamma-aminobutyric acid from inhibitory nerves of lobster. Proc. nat. Acad. Sci. (Washington) *56*, 1110—1115 (1966).

Otsuka, M., K. Obata, Y. Miyata, and O. *Tanaka*: Measurement of γ-aminobutyric acid in isolated nerve cells of cat central nervous system. J. Neurochem. *18*, 287—295 (1971).

Phillis, J. W.: Cholinergic mechanisms in the cerebellum. Brit. med. Bull. *21*, 26—29 (1965).

Precht, W., and *M. Yoshida*: Blockage of caudate-evoked inhibition of neurons in the substantia nigra by picrotoxin. Brain Res. *32*, 229—233 (1971).

Ramón y Cajal, S.: The Structure of Ammon's Horn (translated by *Kraft, L. M.*). Springfield, Ill.: Thomas. 1968.

Roberts, E., and *D. G. Simonsen*: Some properties of L-glutamic decarboxylase in mouse brain. Biochem. Pharmacol. *12*, 113—134 (1963).

Roberts, E., *J. Wein*, and *D. G. Simonsen*: γ-Aminobutyric acid (γABA), vitamin B_6, and neuronal function.—A speculative synthesis. Vitamins and Hormones *22*, 503—559 (1964).

Storm-Mathisen, J.: Glutamate decarboxylase in the rat hippocampal region after lesions of the afferent fibre systems. Evidence that the enzyme is localized in intrinsic neurones. Brain Res. *40*, 215—235 (1972).

Storm-Mathisen, J., and *F. Fonnum*: Quantitative histochemistry of glutamate decarboxylase in the rat hippocampal region. J. Neurochem. *18*, 1105—1111 (1971).

Sze, P. Y., and *R. A. Lovell*: A reexamination of the effect of thiosemicarbazide on brain GABA and glutamic decarboxylase *in vivo*. Life Sci. Part *I*, 9, 889—899 (1970).

Walberg, F., and *J. Jansen*: Cerebellar corticovestibular fibres in the cat. Exp. Neurol. *3*, 32—52 (1961).

Walberg, F., and *J. Jansen*: Cerebellar corticonuclear projection studied experimentally with silver impregnation methods. J. Hirnforsch. *6*, 338—354 (1964).

Wofsey, A. R., *M. J. Kuhar*, and *S. Snyder*: A unique synaptosomal fraction, which accumulates glutamic and aspartic acids, in brain tissue. Proc. nat. Acad. Sci. (Washington) *68*, 1102—1106 (1971).

Woodward, D. J., *B. J. Hoffer*, *G. R. Siggins*, and *A. P. Oliver*: Inhibition of Purkinje cells in the frog cerebellum. II. Evidence for GABA as the inhibitory transmitter. Brain Res. *33*, 91—100 (1971).

Yoshida, M., and *W. Precht*: Monosynaptic inhibition of neurons of the substantia nigra by caudate-nigral fibres. Brain Res. *32*, 225—228 (1971).

Authors' address: Dr. *J. Storm-Mathisen*, Norwegian Defence Research Establishment, Division for Toxicology, P. O. Box 25, N-2007 Kjeller, Norway.

Discussion

Szentágothai: The relatively low GAD activity in the cerebellar cortex as compared to that in the nuclei might be explained by the large volume fraction taken up by the excitatory elements such as mossy terminals, granule cells and parallel fibres.

Storm-Mathisen: I admit such a possibility, though in the dorsal part of Deiter's nucleus the volume fraction taken up by the Purkinje terminals, believed to contain most of the GAD in this location, is also rather small. For the interpretation of biochemical localization studies it is of the utmost importance to have quantitative morphological data on the different tissue components, such as those reported by Szentágothai with regard to the cerebellar cortex.

Journal of Neural Transmission, Suppl. XI, 255—280 (1974)

The Clinical Relevance of Neurohistological Investigations of Intestinal Biopsies

G. Lassmann

Neurological Institute, University of Vienna, Austria
(Head: Prof. Dr. *F. Seitelberger*)

With 11 Figures

Summary

The results of neurohistological investigations of the intestine, performed during the last 2 1/2 years are presented. The material is based on 98 rectum biopsies (including 43 showing a normal innervation, 28 of morbus Hirschsprung, 6 of atresia ani, 9 of morbus Down, one of Niemann-Pick, 5 of which cases were accompanied by different cerebral symptoms), 8 duodenal biopsies accompanied by different clinical symptoms, and one surgical specimen of the appendix. The clinical relevance and practical diagnostic value of neurohistological investigations on bioptic material of the intestine is discussed. Besides, the progress in research due to more specific techniques now available is demonstrated. Unsettled problems concerning the local organization of the intramural nervous system (nerve fibres, ganglion cells and enterochromaffin cells) and the neural control of bowel movement are discussed.

During a long period, neurohistological investigations of the intestine were only performed on autoptic material or surgical specimens, leading to the description of degenerative, or hyperplastic alterations (local neuromata, ganglioneuromas or generalized hyperplasias). Besides, the reaction of argentaffin cells to pathologic conditions was studied (Bourgonnement-Masson, argentaffin cell tumors, carcinoid). The extraction and analysis of the specific substance produced by this tumor helped to explore the physiological function of these cells as serotonin producing and storing elements.

Even up to now, morphological studies were performed using

unspecific and capricious methods often leading to divergent results. One example for this is the long-lasting struggle between neuronists and reticularists. Using the new techniques we are now able to differentiate between adrenergic and cholinergic fibres or serotonin and histamine containing cells, a fact which will help us to obtain more exact statements about pathological alterations.

A further approach was obtained by studies of autoptic material of the rectum in verified cases of storage diseases of the central nervous system made by *Bodian* and *Lake* (1962). These authors showed that in all cases in which central nervous formations were involved, the same alterations could also be found in the peripheral innervation of the intestine.

Today, neurohistological investigations on rectum biopsies deal with numerous problems, as for example:

1. primary and secondary induced degenerative changes in the cause of local inflammation, local or traumatic lesions of the spinal cord at different levels;
2. genetically determined aplasia or dysplasia, local or generalized hyperplasias in the course of alterations of different origin;
3. the different reactions of cholinergic and adrenergic formations in the above mentioned alterations;
4. the morphology and function of different types of enterochromaffin cells in various conditions, for example after oral administration of L-Dopa;
5. storage diseases involving nervous or other tissue elements of the intestine.

Some of these problems will be discussed presenting our own clinical material, but first it is useful to discuss some basic problems.

The use of improved specific histological and histochemical techniques allowed to make some precise statements about morphological alterations in the intestine. Since in all these methods cryostat sections can be used, we can now use more different methods for routine examinations. Especially the demonstration of cholinergic and adrenergic nerve fibres also enables us to differentiate between sympathetic and parasympathetic nerve cells and to determine the content of serotonin-, catecholamine- and histamine-producing or -containing cells.

For solving special problems, some of these methods (as for instance the demonstration of specific cholinesterase) have been employed a long time ago. By means of this technique, for example, we detected the early state of neuromas in the human appendix and we were enabled to explore more functional data than by using the less specific techniques of silver impregnation (Fig. 1).

Much more difficult was the application of the method of formaldehyde-induced fluorescence for the demonstration of catecholamines (*Eränkö*, 1955; *Hillarp*, 1962; *Falck*, 1962; *Sakharov*, 1971). The modification by *Csillik* (1967) and *Sakharov* using cryostat sections enabled us to use these methods in combination with other histological techniques on the same material and to exclude methodical errors as far as possible.

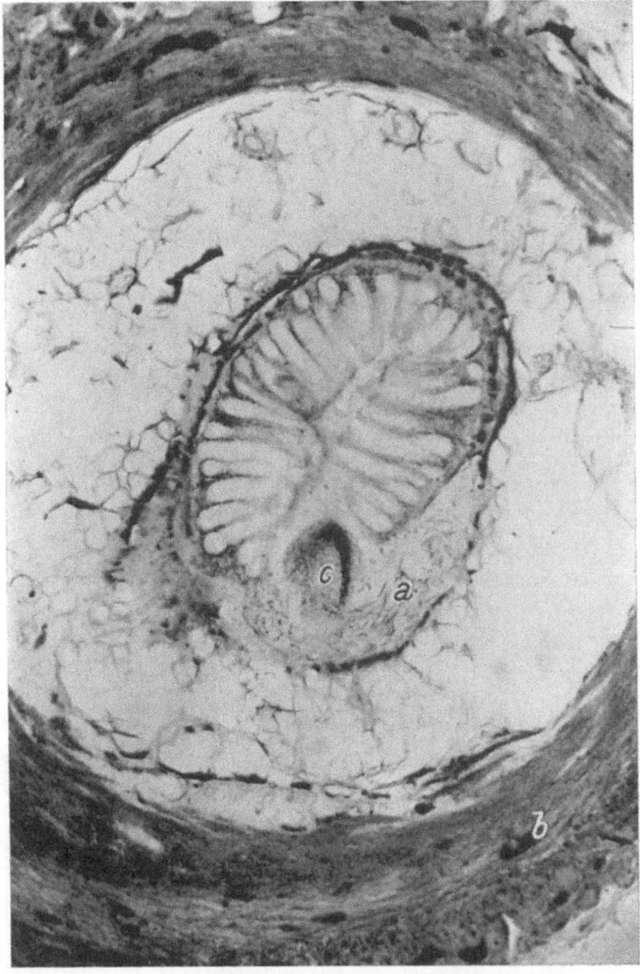

Fig. 1. VP: 423/59. Appendicité neurogène (human material). Local formation of a neuroma in the basal layer of the mucosa (a). b) Auerbach's plexus; c) lymph node. Specific cholinesterase technique (Koelle). ×35

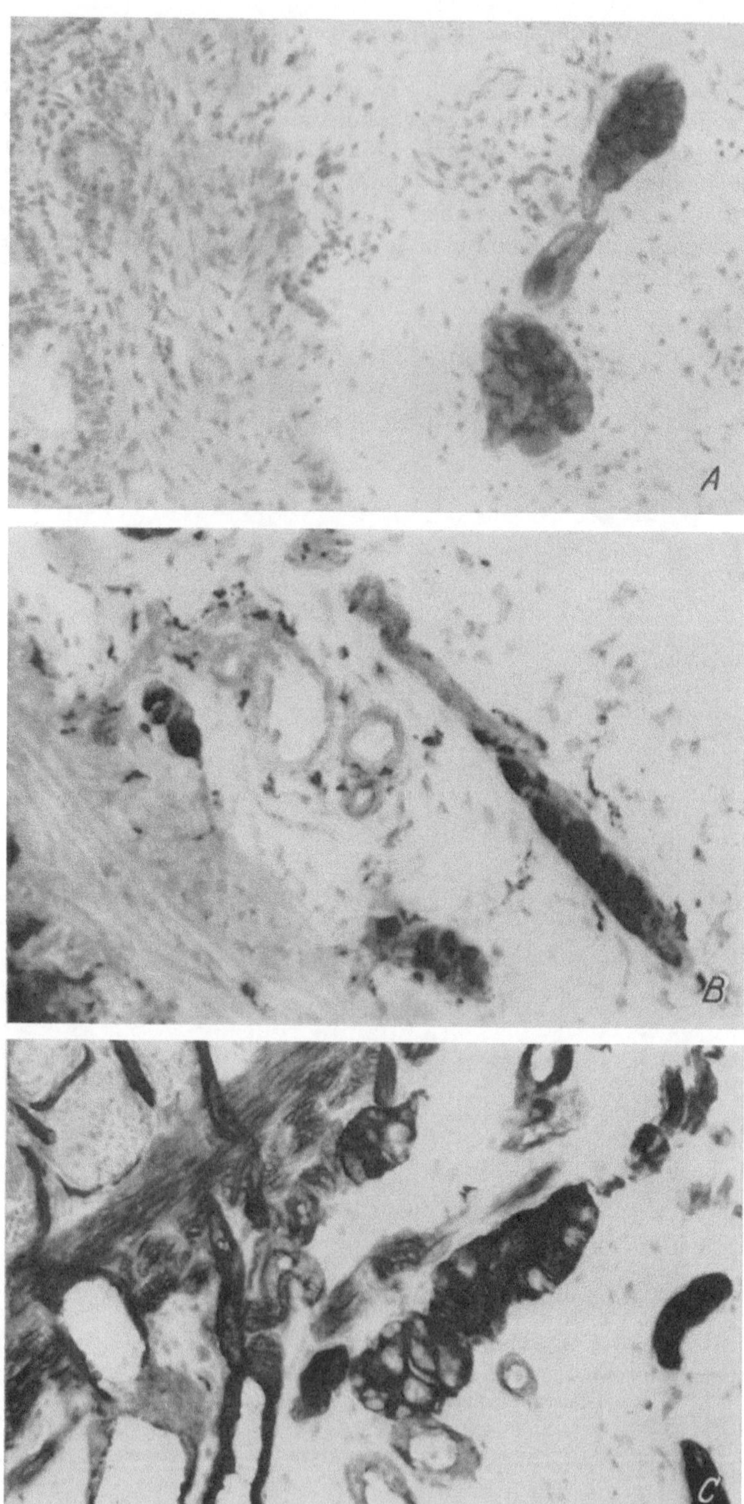
A
B
C

The different reactions of intramural ganglion cells and ganglion cells of the sympathetic trunc in different methods is of practical importance. Intramural ganglion cells—generally considered to be of parasympathetic origin—can easily be demonstrated by the methods for staining specific cholinesterase, but in contrast to sympathic ganglion cells they show no fluorescence using methods for demonstration of catecholamines. This reaction of the intramural ganglion cells offers, so far known, the only evidence for these cells being of parasympathetic origin, as embryologic studies (*Andrew,* 1970, 1971) gave only poor evidence of this hypothesis.

Both methods do not allow the differentiation between the two types of ganglion cells in the Auerbach and Meissner plexus of the intestine. For this we have to use silver impregnation or vital staining methods, but the small amount of bioptic material available invalidates the use of these techniques. For studying functional problems, the selective staining of these two cell types is necessary. Therefore, we tried to use histochemical methods for identification of the two cell types as described by *Korotchkin* (1966), but, so far, we did not succeed (Fig. 2).

Another problem concerns the "interstitial cells" first described by *Cajal* and *Lawrentiev* (cited by *Taxi,* 1965). The recent work of *Taxi* (1965) and the brilliant illustrations demonstrated in this meeting by *Stach* (1971, 1972) give further evidence of their existence, but leave the problem of their origin and functional significance unsolved. With the methods used for routine investigations it has been, so far, impossible to identify interstitial cells in our material. Electron microscopic investigations just started in our laboratory might furnish further evidence.

Other basic problems concerning the normal nervous organization of the intestinal tract are also unsettled. One of these, the visualization and distribution of afferent endings and their involvement in the neural control of bowel movement in normal and pathological conditions, will be briefly discussed later.

On the other hand, using modern methods it is now not only possible to examine the presence of specific structures but also their functional state. Silver impregnation, for example, stains the protein fraction of the dense core vesicles of enterochromaffin cells but does not give evidence about their content of biogenic amines as this is possible with fluorescent histochemical methods.

Fig. 2. L 54/71 male, 1a. M. Hirschsprung. Biops.colostomy. Ganglion cells of Meissner's plexus. Type I and II without distinction. A: spec. cholinesterase; B: acid phosphatase; C: alcalic phosphatase. ×460

Pharmacological alterations in the enterochromaffin-like-cells, following oral administration of certain drugs, such as L-dopa, can also be demonstrated with the fluorescence method. Therefore, this may be used for studying the therapeutic effect of oral application of these drugs, which may be useful for solving clinical problems. At present, our material is not large enough to make exact statements about the distribution and number of cells and nerves which produce and/or contain different biogenic amines. This fact should be specially kept in mind when it is necessary to differentiate between pathological and normal states. In cases in which there is complete deficiency of specific structures, this differentiation is somewhat easier, but we are mostly confronted with transitional states from normal to pathological.

The present paper deals with results obtained by examining our own material (98 rectum biopsies, 8 biopsies of the small intestine and 1 appendix).

Materials and Methods

Rectum biopsy is performed 4—5 centimeters oral to the skin-mucosa line of the anal region under direct visual control. Often it was, however, necessary to take several biopsies at different distances to explore the length of aganglionic segments. This can be done without general anaesthesia. The only complication may consist of small hemorrhages, which can easily be controlled by using a tampon. The material is frozen in isopentane, cooled by liquid nitrogen or dry ice; small pieces according to the method described by *Sakharov* are fixed in ice-cooled 4 % formaldehyde-Ringer-solution. Besides the routine methods, histochemical techniques are used on cyrostat sections (PAS, SSB, and other methods). For the demonstration of specific cholinesterase and catecholamines the methods described by *Karnowsky-Root* (1964) and *Csillik* are used.

If the material has to be carried over a far distance by car or train, the biopsy is placed on gauze, incubated in physiological saline solution in a petri dish, cooled by ice in a thermocontainer. This method, used in Switzerland for this purpose, allows the demonstration of specific cholinesterase even after 2 hours. At the moment, however, we have no personal experience about the possibility to demonstrate biogenic amines in material handled in this way.

At first some problems concerning *megacolon congenitum* (morbus Hirschsprung) will be discussed. This genetic disorder is much more common than is generally believed. In the USA there are about 700 new cases every year, which means a frequency of 1 on 5000 births. This number may increase during the next years, since the first children of patients, successfully treated by surgery, can be soon expected, if no eugenic advice is given. In Austria this problem will also become acute soon. In former years a prevalence of males (1 : 4) was supposed to occur, recent investigations, however, seem to prove that cases with a long aganglionic segment show no sex preference. If this genetic disorder is recognized early enough, surgical treatment may be followed by complete clinical recovery. Therefore, an exact histological examination of the biopsy is necessary. The resected intestine should also be examined with the methods mentioned above to solve unsettled problems on the nervous regulation of bowel movement.

The classical form of the disease with its typical clinical symptoms —chronic or intermittent constipation, abdominal distension, vomiting and—if not treated—even exitus—, is characterized by an aganglionic segment over a differing extent.

Recent investigations, however, described cases with identical clinical disturbances showing a local increase of ganglion cells and even other cases without any neurohistological alteration (*Netzeloff et al.*, 1970; *Ehrenpreis*, 1972; *Hillemand et al.*, 1971).

Today the a- or hypoganglionic segment is considered to be the result of a disturbance of the migration of ganglionic cells from oral into a caudal direction along the intestine. This migration seems to stop before the rectum is reached or even at the level of the proximal part of the colon descendens. *Mayer-Ruge* (1968, 1972) described a more or less distinct hypoganglionic segment of various length proximal to the aganglionic rectal part of the colon descendens. This finding seems to confirm the considerations about the development of this gentic disturbance. Nevertheless, *Hüther* (1954) described an ingrowth of intramural nerve cells in the rectum from caudal to oral in human embryos and *Cantino* (1970) observed the same in rats and guinea-pigs. This is in contradiction to the theory mentioned above. Therefore, this matter needs some fundamental prove.

A pathophysiological basis of the functional state of the "*narrow*" segment may be the absence of the ganglion cells in the plexus myentericus and submucosus. Therefore, the reflex mechanism of the intestinal musculature is interrupted, while the inhibiting effect of the adrenergic nerves influencing the intramural muscle and ganglion cells is also missing. This explanation cannot be employed in cases combined with nerve cell hyperplasie or absence of neurohistological alterations. In these cases one could think of a, still hypothetical, enzyme deficiency.

It is of interest that experimental removal of ganglion cells from the Auerbach plexus in the lower colon of dogs is not followed by obstipation or a so-called "narrow segment". To the contrary, a dilatation of the aganglionic part is observed (*Hovnanian et al.*, 1971). In our opinion, experiments destroying only the ganglion cells of the Auerbach plexus leaving the cells of the Meissner plexus intact, may not lead to the same clinical effects as present in morbus Hirschsprung. Especially, we cannot agree with the opinion that after removal of the intramural ganglion cells by dissecting the whole muscularis propria, a comparison with morbus Hirschsprung can be

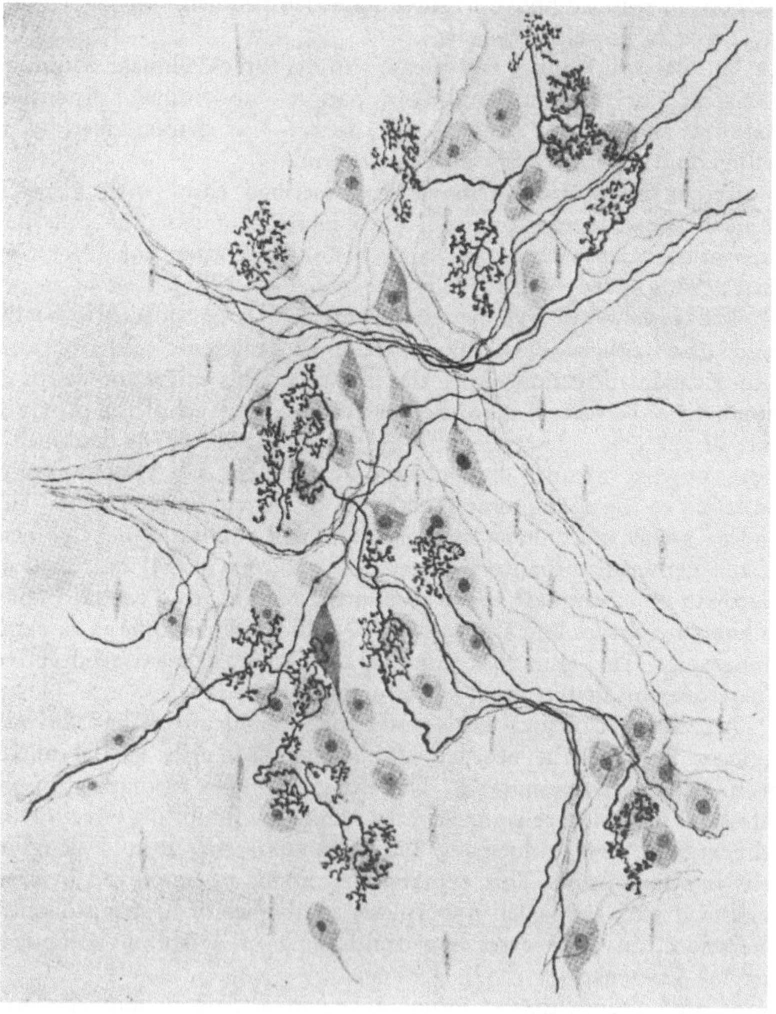

Fig. 3. Endformation of afferent fibres in the intramural ganglion of the cat's small intestine (from *Kolossow* and *Milochin*, 1963)

made. Electronmicroscopic studies of myenteric nerves in this disease and normal bowel (*Howard* and *Carret,* 1970) confirm the findings already known from light microscopic investigations. Normal contacts between nerve fibres and muscle cells were found and seemed to be generally present. This demonstrates that sacral parasympathetic nerve fibres have also contacts with muscle cells. This may explain the fact that reduced motor activity of the aganglionic segment can be abolished by spinal anaesthesia. So it seems that the constant activity exerted by these fibres on the muscle cells, which is not regulated by intramural ganglionic centres or inhibited by the interaction of noradrenergic nerves on the ganglion cells, is at least one of the causes of the so-called "narrow segment" in megacolon congenitum.

The presence of myelinated nerve fibres in the normal and in cases of Hirschsprung's disease proves the existence of afferent endings. Differences in bowel regulation, also in long aganglionic segments, due to a segmental arrangement of afferent endings, have, so far, not been described. It seems, therefore, that differences exist in extramural innervation patterns. At the moment this is, of course, a hypothetical statement, but it should inspire future investigations into two problems:

1. *Investigations concerning the visualization and distribution of afferent endings.* A correlation of difference in type of endings with difference in function such as slowly adapting mechanoreceptors in series location, or rapidly adapting mechanoreceptors (probably all kinds of encapsulated endings), is necessary as well as knowledge of their pathological changes under different conditions. Tension receptors present in series in smooth muscle cells have been suggested by *Leek* (1972) to take the form of flattened stars, their points making contact with muscle cells. In *Leek*'s opinion there is only a poor correlation between the structures described by histologists and their functional state, except for the endings embedded in collagen whorles in ruminal muscle layers as described by *Steven* and *Marshall* (1970) which may be tension receptors. Looking through the references of *Leek*'s paper it is noteworthy that all Russian, European and Japanese authors working in this field are omitted. In our opinion at least some of the arborated receptors found in the muscle layer and between the ganglion cells of Auerbach's plexus (*Kolossow et al.*, 1963, 1965; *Kadanoff et al.*, 1963; *Milochin*, 1953, 1958, 1959, 1960, 1963; *Seto,* 1963), fulfil the condition of star-shaped endings as postulated by *Leek* (Fig. 3). Further investigations concerning these problems seem to be a matter of great importance. They could also be performed in bioptic material.

2. *Differentiation between Type I and II ganglion cells in the intestine*, to elucidate their different role in the regulation of bowel movement. This can be done only with silver impregnation methods on autoptic material or by vital staining with methylene blue. All other methods including histochemical ones, as for example the acid phosphatase technique, failed (*Korotchkin*, 1966). Having solved these pıoblems we may probably gain more knowledge about functional disturbances of the intestine.

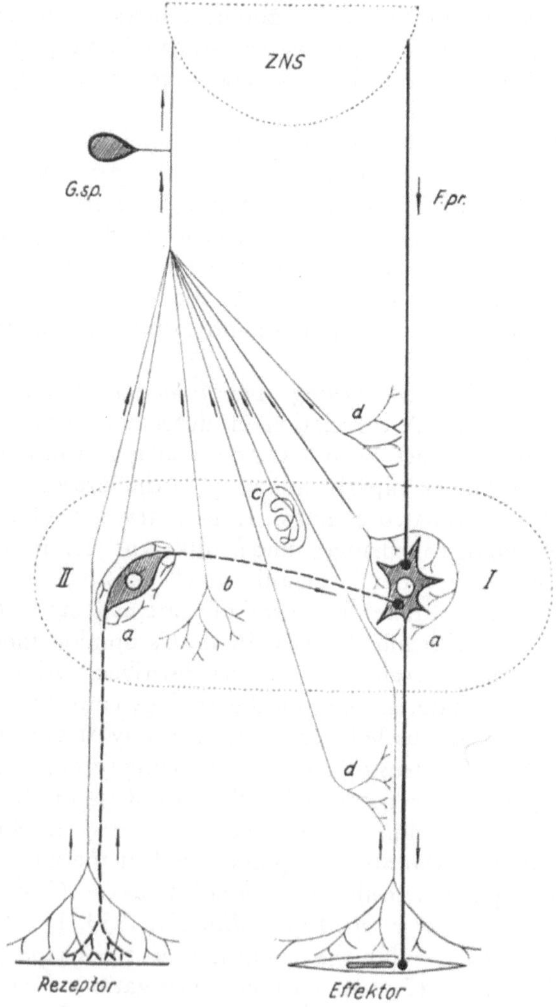

Fig. 4. Diagram of intraneuronal connections in vegetative ganglions (from *Kolossow* and *Milochin*, 1963)

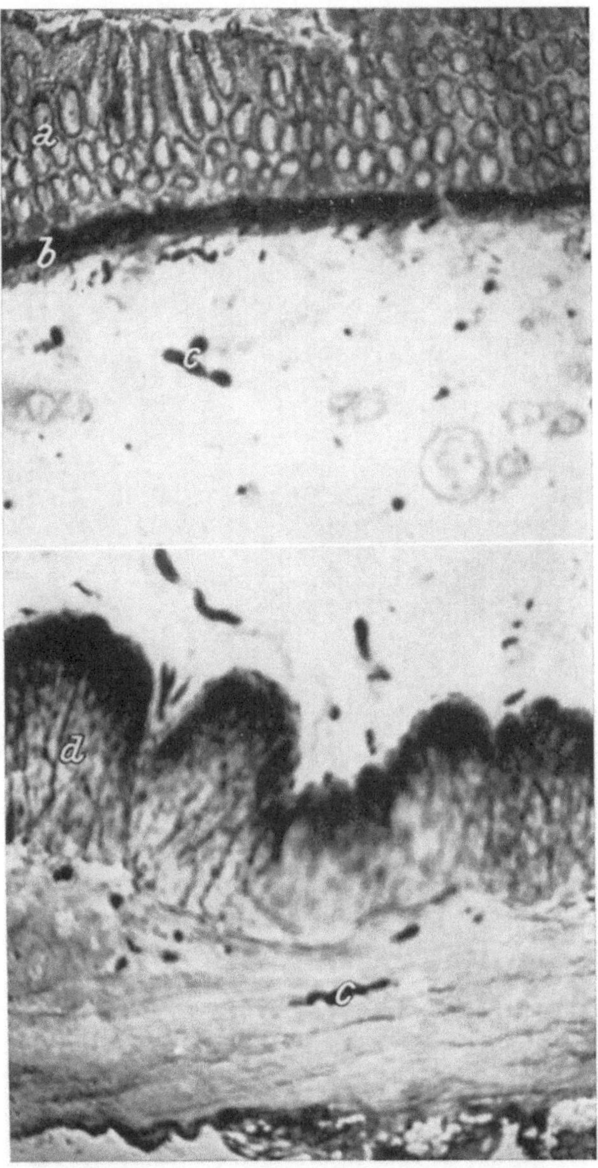

Fig. 5. L 32/69, male 8 month. Morbus Hirschsprung. Rect.biop. Aganglionic segment. a) mucosa; b) muscularis mucosae; c) nerve fibres; d) contracted circular layer of the muscularis propria. Specific cholinesterase technique Karnovsky-Root. ×40

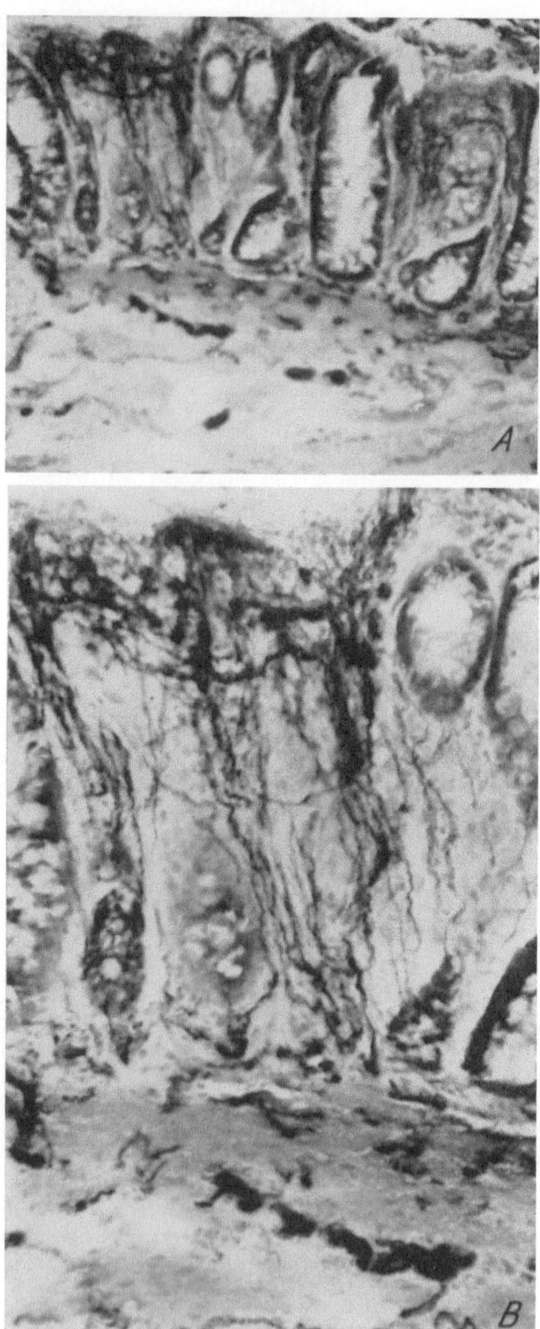
A
B

Present knowledge about the functional qualities of these two types of ganglion cells is poor. By *Kolossow* (1965), the facts already known and some hypothetical statements have been summarized in a diagram (Fig. 4). Recent investigations by *Zlatinskai et al.* (1968) seem to verify some of *Kolossow*'s statements.

Using the method for demonstration of specific cholinesterase, it is easy to differentiate between ganglionic and aganglionic segments even if the bioptical material contains only parts of the submucosa. Besides the absence of nerve cells, there is always some hypertrophic change in nerve elements in the mucosa and submucosa. Sometimes, even neurinoma-like structures are formed in the upper part of the mucosa (Figs. 5, 6).

In those cases where only the number of nerve cells is diminished, these hypertrophic changes are absent and, therefore, the diagnosis may be much more difficult and only possible by the determination of the numerical distribution of ganglionic cells in Meissner's plexus. Exact counts of nerve cells in Auerbach's plexus in normal and pathological conditions can be found in the paper by *Mayer-Ruge* (1968) and in the plexus of Meissner in the paper by *Aldridge* and *Campell* (1968). As the number of nerve cells in the most caudal 3—4 cm of the rectum is significantly diminished, it is important to take biopsies from more proximal parts and to examine a large amount of slides in order to gain an effective diagnosis. It should be stressed that in paraffin sections containing only mucosa and parts of the submucosa, this cannot be properly done.

Alterations regarding the content and distribution of noradrenergic nerves in Hirschsprung's disease cannot directly be used for diagnosis. Of course, if there is a complete loss of nerve cells in Auerbach's or Meissner's plexuses the noradrenergic plexus surrounding these cells is also absent. Alterations in the distribution of adrenergic nerve fibres around the blood vessels could neither be demonstrated by *Ehrenpreis* (1971) and *Gamon et al.* (1968), nor by us. Perhaps by checking larger amounts of material from several cases we will gain more insight into this problem. The same can be said about the alterations regarding the enterochromaffin cells, since, so far, there are no exact measurements of these cells in normal and pathologic conditions.

Up to now we have found no case of Hirschsprung's disease in our material showing an increase of intramural nerve cells as described

Fig. 6. L 5/70, male 3a. Morbus Hirschsprung: aganglionic segment. Rect. biop. A: local neural hyperplasia in the mucosa; B: detail of A. Specific cholinesterase technique Karnovsky-Root. A: ×100; B: ×240

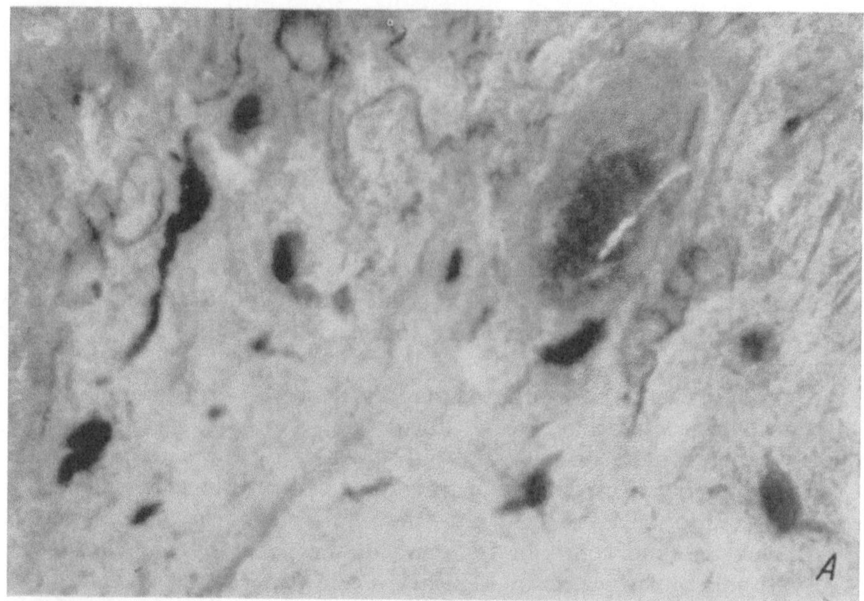

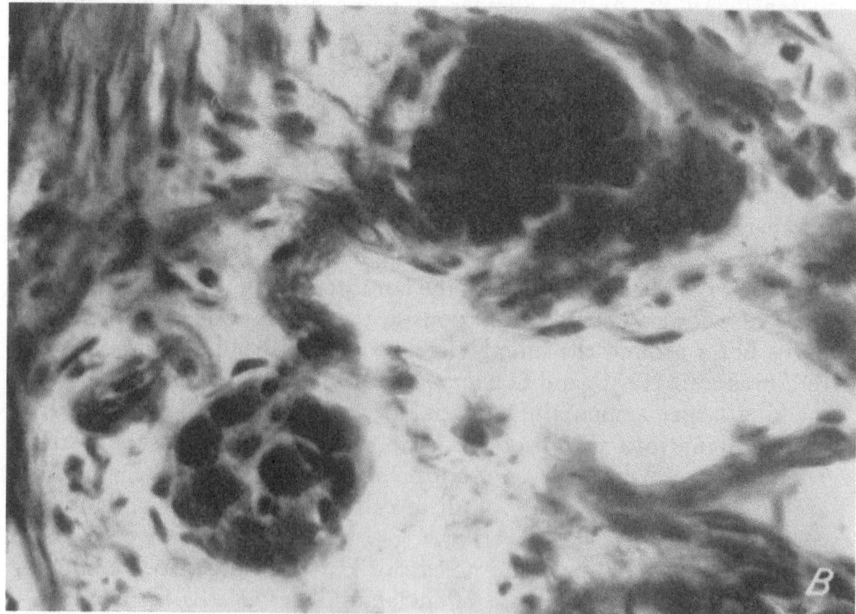

Fig. 7. L 41/71, male 1a. Morbus Hirschsprung: aganglionic segment. Rect.biop. Local accumulation of ganglions in Meissner's plexus in the intermediate zone between the aganglionic segment and the normally innervated part of the rectum. A: Cholinesterase, ×110; B: silverimpregnation (Jabonero) ×400

by *Mayer-Ruge* (1972) and *Netzeloff* (1970). In two cases only we were able to demonstrate a local increase of nerve cells in the transitional part between the aganglionic segment and the part of the colon which was normally innervated. This fact could easily lead to a diagnostic misinterpretation if only one suction biopsy would be taken from this site (*Lassmann* and *Wurnig*, 1973; Fig. 7).

In the following, the history of one of our patients will be reported in order to demonstrate how difficult it is to explain the different clinical symptoms with our theoretical knowledge.

This patient, a boy of 15 years old, was said to have normal action of the bowels during the first two years. In the following time, up to his 9th year, he suffered from obstipation with increasing intensity. During the next years the action of the bowels was normal again and only half a year before surgery he again suffered from increasing obstipation, which could not be treated by medicaments any longer.

Usual pathohistologic examination of the large intestine after resection showed an aganglionic segment of 15 cm in length; the adjacent intestine, up to 40—50 cm in an oral direction, was balloon-like distented (Fig. 8). Besides shorter aganglionic segments we were able to demonstrate in this part also degenerative changes of nerve cells, which may have been caused by pressure atrophy; the musculature was also extraordinarily thin in these parts.

Alternating periods of obstipation and normal action of the bowels have been also described by other authors, but till now we cannot give a satisfactory explanation of this phenomenon. It is possible that alteration of nerve elements in the distented part of the colon such as a primary insufficient grouping or spot-like secondary degeneration, is one of the causes of the different peristaltic activity observed during longer periods. The present case demonstrates, as did also the 4 cases described by *Ponky et al.* (1972), that adequate histological examination should be performed of every case of clinically not explorable obstipation to prevent in the future assignments such as the following: "This child was sent for classification of his chronic obstipations as it was impossible to get him continent in spite of thrashing and other educational treatments."

In concluding I want to point out again the urgent problem of eugenic advice, although this is not very popular in Europe, in referring to case 4 of *Ponky et al.* (1972). This is a case of a male suffering of constipation during lifetime caused by an aganglionic segment that was successfully treated at the age of 16 years. Later he married and got two children, both of them suffering from Hirschsprung's disease. One girl died early. The diagnosis was verified on autoptic material. The other child, also a girl, was found to have a

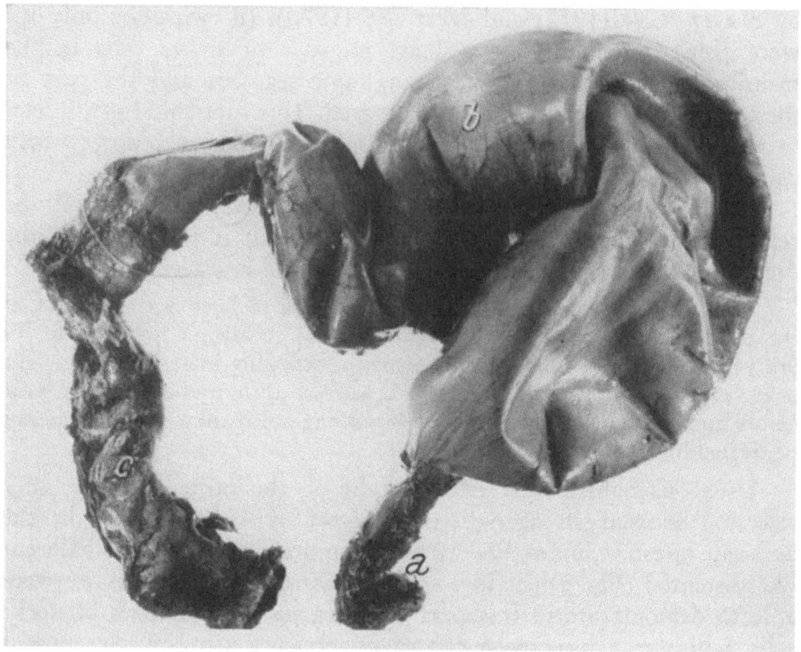

Fig. 8. L 24/72, male 16a. Morbus Hirschsprung. Resected large intestine. a) Aganglionic segment of the rectum; b) balloon-like distended part; c) colon transversum

total lack of nerve cells in the colon and the distal 10 cm of the terminal part of the ileum. These two cases, we feel, demonstrate the necessity of eugenic advice as, so far, we have no other method to stop the propagation of the disease.

Sometimes it is useful to try to solve unsettled problems of general interest by exploring organs in which, for example, the vascular system contains locally all the specific structures in physiological conditions that are otherwise scattered throughout the whole organism, as for instance, in the corpora cavernosa of the penis (*Lassmann*, 1965). Another possibility is to study the alterations in organs in which normally one part of the neural structure is missing, as *Csillik* (1972) did in the epiphysis cerebri. The same concept may sometimes be used in genetical disturbances and, therefore, we have started to examine rectum biopsies of cases of morbus Down, considering four main theories:

1. An incident rate of 3.4 % of Mongolism associated with Hirschsprung's disease is reported by *Gravier* and *Sieber* (1966). These authors supposed that the genetic factor of Hirschsprung's disease is carried by chromosome N 21. On the other hand, *Wetter-*

berg et al. (1972) reported that in children with Down's syndrome, the domapine beta-hydroxylase activity in serum is lower than in other normal children or in mentally retarded patients not suffering from morbus Down. The low plasma activity does not seem to be due to the presence of an endogenous enzyme inhibitor. The fact that children with Down's syndrome, at the age of 5—15 years, had a plasma DBH activity as low as newborns, whereas adults with Down's syndrome had nearly normal plasma levels indicates that the maturation of the autonomic nervous system may be slower in patients with Down's disease. This hypothetical statement should inspire morphological and histochemical investigations into Down's disease and other mentally retarded children as, possibly, a disturbance occurs in the synthesis or release of noradrenaline as reflected by the low plasma DBH activity found in these patients.

2. With regard to the short life-span of these patients there may be changes in the nerve cells, which could be detected at any early stage.

3. The diminished serotonin level of the serum may be correlated with a change of the number or of the functional state of the enterochromaffin cells. If so, this fact could give some evidence for the failing of serotonin treatment in these cases (*Partington et al.*, 1971; *Hakan*, 1969).

4. There may be some direct significant changes in the intramural innervation as constipation is very often seen in this disease.

Up to now the only alterations we were able to find in our small material were a local discrete hyperplasia of nerve fibres in the mucosa of the rectum, a small amount of nerve cells in the basal part of the stroma, which is never found normally in this location, and a larger amount of muscle cells, also in the basal parts (Fig. 9). Sometimes, encapsulated afferent endings in the muscularis mucosae could at least be observed (Fig. 10). A statement about the amount of noradrenaline in nerve fibres is difficult as there seems to be no discrete or even a complete loss of them as compared with the normal arrangement. Nevertheless, the material is not yet large enough for a final statement.

Two cases of atresia ani, treated with sphincterotomy, showed some differences in adrenergic innervation of the resected intestinal muscle. The content of cholinesterase-positive nerve fibres was normal, whereas the specific adrenergic innervation of the internal sphincter was extraordinarily diminished in one case being even lower than in the more proximal parts of the rectum. The other case showed a normal noradrenergic innervation. More numerous investigations will indicate to what extent these results may be of clinical relevance.

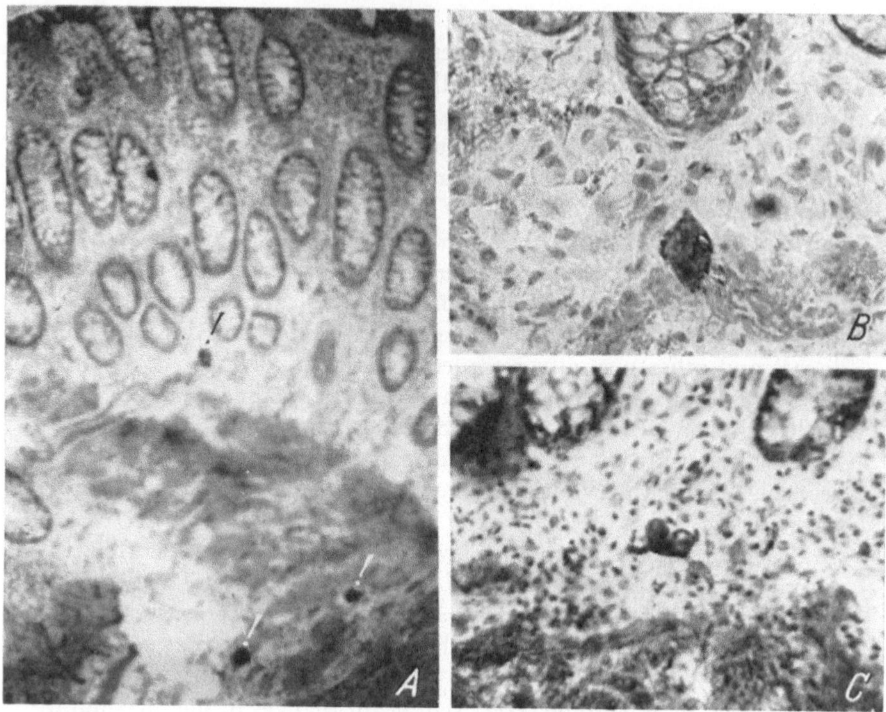

Fig. 9. A: L 78/71, male. Morbus Down, rect.biop. Ganglion cells in the basal layer of the mucosa, in the muscularis mucosae, and in typical position (!). B: L 79/71, girl 7a. Morbus Down, rect.biop. Ganglion cells in the basal layer of the mucosa. C: L 16/72, girl 14a. Morbus Down, rect.biop. Ganglion cells in the basal layer of the mucosa. Specific cholinesterase. Technique: Karnovsky-Root. A: ×100, B: ×240, C: ×200

Nevertheless, using the histological methods mentioned, some more insight in the functional aspects of these disturbances will be gained than was possible with the unspecific methods for demonstration of nerve elements.

A further problem to be briefly discussed here concerns the practical value of rectal biopsies for the diagnosis of storage diseases. No difficulties arise in the diagnosis of neuronal storage diseases and leucodystrophies. It should, however, be considered that, besides of positive results in storage diseases as, *e.g.*, in one case of Niemann-Pick's disease, storing cells in the mucosa have been also found in other conditions showing no clinical symptoms at all such as, *e.g.*, PAS-positive cells called muciphages. We could demonstrate that the occurrence of such cells is much more frequent in children than has been believed before (*Azzopardi* and *Evans*, 1966). Next to these cells we also found other lipid storing cells showing no positive PAS reaction. The clinical significance and frequency of these cells are, so far, obscure.

The number of enterochromaffin cells also differs in our cases. Further investigations will have to explore whether the difference in number of these cells is based only on a different functional state or on a real decrease in their number.

Finally some remarks about the problem of the "appendicité neurogène" should be made. This kind of disorder of the appendix is characterized by a neuromaformation in the mucosa of the organ leading to a total obstruction of the lumen without any signs of inflammation. The alteration is accompanied by clinical symptoms such as local appendicular attacks and less distinct general symptoms. But, not infrequently, there is also a mental disturbance in the form of depression.

The neuromaformation was first described by *Maresch* (1921) and *Masson* (1924). As this was the first classical observation of a local

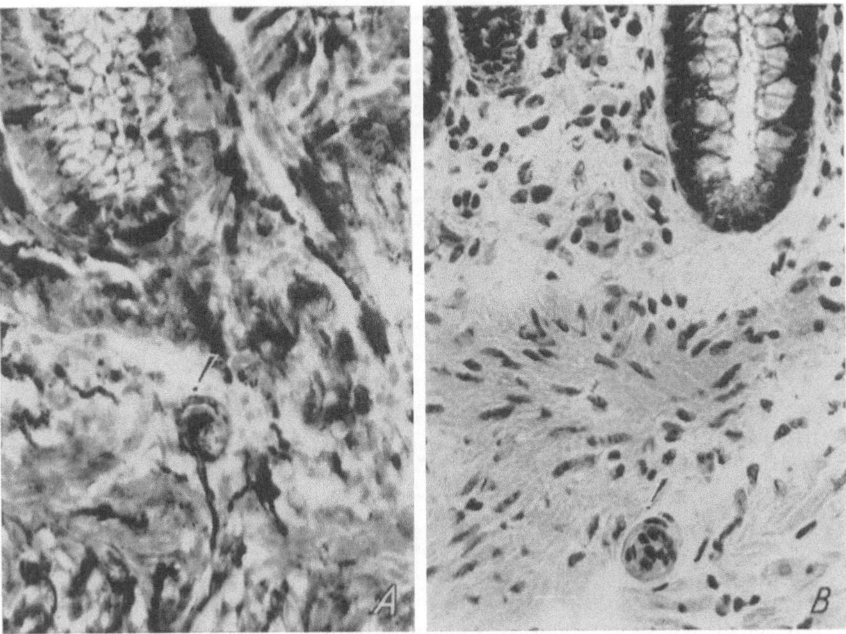

Fig. 10. A: L 61/70, male 3a. Constipation. Rect.biop. Normal innervation. Encapsulated ending in the basal layer of the mucosa. B: L 79/71, girl 7a. Morbus Down, rect.biop. Corpuscular nerve ending in the muscularis mucosae (!). A: Osmium zinc jodide, ×400; B: hematoxylin-eosin, ×400

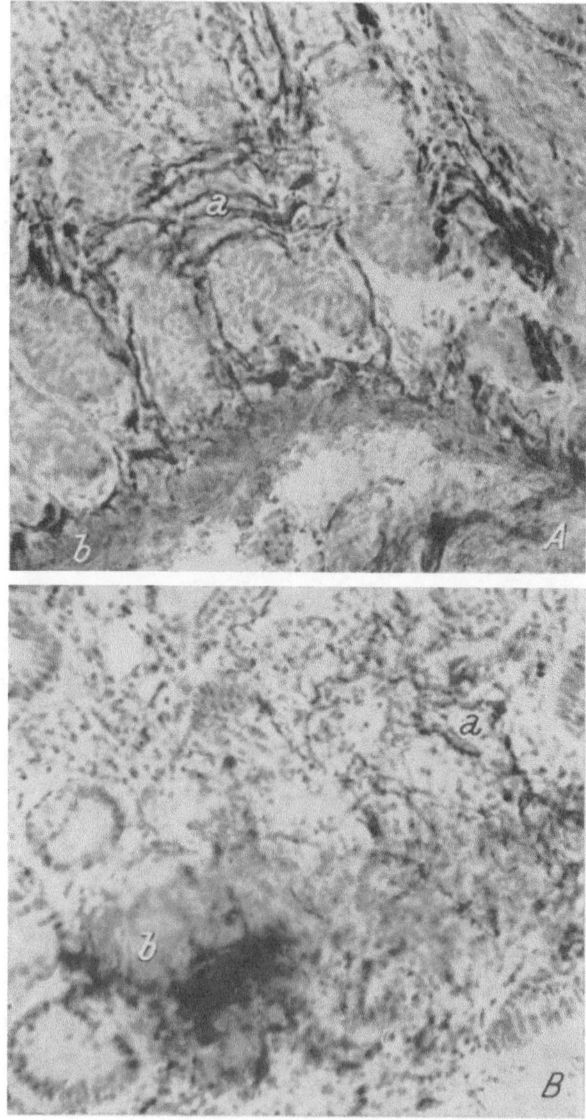

Fig. 11. A: L 16/71, male 49a. Hypertony. Duod.biop. Local hyperplasy in the basal layer of the mucosa (a). Muscularis mucosae (b). B: L 31/71, female 70a. Chronic enteritis. Duod.biop. Local hyperplasy of nerve elements (a), neuroma in the basal layer of the mucosa (b). Specific cholinesterase. Technique: Karnovsky-Root. ×200

neuroma in the intestines it attracted the attention of many authors (*e.g. Feyerter*, 1951; *Coronini et al.*, 1957; *Lassmann*, 1949, 1950, 1961, 1962). Besides the description of the alteration in the organ, some more detailed clinical observations were also mentioned. Later, in 1951, *Triska* reported the first case of generalized neuroma-formations in the upper intestine which caused several chirurgical interventions because of different intestinal symptoms. These interventions failed to be effective and I would like to conclude my report with a piece of advice on this point. We are now able to confirm the existence of even small initial neuromas or local hypertrophies of the local nervous system in biopsies of the small intestine as shown in figure 11. The importance of this is the early detection of the few cases of generalized neuromas in the upper intestine, when in spite of typical clinical symptoms surgical treatment is unadvisable.

In concluding it was our intention to demonstrate some clinically interesting results of our neurohistological and histochemical investigations made on biopsies of the rectum and intestine, showing also that the use of modern methods enables us to make more specific statements about the changes in the nerve elements of an organ. Using the method of rectum biopsy, which is only a minor surgical intervention, it is possible to examine a wide spectrum of clinical diseases. This may prevent much more difficult surgical interventions, such as for instance brain biopsies, to explore storage diseases.

This may also be of value in the course of intestinal diseases as was just shown in Krohn's disease, which can also be diagnosed by rectum biopsy: a much better way than the more complicated intestinal biopsies.

References

Aldridge, R. T., and *P. E. Campell*: Ganglion cell distribution in the normal rectum and anal canal. A basis for the diagnosis of Hirschsprung's disease by ano-rectal biopsy. J. Ped. Surg. *3*, 475—490 (1968).

Andrew, A.: The origin of intramural ganglia. III. The vagal scoure of enteric ganglion cells. J. Anat. *107*, 327—336 (1970).

Andrew, A.: The origin of the intramural ganglia. IV. The origin of enteric ganglia, a critical review and discussion of the present state of the problem. J. Anat. *108*, 169—184 (1971).

Azzopardi, J. G., and *D. J. Evans*: Mucoprotein-containing histiocytes (muciphages) in the rectum. J. clin. Pathol. *19*, 368—374 (1966).

Bodian, M., and *B. D. Lake*: The rectal approach to neuropathology. Brit. J. Surg. *50*, 702—714 (1962).

Cantino, D.: An histochemical study of the nerve supply in the developping alimentary tract. Experentia *25*, 766—767 (1970).

Coronini, C., W. Kovac, and *G. Lassmann*: Die neurogene Appendicopathie; ein vegetativer Test für Klinik und Pathologie. Acta Neuroveg. (Wien) *16*, 250—278 (1957).

Coronini, C., W. Kovac, G. Lassmann, and *J. Smereker*: Untersuchungen über die periphere Neurosekretion in der Appendix. Acta Neuroveg. (Wien) *16*, 279—293 (1957).

Csillik, B., and *E. Knyihar*: Structural organization of the peripheral anatomic innervation apparatus. Recent Developments of Neurology in Hungary III (*Lissak, K.*, ed.). Budapest: Akadémiai Kiadó. 1972.

Csillik, B., and *G. Kalman*: Vacuumless freezing drying: its application in catecholamine histochemistry. Histochemie *9*, 275—280 (1967).

Ehrenpreis, Th.: Hirschsprung's disease. Am. J. Dig. Dis. *16*, 1032—1052 (1971).

Ehrenpreis, Th.: More comments on Hirschsprung's disease. Letter to the Edit. Dis. Col. Rect. *15*, 311 (1972).

Eränkö, O.: The histochemical demonstration of noradrenaline in the adrenal medulla of rats and mice. J. Histochem. Cytochem. *4*, 11—13 (1956).

Falck, B.: Observations on the possibilities of the cellular localisation of monoamines by a fluorescence method. Acta physiol. scand. *56*, Suppl. 197, 1—25 (1962).

Falck, B., N. A. Hillarp, G. Thieme, and *A. Torp*: Fluorescence of catecholamines and related compounds condensed with formaldehyde. J. Histochem. Cytochem. *10*, 348—554 (1962).

Feyrter, F.: Über die Pathologie der vegetativen nervösen Peripherie und ihrer ganglionären Regulationsstätten. Wien: W. Maudrich. 1951.

Gannon, B. J., H. R. Noblett, and *G. Burnstock*: Adrenergic innervation of the bowel in Hirschsprung's disease. Brit. Med. J. *3*, 338—340 (1968).

Gravier, L., and *W. K. Sieber*: Hirschsprung's disease and Mongolism. Surg. *60*, 458—461 (1966).

Hakan, S.: Behandling med 5 Hydroxitrypthofan vid Morbus Down. Sartr. u. Nord. Med. *82*, 952—954 (1969).

Hillemand, M. M. P., J. Mureau, Ph. Delaviere, and *L. P. Burdois*: Les megacolons congénitaux de l'enfant sans aganglionose. Arch. Franc. Mal. d. l'Appareil Digestive (Paris) *60*, 319—330 (1971).

Hopwood, D.: The histochemistry and electron histochemistry of chromaffin tissue. Progr. Histochem. Cytochem. *3*, No. 1 (1971).

Hovnanian, A. P., and *R. Golkar*: Hirschsprung's disease revised, Part I. Preliminary report. Dis Col. Rect. *14*, 108—116 (1971).

Howard, E. R., and *J. R. Garret*: Electron microscopy of myenteric nerves in Hirschsprung's disease and in normal bowel. Gut *11*, 1007—1014 (1970).

Hüther, W.: Die Hirschsprung'sche Krankheit als Folge einer Entwicklungsstörung der intramuralen Ganglien. Beitr. path. Anat. *114*, 161—191 (1954).

Kadanoff, D., and *A. Gürowsky*: Morphologie der Rezeptoren des Atmungs- und Verdauungssystems. Jena: VEB Gustav Fischer. 1963.

Karnovsky, M. J., and *L. Roots*: A direct-coloring thiocholin technique for cholinesterases. J. Histochem. Cytochem. *12*, 219—221 (1964).

Kolossow, N. G.: On afferent neurons of the autonomic nervous system. Dokl. Akad. Nauk. SSSR *161*, 490—492 (1965).

Kolossow, N. G., and *A. A. Milochin*: Die afferente Innervation der Ganglien des vegetativen Nervensystems. Z. mikr.-anat. Forsch. *70*, 427—463 (1963).

Korotchkin, L. I.: Morphological and cytochemical divergency in neurocytogenesis. Z. mikr.-anat. Forsch. *75*, 1—19 (1966).

Lassmann, G.: Zum Problem der neurogenen Appendicitis. Mikroskopie *4*, 257—320 (1949).

Lassmann, G.: Sull' appendicitis neurogena nel quadro delle malattie neurovegetative allergiche. Minerva Med. *41*, 1—12 (1950).

Lassmann, G.: Histochemische Darstellung der spezifischen Cholinesterase (Azetylcholinesterase) in den nervösen Formationen bei einem Fall von neurogener Appendicopathie mit zentraler Neurombildung. Acta Neuropathol. *1*, 308—310 (1961).

Lassmann, G.: Die Darstellung der spezifischen Cholinesterase in den nervösen Formationen der menschlichen Appendix. Acta Histochem. *13*, 113—122 (1962).

Lassmann, G.: Neue Befunde über die Nervenversorgung des Gefäßsystems. Acta neuroveg. *27*, 546—599 (1965).

Lassmann, G., and *P. Wurnig*: Lokale Ganglienzellhyperplasie in der Submucosa am oralen Ende des aganglionären Segmentes bei Morbus Hirschsprung. Z. Kinderchir. *12*, 236—243 (1973).

Leek, B. F.: Abdominal visceral receptors. Handbook of Sensory Physiology, Vol. III/1, pp. 113—160. Berlin-Heidelberg-New York: Springer. 1972.

Lukashin, V. G.: Receptorendings of dendrites of Dogiel cells Type II in the large intestine of the cat. Arkh. Anat. *56*, 63—66 (1969).

Lukashin, V. G.: Sensory innervation of autonomic ganglia. Dokl. Akad. Nauk. SSSR *189*, 909—912 (1969).

Maresch, R.: Über das Vorkommen neuromartiger Bildungen in obliterierten Wurmfortsätzen. Wien. klin. Wschr. *34*, 181—182 (1921).

Masson, P.: Les nevromes sympathiques et l'appendicite oblitérante. Lyon chir. *1*, 1—15 (1921).

Masson, P.: Appendicite neurogène et carionoides. Ann. d'Anat. pathol. *1*, 3—15 (1924).

Masson, P.: Carcinoids (argentaffine cell tumors) and nerve hyperplasia of the appendicular mucosa. Am. J. Path. *4*, 181—189 (1928).

Masson, P.: Contribution to the study of sympathic nerves of the appendix. The musculo-nervous complex of the submucosa. A. J. Path. *6*, 217—227 (1930).

Meier-Ruge, W.: Morbus Hirschsprung — Neuere Probleme der Diagnose. In: Morbus Hirschsprung — Neuere Probleme der Diagnose. Inkontinenzbehandlung im Kindesalter (*Wurnig, P.*, ed.). (Pädiatrie und Pädologie, Supplementum II.) Wien-New York: Springer. 1972.

Meier-Ruge, W.: Das Megacolon. Virch. Arch. Pathol. *344*, 67—85 (1968).

Milochin, A. A.: Zur afferenten Innervation des Verdauungstractus. Sammelwerk über Fragen der Morphologie der Rezeptoren der inneren Organe und des Gefäßsystems, S. 77. Moskau-Leningrad: Akad. Wiss. UdSSR. 1953.

Milochin, A. A.: Über eigenrezeptorische Neurone des vegetativen Nervensystems. Z. mikrosk.-anat. Forsch. *63*, 497—517 (1958).

Milochin, A. A.: Die afferente Innervation des Verdauungstraktes bei einigen niederen Wirbeltieren. Z. mikrosk.-anat. Forsch. *65*, 402—412 (1959).

Milochin, A. A.: Über die afferente Innervation der peripheren vegetativen Neuronen. Z. mikrosk.-anat. Forsch. *66*, 483—488 (1960).

Milochin, A. A.: Morphologischer Nachweis der afferenten Innervation des vegetativen Nervensystems. Z. mikrosk.-anat. Forsch. *69*, 615—629 (1963).

Netzelof, C., *D. Guy-Grand*, and *E. Thomine*: Les megacolons avec hyperplasie des plexus myentériques. (Une entité anatomo- clinique, a propos de 3 cas.) La Presse Medic. *78*, 11—18 (1970).

Partington, M. W., and *J. B. McDonald*: 5 Hydroxytryptophan (5 HTP) in Down's Syndrom. Devel. Med. Child Neurol. *13*, 362—372 (1971).

Ponky, J. L., *C. Grodinsky*, and *E. Brush*: Megacolon in teen-aged and adult patients. Dis. Col. Rect. *15*, 14—22 (1972).

Sakharova, A. V., and *D. A. Sakharov*: Visualization of intraneuronal monoamines by treatment with formalin solutions. Histochemistry of nervous transmission. Progr. Brain Res. *34*, 11—25 (1971).

Seto, H.: Studies on the Sensory Innervation. In: Human Sensibility (*Igaka Shoin*, ed.). Springfield, Ill.: Charles C Thomas. 1963.

Stach, W.: Über die in der Dickdarmwand aszendierenden Nerven des Plexus pelvinus und die Grenze der vagalen und sakral-parasympathischen Innervation. Z. mikr. Anat. *84*, 65—90 (1971).

Stach, W.: Der Plexus entericus extremus des Dickdarms und seine Beziehungen zu den interstitiellen Zellen (Cajal). Z. mikr. Anat. *85*, 245—272 (1972).

Steven, D. H., and *A. B. Marshall*: Cited by *L. F. Leek*.

Stovicek, G. V.: Several problems on the neural morphology of Dogiel Typ II, cells. Arkh. Anat. *54*, 90—92 (1968).

Taxi, J.: Thesis. Annal. d. Scienc. Nat. Zool. Paris, 12e Série, Tome 7, 413—674 (1965).

Tirkhanowa, L. P.: On the receptive neurons in the ganglia of the intramural nervous plexus of the cat's large intestine. Arkh. Anat. *54*, 98—102 (1968).

Triska, H.: Spätresultate nach Appendektomien mit besonderer Berücksichtigung der Appendicite neurogene. Wien. Med. Wschr. *101*, 611—613 (1951).

Wetterberg, L., *K. H. Gustavson*, *M. Backström*, *S. B. Roos*, and *Ö. Fröden*: Low dopamine-beta-hydroxylase activity in Down's syndrome. Clinic. Genetics *3*, 152—153 (1972).

Zlatiskai, N. N.: On the sensory innervation of the small intestine of the cat. Arkh. Anat. *54*, 102—105 (1968).

Author's address: Doz. Dr. *G. Lassmann*, Neurological Institute, University of Vienna, Schwarzspanierstraße 17, A-1090 Wien, Austria.

Discussion

Shimoda: I want to mention two kinds of cardiospasm, clinically described in Japan. The first type is characterized by a brain stem epilepsy with an abnormal EEG; in this type, Auerbach's and Meissner's plexuses do not show any abnormality. The EEG of the second type is normal, but malformations of Auerbach's and Meissner's plexuses are apparent. Is this in analogy with Hirschsprung's disease?

Lassmann: Similar findings were not reported with respect to morbus Hirschsprung until now.

Stach: I regard the demonstration of ganglion cells by means of the cholinesterase technique as rather problematic. We found in stretch specimens well-developed Auerbach's and Meissner's plexuses in so-called "aganglionic" segments; the innervation of the mucosa appeared also to be normal. Accordingly, I am still quite sceptic with regard to the diagnosis on the basis of biopsy material.

Lassmann: In our material we examined not only the biopsies but in several cases also the resected parts of the large intestine. This enabled us to proof the length of the aganglionic segment. In all cases we could proove the absence of ganglion cells in all parts of the intramural nervous system throughout the narrow segment. Our findings are in accordance with the results of other investigators. We are quite sure that for diagnostic purposes the inspection of bioptic material including only the mucosa and parts of the submucosa is sufficient in every case. It seems, therefore, possible that the case you examined is one of a megacolon without defects in the nerve supply. In these very rare cases an enzyme defect is discussed.

Szentágothai: A low segmental origin of ganglion cells in the lower part of the rectum could be expected only if it were true what was claimed earlier, but was not generally accepted by contemporary neuroembryologists, namely that the parasympathetic ganglionic neurones emerge from the medullary tube with the growing tips of preganglionic fibres.

Stach: In this context I want to mention that no such significant differences between ganglion cells in the small and large intestine are present that would support any idea of a different origin of neurones of the colon, colon sigmoideum and rectum.

Lassmann: The problem of the ingrowth of ganglion cells in the rectum from anal to oral is of practical importance because narrow segments in more oral regions are situated in between normal innervated parts of the intestine. I think this problem should be examined again.

Sakharov: I want to ask whether it is not possible, in principle, to treat intestinal diseases concerned with lack or deficiency of catecholamines with L-Dopa.

Lassmann: Up to now, as you know, L-Dopa treatment is predominantly used in Parkinson's disease. Recently, the value of this therapy in liver diseases is discussed, but there is not much further experience. I agree that your question deals with a very interesting problem. My intention is to study the effect of L-Dopa therapy in rectum biopsies in Parkinson's disease before, during, and after the therapeutic use of the substance.

One problem to be solved is to solve the question whether the larger amount of enterochromaffin cells in the rectum corresponds with a higher number of enterochromaffin-like cells in this region. In this way, biopsies of this region can be used to control the effect of the L-Dopa therapy. On the other hand, rectum biopsies would be very convenient because of harmless handling.

Csillik: I propose treatment of patients, in whom the intestinal nerve plexuses are failing, with the nerve growth factor of Levi-Montalcini.

Lassmann: In our opinion an experiment using nerve growth factor treatment, if this is possible at all, would seem to be successful only in cases with lack of differentiation and not in cases with total absence of ganglion cells.

Journal of Neural Transmission, Suppl. XI, 281—283 (1974)

A Simple Microscope for General Routine Diagnosis in Neurohistology

F. Herzog and **G. Lassmann**

Entwicklungslabor der C. Reichert Optische Werke AG., and Neurological Institute, University of Vienna, Austria
(Head: Prof. Dr. *F. Seitelberger*)

Summary

A microscope, capable of working with ordinary transmitted light and incident light fluorescence will be described. Filter systems and optics are chosen so that they fit best to the requirements of general routine diagnosis in neurohistology.

The introduction of fluorescence microscopical methods for the detection of aromatic monoamines in tissues has opened a wide field for clinical diagnoses by means of neurohistology. On the other hand, methods of classical light microscopy have to be used for special observations.

In the following a microscope is described which can be used in both, fluorescence microscopy and light microscopy.

A. Fluorescence Microscopy

1. Incident illumination. According to suggestions made by *Ploem* (1969) for the use of vertical illumination for observations of fluorescent aromatic monoamines, we used an incident light fluorescence equipment with a mercury burner HBO 50, for Reichert "Fluorpan".

2. Filters. The use of suitable filters for fluorescence microscopy is the most important requirement for efficient work. *Corrodi* and

Jonsson (1967) have shown that the absorption maxima of noradrenaline and serotonin lie in the range of 410 nm. Therefore some trials were made to select the 405 nm emission peak for the mercury spectrum (*Ploem*, 1971). Two types of interference filters have been applied. A Balzers KP 425 in connection with a Schott BG 25 and an interference filter $\lambda = 405$ nm by Reichert.

The high brightness of the microscopical image is due to the higher transmission of the KP 425 + BG 25. The use of a coloured barrier filter 2 mm GG 9 (Schott) causes a lower degree of discrimination between the bluish fluorescence of noradrenaline (peak 480 nm) and the greenish one of serotonin (peak around 525 nm). Table 1 B illustrates the transmission curves.

The interference filter $\lambda = 405$ nm has a lower band width. Therefore it permits a very good selection between the fluorescence colours of noradrenaline and serotonin since a colourless barrier filter KV 418 (Schott) is sufficient. Table 1 A shows the transmission curves. The brightness of the images is lower than with KP 425 + BG 25.

Table 1

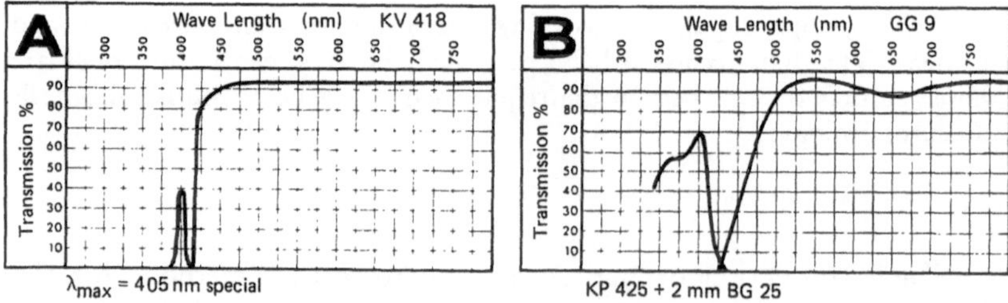

3. Optics. The recently fabricated lenses Fluorite 16/0.50 and SPl 40/0.90 are very powerful accessories because of their high numerical aperture. In some special cases, an immersion lens glycerol 60/0.95 has been applied. To obtain high brightness plano-compensating eyepieces are prefered.

B. Light Microscopy

1. Brightfield. For the usual methods of histochemistry, a brightfield condensor and an quartziodine lamp has been used.

2. Darkfield. A very important help is a simple darkfield diaphragm, which fits in the condensor. It permits the prescreening of tissue sections for the fluorescence microscopical observation with the lenses 16/0.50 and 40/0.90.

References

Corrodi, H., and *G. Jonsson*: The formaldehyde fluorescence method for the histochemical demonstration of biogenic monoamines. A review on the methodology. J. Histochem. Cytochem. *15*, 65—78 (1967).

Ploem, J. S.: A new microscopic method for the visualization of blue formaldehyde-induced catecholamine fluorescence. Arch. int. Pharmacodyn. *182*, 421—424 (1969).

Ploem, J. S.: The microscopic differentiation of the colour of formaldehyde-induced fluorescence. Progr. Brain Research *34*, 27—37 (1971).

Author's address: Mag. *F. Herzog*, C. Reichert Optische Werke AG., Hernalser Hauptstraße 219, A-1170 Wien, Austria.

Discussion

Taxi: Why gives noradrenaline a blue or blue-green colour in this fluorescence microscope, in spite of the fact that, according to Falck and Hillarp, the fluorescence colour of noradrenaline is green to green-yellow?

Herzog: The filter system is of much importance. If the exciting light is blue, the ocular filter must be orange that suppresses the blue band of noradrenaline fluorescence. Accordingly, what remains is green-yellow. However, if excitation is performed with violet light, a colourless ocular (barrier) filter should be used which provides a good reproduction of the bluish fluorescence of noradrenaline.

Journal of Neural Transmission, Suppl. XI, 285—298 (1974)

On the Problem of the Origin of Nerve Terminals in the Sympathetic Ganglia

F. Joó

Electron Microscope Laboratory of the Biophysical Institute of the Biological Research Centre of the Hungarian Academy of Sciences, Szeged, Hungary

With 16 Figures

Summary

According to the current concepts of nerve transmission in sympathetic ganglia, certain adrenergic cellular elements located around or among ganglion cells are held to be responsible for the intraganglionic inhibition of the effect of preganglionic impulses.

A review of results, obtained from electronmicroscopic studies of *in vitro* cultured sympathetic ganglia from chick embryos and of preganglionally denervated superior cervical ganglia of adult rats, suggests the existence of well-defined intraganglionic axon terminals showing acetylcholinesterase activity, which could originate as recurrent collaterals from cholinergic ganglion cells or as axonal processes of acetylcholinesterase-positive interneurones.

Introduction

In recent years attention has been paid by several investigators (*Lawrentjew*, 1924; *Grillo*, 1966; *Williams* and *Palay*, 1969) to the possible occurrence of interneurones in sympathetic ganglia. Following the electron microscopic description (*Elfvin*, 1963) of the principal structural relationships between the pre- and postganglionic elements, adrenergic synaptic terminals of unknown origin have recently been reported to exist in addition to the typical cholinergic terminals of preganglionic nerve trunks in various sympathetic ganglia (*Hamberger* and *Norberg*, 1965; *Csillik et al.*, 1967). On the other hand, *Norberg* and *Hamberger* (1964) described sparsely scattered groups of small, chromaffin-positive cells in ganglia. In addition, preganglionic synaptic endings, as evidenced by their degeneration after preganglionectomy, have been found to terminate on the surface of these cells (*Grillo*, 1966; *Williams* and *Palay*, 1969). *Eccles* and *Libet* (1961) were of the opinion that preganglionic stimulation causes the chromaffin cell to liberate catecholamines which, after diffusing to a site on a ganglion cell, hyperpolarize its membrane and so modify the response of this cell to preganglionic signals. *Matthews* and *Raisman* (1969) revealed contacts under the electron microscope between chromaffin-like cells and dendrites from ganglion cells in the superior cervical ganglion of rat and concluded that these contacts would be the sites where the chromaffin cells would influence the ganglion cells.

In many ways it is now confirmed by pharmacological studies (*Marrazzi*, 1939; *Knoll* and *Vizi*, 1970; *De Groat* and *Saum*, 1971) that the inhibition of ganglion cells is effected by catecholamines, but it still remains to be elucidated whether these are released from the intraganglionic nerve endings or from the chromaffin cells.

Although several experimental morphological studies have been made since the first assumption (*Dogiel*, 1896) of the existence of such an interneurone interposed between the pre- and postganglionic element, the exact nature of the intraganglionic connections of cell processes have, so far, not been fully revealed.

The present study, certain parts of which have recently been published in detail (*Presley et al.*, 1971; *Joó et al.*, 1971), was undertaken in order to obtain a better insight into the intraganglionic nerve connections of sympathetic ganglia cultured *in vitro* from chick embryos and into the nature and relationships of axons which may be derived from other than preganglionic sources.

Methods

Tissue Culture of Sympathetic Ganglia and Processing of the Cultures for Microscopic Investigations

Lumbar sympathetic ganglia were removed from 11 to 13-day old chick embryos and subdivided in a growth medium. From a pool of such fragments cultures comprising two or three pieces of ganglion were contained in a central drop of about 0.2 ml of medium in a Maximow double coverslip preparation. In this assembly, the cultures rested on a Melinex surface. Basic growth medium consisted of four parts Medium 199, 1 part foetal bovine serum, 1 part chick embryo extract. Cultures were maintained at 37° C for periods up to 10 days.

A modification (*Lever et al.*, 1968) of the original fluorescent method of *Falck et al.* (1962) was used to demonstrate the presence and cellular

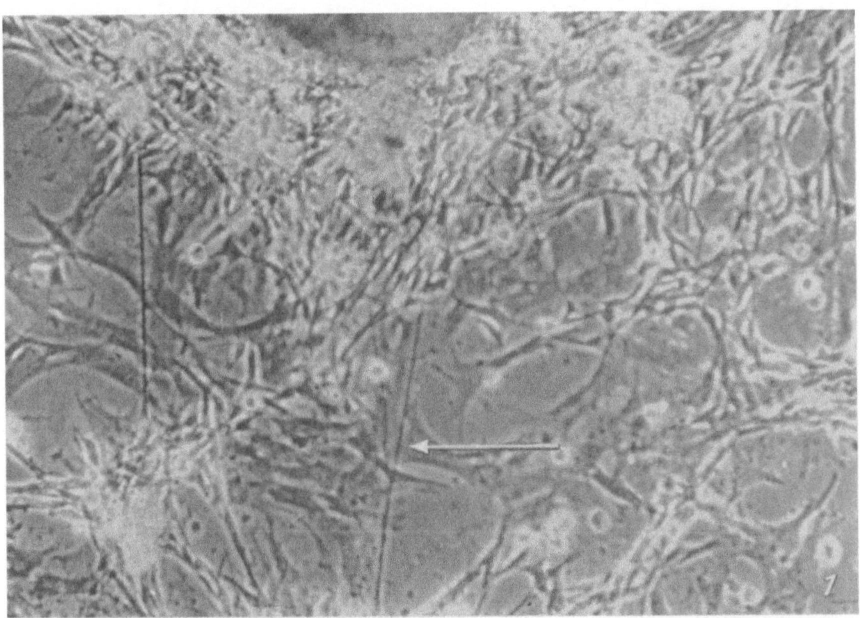

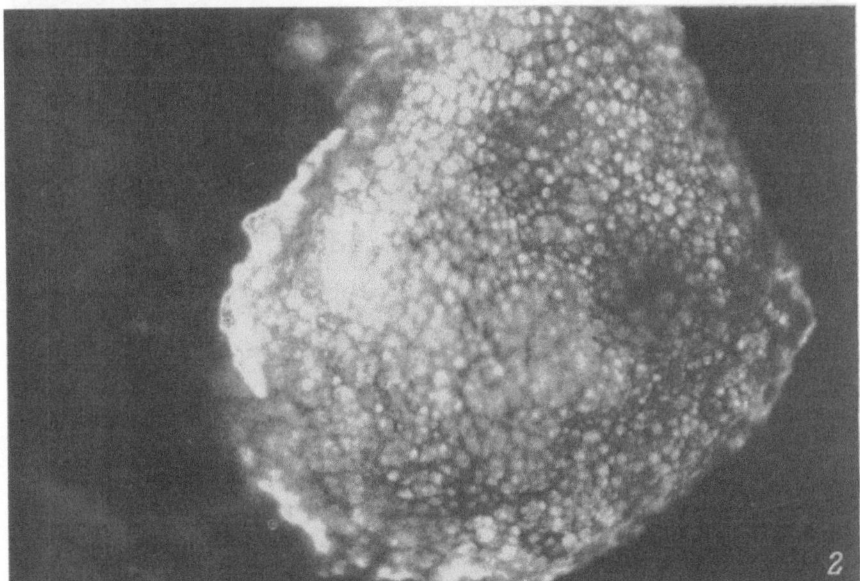

Fig. 1. Phase contrast microscopic picture of chick sympathetic ganglion on the 10th day of *in vitro* culturing. The arrow points to one of the axonal outgrowths. ×160

Fig. 2. Noradrenaline fluorescence of ganglion cells in a 3-day-old culture. ×140

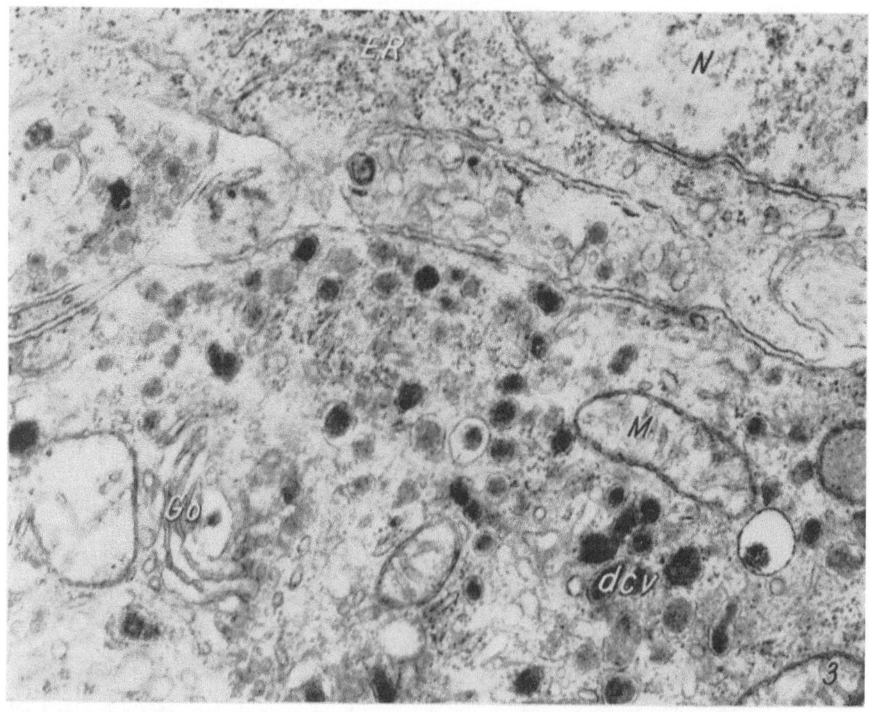

Fig. 3. Appearance of a chromaffin cell containing large granules (dcv) and of a principal ganglion cell (ER, ergastoplasm; N, nucleus) in 5-day-old tissue culture M, mitochondrion; Go, Golgi apparatus. ×50,000

distribution of noradrenaline in the *in vitro* cultured sympathetic ganglia. Some of the cultures were prefixed in 2.5 per cent buffered glutaraldehyde solution for 20 min prior to fixation in osmium tetroxide. Samples were then dehydrated in ascending series of alcohol and embedded in Araldite.

Investigations on Denervated and Undenervated Superior Cervical Ganglia of Adult Rats

Unilateral preganglionic sympathectomy was aseptically performed on 6 adult male white Wistar rats under Nembutal anaesthesia. In each case a short section of the sympathetic chain immediately caudal to the superior cervical ganglion was excised. The animals were sacrificed either at 24 h (acute series, 3 rats) or at 5 weeks (chronic series, 3 rats) after operation.

Fig. 4. Axodendritic synapse in a 7-day-old culture. Sv, synaptic vesicles; D, dendrite. The arrow indicates the site of synaptic contact. ×65,000

Fig. 5. Axosomatic synapse on a chromaffin cell. Sv, synaptic vesicles; M, mitochondrion. The arrow points at the postsynaptic membrane thickening. ×65,000

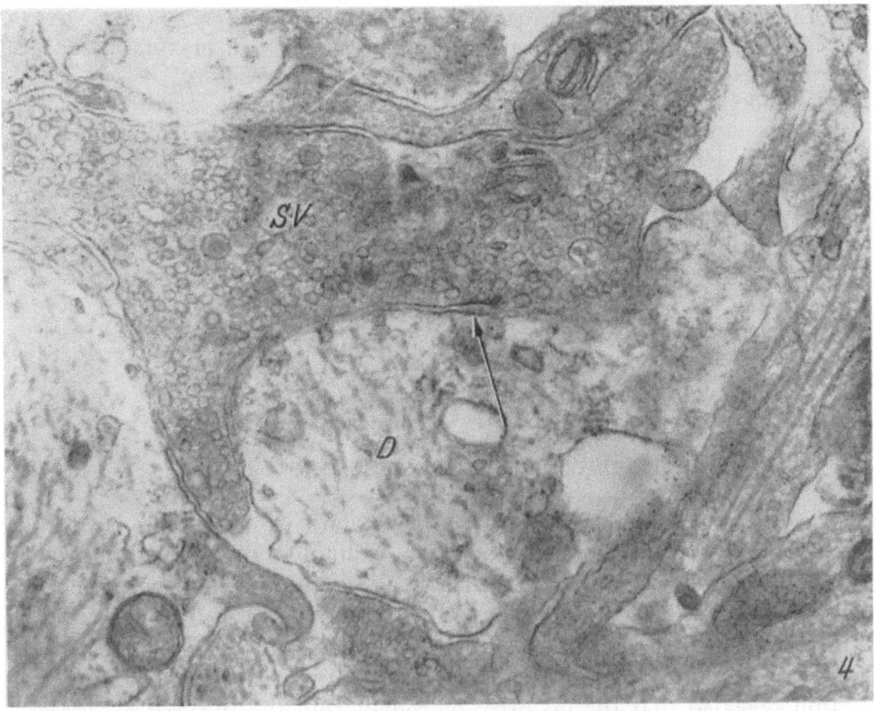

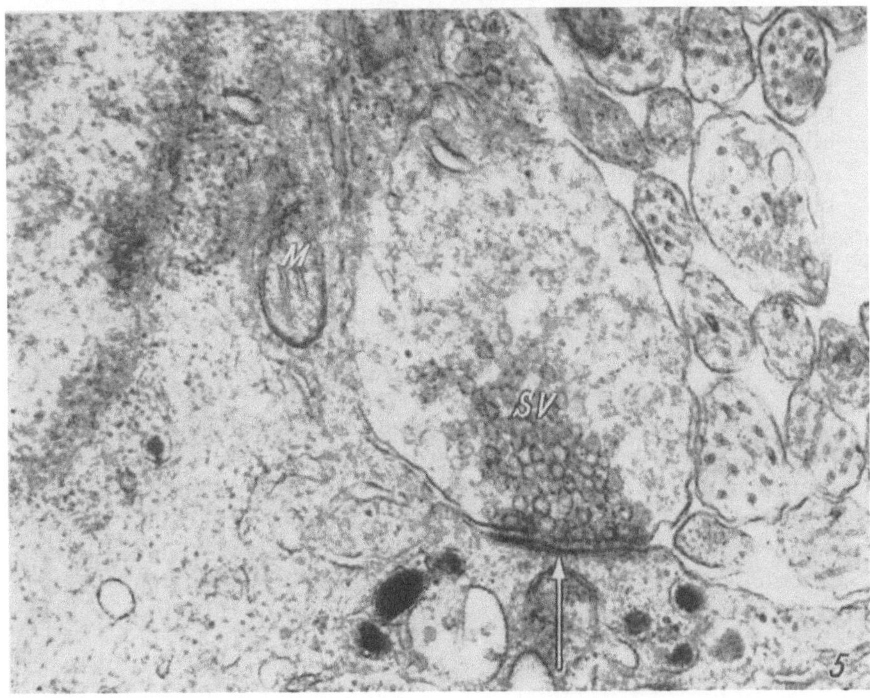
M
SV
5

Denervated and undenervated (control) superior cervical ganglia were removed and in each instance bisected longitudinally to provide material for routine electron microscopy and for electron histochemistry.

Ganglion specimens were treated in two alternative ways: (1) fixed directly in buffered (pH 7.4) 1 % osmium tetroxide, or (2) prefixed for 4 h in 2.5 % buffered (pH 7.2) glutaraldehyde and then processed for the demonstration of acetylcholinesterase (AChE) activity in the presence of ethopropazine (2×10^{-4} M) by a thiocholine technique (*Lewis* and *Shute*, 1966) prior to fixation in osmium tetroxide. All material was finally Araldite-embedded and, after staining by the *Reynolds'* lead citrate method (1963), fine sections were examined in a Siemens Elmiskop 1 electron microscope.

Results

Observations on the Sympathetic Neurone in Vitro

From the third day on successful explants exhibited cellular and axonal centrifugal outgrowth when examined by phase contrast microscopy (Fig. 1). After formol gassing a strong fluorescence, characteristic of catecholamines, was observed in axons and in presumptive nerve cells at the periphery of the cultures as well as within the cells of the explants (Fig. 2). The electron microscopic investigation revealed that the majority of cells within the cultured explants were typical sympathetic neurones, but small granulated cells of chromaffin appearance were also present (Fig. 3). On dendrites of principal ganglion cells (Fig. 4) and at the perikarya of the small granulated cells (Fig. 5) characteristic nerve terminals were occasionally found containing clear synaptic vesicles of a 40—60 nm diameter range. Intact axodendritic and axosomatic synapses were frequently observed in 3, 5 and 10 day-old cultures as well, in which, axons deriving either from the pre- or postganglionic nerve trunk should have undergone various stages of nerve degeneration.

Observations on Preganglionally Denervated Rat Superior Cervical Ganglia in the Acute Series

Surviving axons were seen in relation both to ganglion cell bodies and their dendrites. These axons contained axoplasmic components characteristic of cholinergic terminals—concentrations of clear vesicles (30—50 nm in diameter) and a few larger (100nm) dense-cored vesicles—besides being partly or completely surrounded by a reaction product indicative of the presence of acetylcholinesterase (Figs. 6—8).

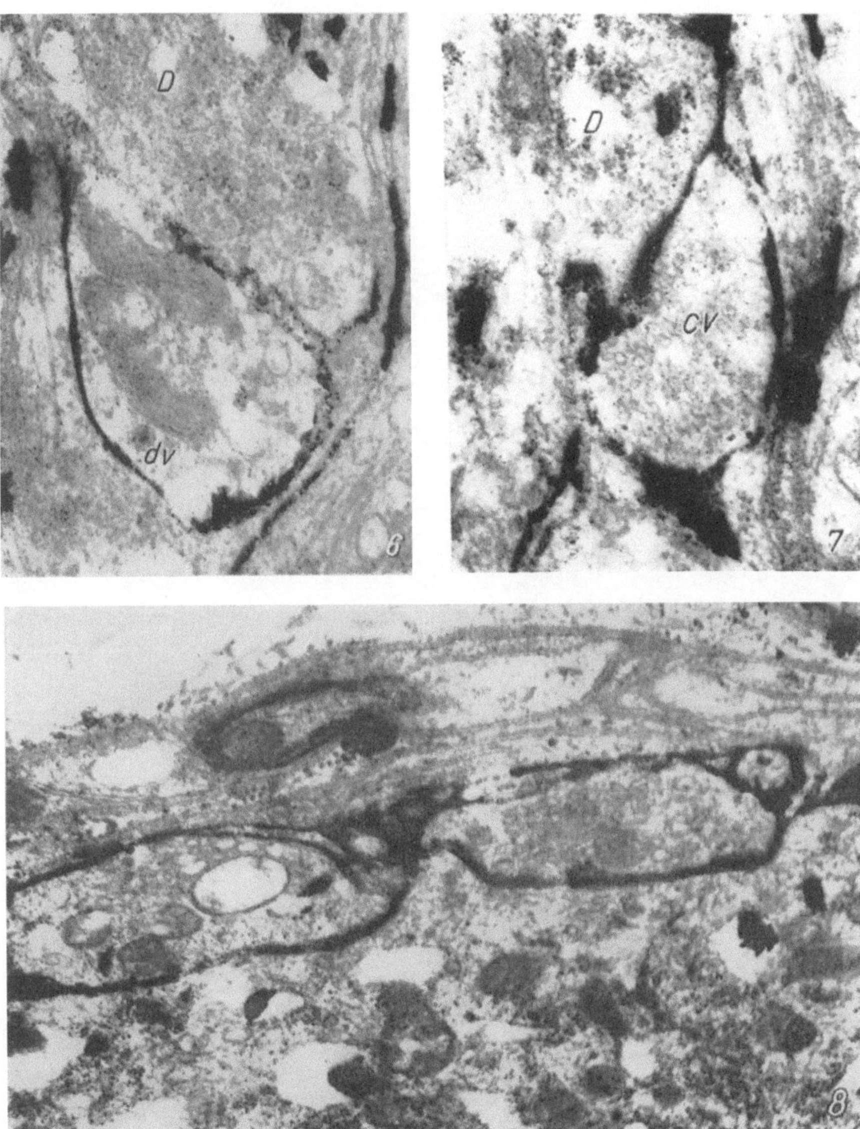

Figs. 6—8. Intact axon terminals in the rat superior cervical ganglion 24 h after preganglionic sympathectomy. The presence of acetylcholinesterase (AChE) is indicated by dense reaction product after acetylthiocholine incubation. Figs. 6 and 7 show axodendritic synapses. Note dendrite (D), 100 nm dense-cored vesicles (dv), 50 nm clear vesicles (cv). Fig. 8 shows surviving axosomatic synapses. Fig. 6: ×52,000; Fig. 7: ×57,000; Fig. 8: ×40,000

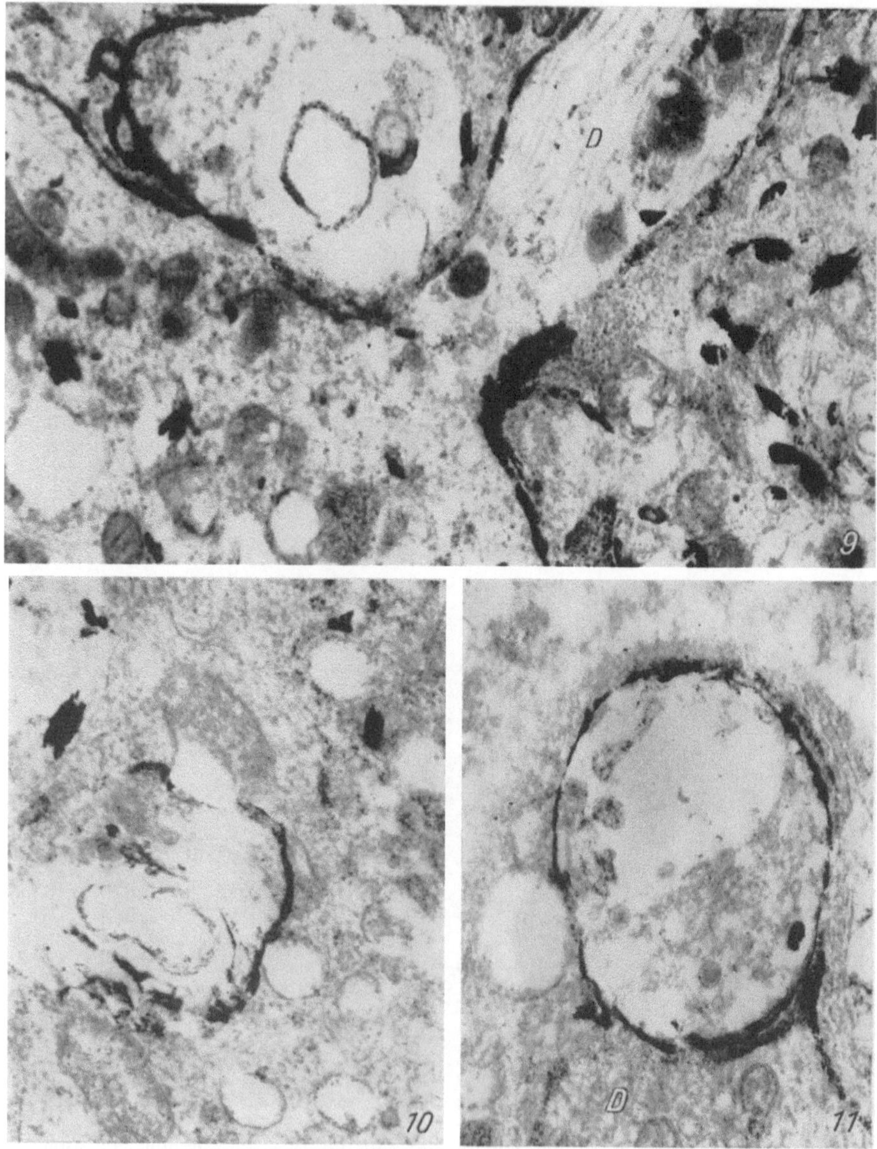

Figs. 9—11. Degenerating (AChE-positive) axon terminals in the rat superior cervical ganglion 24 h after preganglionic sympathectomy. Degenerating axon terminals may be axosomatic (Figs. 9, 10) or axodendritic (Fig. 11). Note dendrites (D). Fig. 9: ×40,000; Fig. 10: ×65,000; Fig. 11: ×40,000

Axon terminals undergoing degeneration were also found in relation to ganglion cell bodies (Figs. 9, 10) and to their dendrites (Fig. 11). All such terminals were AChE-positive. Degenerating axons were identified by one or both of the following features (*Hámori et al.*, 1968): (1) axoplasmic clumping (Fig. 10) with increased osmiophilia, (2) axoplasmic vacuolation (Fig. 11), sometimes associated with obvious axonal swelling.

Observations on Preganglionally Denervated Rat Superior Cervical Ganglia in the Chronic Series

Five weeks after preganglionic sympathectomy both axodendritic and axosomatic nerve terminals were found in close relationship to the postganglionic neurones of the superior cervical ganglion (Figs. 12 to 16). From these surviving terminals 17 were AChE-positive (Figs. 13, 15) and one was AChE-negative (Fig. 16). This AChE-negative axodendritic terminal contained a concentration of small (30—50 nm) vesicles, some of which were dense-cored. In every AChE-positive terminal all the small vesicles were clear.

Conclusion

The demonstration of surviving axodendritic and axosomatic terminal contacts, both in *in vitro* cultured tissue of the sympathetic chain and in ganglia divested of their preganglionic nerve supply, draws our attention to the existence of a rather complicated system consisting of axon terminals not being of preganglionic origin, which could possibly modulate the cellular functioning between different ganglion cells. The origin of surviving synapses from extraganglionic sources seems to be clearly ruled out in cultured chick sympathetic ganglia, while in denervation experiments performed on adult rats, only their preganglionic derivation could be excluded. Since the majority of these remaining synapses in denervated superior cervical ganglia were AChE-positive, they might be regarded either as recurrent collaterals originating from cholinergic postganglionic axons or as axonal processes of another ganglion cell type—possibly an interneurone. The minority of axon terminals on ganglion cells (remaining after preganglionic neurectomy), being AChE-negative and containing concentrations of small dense-cored vesicles, might be regarded as adrenergic. Their origin is again speculative, but they could conceivably belong to recurrent collaterals from adrenergic

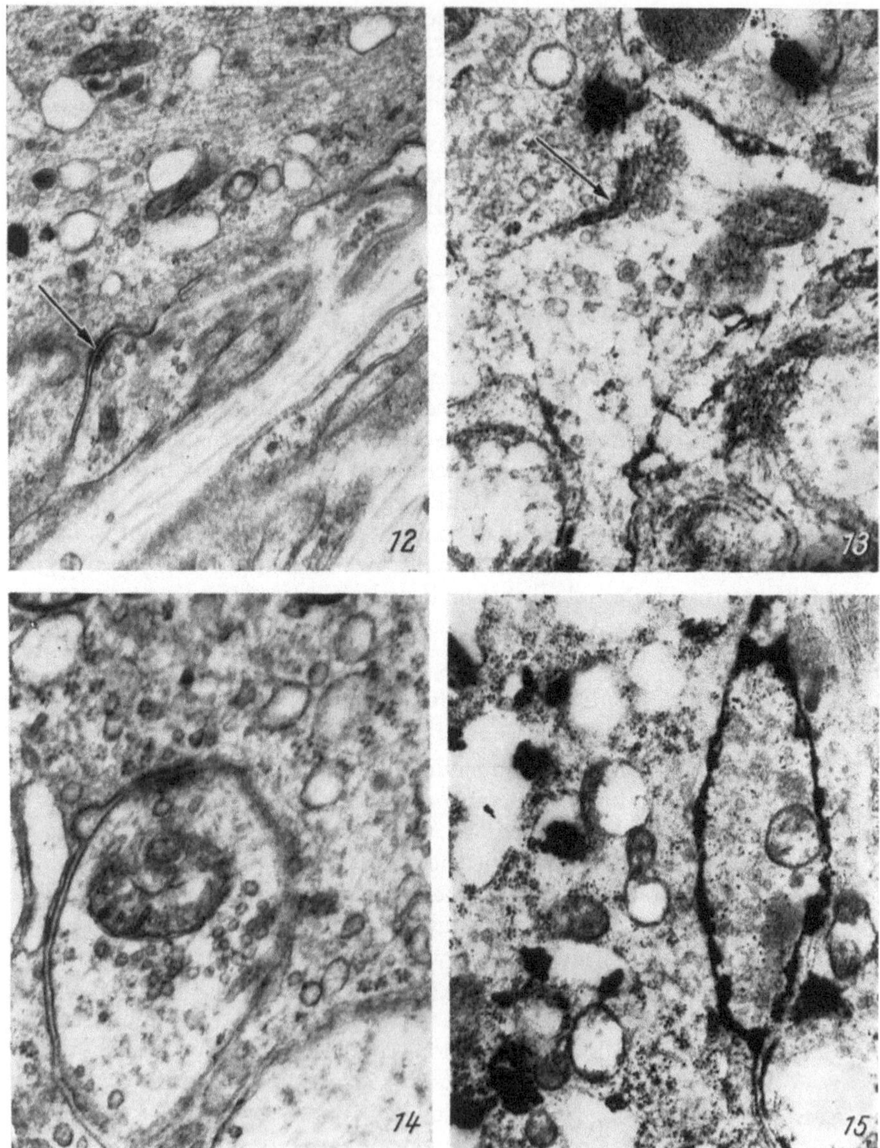

Figs. 12—15. Intact terminal axons in rat superior cervical ganglion 5 weeks after preganglionic sympathectomy. Both axodendritic (Figs. 12, 13) and axosomatic (Figs. 14, 15) contacts are shown. Note synapses with membrane thickening (arrowed). Figs. 13 and 15 were previously processed for the demonstration of AChE. Fig. 12: ×30,000; Fig. 13: ×65,000; Fig. 14: ×50,000; Fig. 15: ×65,000

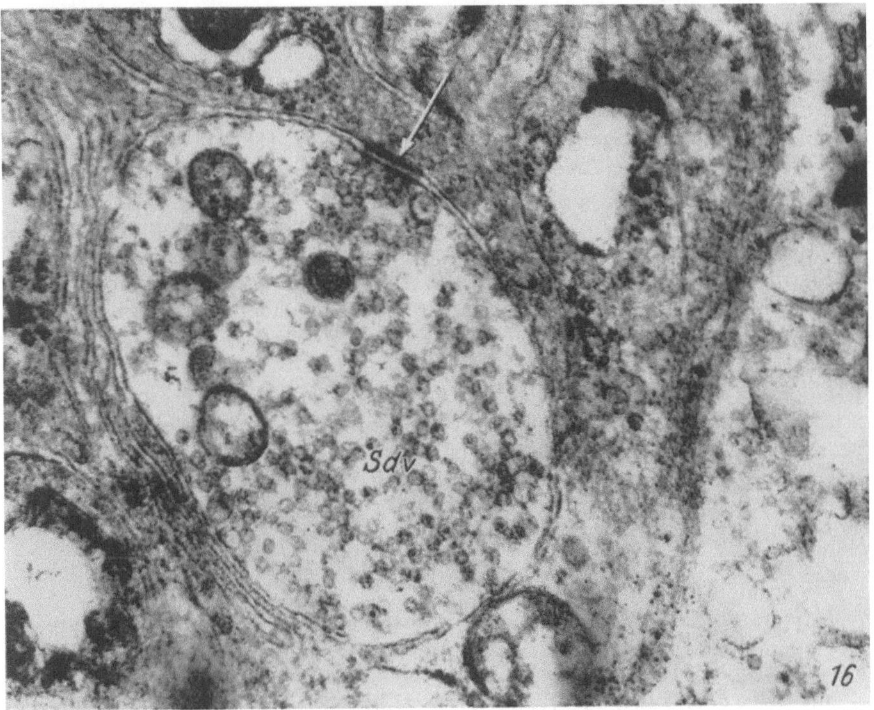

Fig. 16. AChE-negative axodendritic terminal in rat superior cervical ganglion 5 weeks after preganglionic sympathectomy. Note synaptic contact (arrowed) and 50 nm synaptic vesicles, many with dense cores (sdv). ×65,000

postganglionic axons or to axonal processes of an adrenergic interneurone. Since all of the remaining axon terminals, found both in tissue cultures and denervated ganglia, showed fine structural features characteristic either of cholinergic or adrenergic nerves, none of the nerve endings observed can be considered to be cell processes from cells of the chromaffin type.

Acknowledgement

This work was performed in collaboration with Prof. *J. D. Lever* and Dr. *R. Presley* at the Department of Anatomy, University College, Cardiff, Great Britain, during the tenure of a fellowship of the author sponsored by the Wellcome Trust. The author is grateful to Miss *C. Ivens* and Dr. *R. Mottram* for their valuable help and to Mrs. *G. Howells* and Miss *D. Morgan* for expert technical assistance.

References

Csillik, B., Gy. Kálmán, and *E. Knyihár*: Adrenergic nerve endings in the feline cervical superius ganglion. Experientia *23*, 477 (1967).

De Groat, W. C., and *W. R. Saum*: Adrenergic inhibition in mammalian parasympathetic ganglia. Nature New Biol. *231*, 188—189 (1971).

Dogiel, A. S.: Zwei Arten sympathischer Nervenzellen. Anat. Anz. *11*, 679—687 (1896).

Eccles, R. R. M., and *B. Libet*: Origin and blockade of the synaptic responses of curarized sympathetic ganglia. J. Physiol. *157*, 484—503 (1961).

Elfvin, L. G.: The ultrastructure of the superior cervical sympathetic ganglion of the cat. II. The structure of the preganglionic end fibres and the synapses as studied by serial sections. J. Ultrastr. Res. *8*, 441—476 (1963).

Falck, B., N.-A. Hillarp, G. Thieme, and *A. Torp*: Fluorescence of catechol amines and related compounds condensed with formaldehyde. J. Histochem. Cytochem. *10*, 348—354 (1962).

Grillo, M. A.: Electron microscopy of sympathetic tissues. Pharmacol. Rev. *18*, 387—399 (1966).

Hamberger, B., and *K.-A. Norberg*: Studies on some system of adrenergic synaptic terminals in the abdominal ganglia of the cat. Acta Physiol. Scand. *65*, 235—242 (1965).

Hámori, J., E. Láng, and *L. Simon*: Experimental degeneration of the preganglionic fibres in the superior cervical ganglion of the cat. Z. Zellforsch. *90*, 37—52 (1968).

Joó, F., J. D. Lever, C. Ivens, D. R. Mottram, and *R. Presley*: A fine structural and electron histochemical study of axon terminals in the superior cervical ganglion after acute and chronic preganglionic denervation. J. Anat. *110*, 181—189 (1971).

Knoll, J., and *E. S. Vizi*: Presynaptic inhibition of acetylcholine release by endogenous and exogenous noradrenaline at high rate of stimulation. Br. J. Pharmacol. Chemother. *40*, 554—555 P (1970).

Lawrentjew, B. J.: Zur Morphologie des Ganglion cervicale super. Anat. Anz. *58*, 529—539 (1924).

Lever, J. D., T. L. B. Spriggs, and *J. D. P. Graham*: A formol- fluorescence, fine structural and autoradiographic study of the adrenergic innervation of the vascular tree in the intact and sympathectomized pancreas of the cat. J. Anat. *103*, 15—34 (1968).

Lewis, P. R., and *C. C. D. Shute*: The distribution of cholinesterase in cholinergic neurons demonstrated with the electron microscope. J. Cell Sci. *1*, 381—390 (1966).

Marrazzi, A. S.: Adrenergic inhibition at sympathetic synapses. Am. J. Physiol. *127*, 738—744 (1939).

Mattews, M. R., and *G. Raisman*: The ultrastructure and somatic-efferent synapses of small granule-containing cells in the superior cervical ganglion. J. Anat. *105*, 255—282 (1969).

Norberg, K.-A., and *B. Hamberger*: The sympathetic adrenergic neuron. Acta Physiol. Scand. *63*, Suppl. *238*, 1—42 (1964).

Presley, R., J. D. Lever, C. Ivens, F. Joó, and *T. L. B. Spriggs*: Observations on the sympathetic neurone *in vitro*. J. Anat. *108*, 611—612 (1971).
Reynolds, E. S.: The use of lead citrate at high pH as an electron opaque stain in electron microscopy. J. Cell Biol. *17*, 208—212 (1963).
Williams, T. H., and *S. L. Palay*: Ultrastructure of the small neurons in the superior cervical ganglion. Brain Res. *15*, 17—34 (1969).

Author's address: Dr. *F. Joó*, Electron Microscope Laboratory of the Biophysical Institute of the Biological Research Centre of the Hungarian Academy of Sciences, Szeged, Hungary.

Discussion

Vizi: I think this work is a very important contribution to the understanding of the role of noradrenergic nerves in the neurochemical transmission of ganglia. Your findings provide strong evidence that noradrenergic nerves are capable of influencing cholinergic transmission, which was suggested by us (*Paton* and *Vizi*, Br. J. Pharmacol. *35*, 10—28, 1969; *Vizi*, Arch. exp. Path. Pharmak. *259*, 199—200, 1968; *Knoll* and *Vizi*, Br. J. Pharmacol. *42*, 263—272, 1971). Dr. Dawes and I could show that noradrenaline and adrenaline reduce the acetylcholine output from isolated cervical ganglia when the preganglionic fibres were stimulated. I think morphological and physiological findings fit very well. Just a question. Did you use also other types of ganglion? Dr. Csillik already showed that in the ciliary ganglion there are noradrenergic fibres. I hope I am correct quoting this reference.

Joó: Our investigations, as I mentioned already, provide morphological evidence for the existence of axon terminals of the adrenergic type, but the presence of surviving acetylcholinesterase-positive nerve endings in denervated superior cervical ganglia should be emphasized again. It is probable that the noradrenaline containing nerves are indeed involved by some mechanisms in intraganglionic inhibition, but the function of the non-preganglionic, acetylcholinesterase-positive axons revealed still remains to be elucidated. To answer your question, we have not extended our investigations to another type of ganglion, but we will perform studies on ganglia of a parasympathetic nature.

Gerebtzoff: How long after the removal of a lumbar sympathetic ganglion did you find fluorescence in the nerve cells of the explant? I think the centre, at least, of any explant is meant to die off soon.

Joó: The projected slide showed a 3-day-old culture. The fluorescence, indicative of catecholamines in the perikarya of sympathetic neurones of the explants, was observed up to 5—6 days of *in vitro* culturing.

Storm-Mathisen: Do you have any evidence, apart from the existence of acetylcholinesterase-positive nerve terminals, that cholinergic nerve elements exist in these explants? Do they for instance contain cholinacetylase?

Joó: The demonstration of acetylcholinesterase is, at present, widely accepted as a useful histochemical method for vizualizing cholinergic structures. Although some doubts were raised about the validity of identifying structures, mainly in the central nervous system, in this way. In sympathetic ganglia the preganglionic nerve terminals were found in neurophysiological experiments to operate with cholinergic mediation. Therefore, in these nerves, acetylcholinesterase-positivity may signify the cholinergic mediation too.

We have not tried as yet to demonstrate cholinacetylase activity in our material, but it certainly would be of interest to reveal its localization.

Journal of Neural Transmission, Suppl. XI, 299—314 (1974)

The Role of Choline in Maintaining the Fine Structure of Nerve Terminals in the Superior Cervical Ganglion of Cat

Á. Párducz, F. Joó, and **O. Fehér**

Electron Microscope Laboratory, Biophysical Institute of the Biological Research Centre of the Hungarian Academy of Sciences, and Institute of Animal Physiology, József Attila University of Sciences, Szeged, Hungary

With 8 Figures

Summary

The fine structure of synapses was studied in the cat superior cervical ganglion under various experimental conditions. The presence of choline in the performed experiments was found to play an essential role not only in maintaining the intraganglionic nerve transmission, but also in preserving the normal fine structure of presynaptic terminals. Results obtained shed more light on other aspects of the vesicle hypothesis elucidating one of the cellular mechanisms by which the disappearance of synaptic vesicles in stimulated cholinergic nerve terminals could be interpreted.

Introduction

The search for fine structural changes taking place in the terminals of stimulated nerves has long been the subject of electron microscopic investigations. Ultrastructural studies performed until now were, almost without exception, strongly influenced by the currently accepted so-called "vesicle hypothesis", eleborated by *del Castillo* and *Katz* (1956) and confirmed in many ways by *Whittaker* (1970). According to this, the quanta of transmitter are presumed to preexist within the terminal as vesicular acetylcholine.

Although the exact nature of binding of acetylcholine to synaptic vesicles should still be the subject of further studies (*Csillik* and *Joó*, 1967; *Whittaker*, 1971), the fact that changes in the number and distribution of vesicles are occuring during stimulation (*Hubbard* and

Kwanbunbumpen, 1968; *Jones* and *Kwanbunbumpen*, 1970; *Csillik* and *Bense*, 1971; *Perri et al.*, 1972) seems now well established. Most of our knowledge regarding the correlation between changes in number of synaptic vesicles and experimentally enhanced functioning of nerves is based on data derived from studies on peripheral synapses. As regards the cat superior cervical ganglion, no change was found in the number of synaptic vesicles of preganglionic nerve terminals after stimulation if the natural blood supply was maintained, whereas significant decrease was observed in stimulated ganglia, perfused with isotonic Locke solution (*Párducz* and *Fehér*, 1970). Later results of *Friesen* and *Khatter* (1971) and *Birks* (1971), however, indicated a decrease in the number of synaptic vesicles in preganglionic nerve endings of the cat superior cervical ganglion even with normal blood supply.

At present, little is known, however, of those cellular mechanisms by which the formation and breakdown of synaptic vesicles are regulated in nerve terminals at different functional states. The presence of choline has been shown to be essential in maintaining intraganglionic nerve transmission (*Brown* and *Feldberg*, 1936). Later, *Collier* and *MacIntosh* (1969) demonstrated that ganglionic terminals take up labelled choline at rapid rates and that 85 % of the choline content of the ganglion may consist of labelled choline. It was found by *Collier* and *Lang* (1969) that choline was incorporated into phospholipids and phosphorylcholine, too. The intensive and selective incorporation of labelled choline into phospholipids present in the synaptic vesicles was clearly demonstrated by *Bosmann* and *Hemsworth* (1970). Since the crucial importance of choline clearly emerges from these investigations, it seemed to be of interest to correlate the changes in number of synaptic vesicles following stimulation with the choline uptake of intraganglionic nerve terminals in two ways: (1) by adding choline to the Locke's perfusion fluid, and (2) by adding HC-3 to the perfusion fluid, or by giving HC-3 directly to the animals. The effect of exhaustive preganglionic stimulation was recorded by physiological methods and by electron microscopy. Parts of this investigation was recently published (*Párducz et al.*, 1971).

Methods

Cats weighing 1.5—3 kg, anaesthetized with 40 mg/kg sodium pentobarbital were used. The superior cervical ganglia and the cervical sympathetic trunks were exposed and, in some cases, perfused following the method of *Kibjakov* (1933) as modified by *Paton* and *Perry* (1953). Contractions of

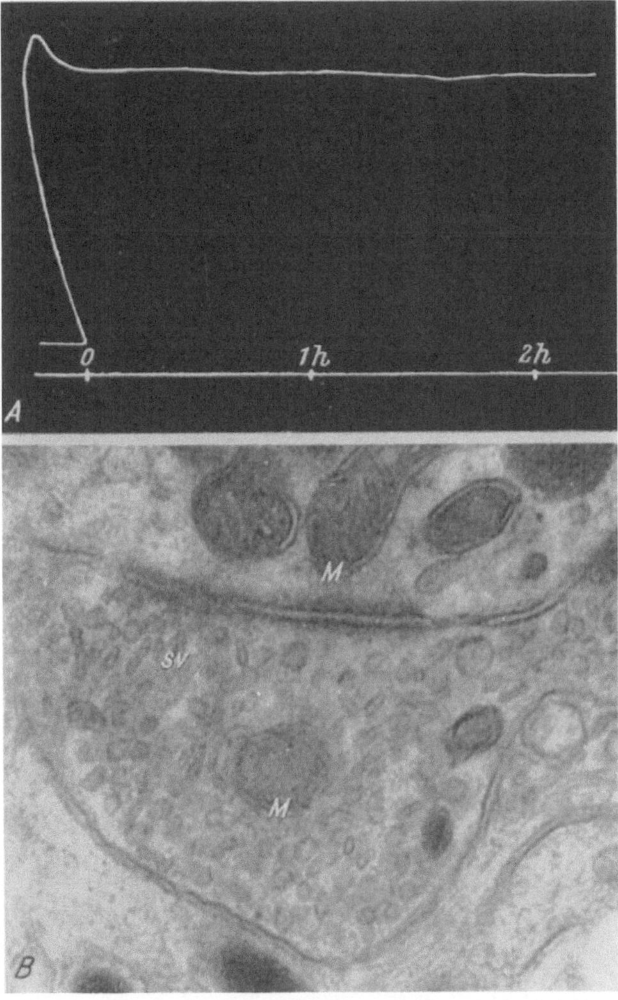

Fig. 1. A: Recording of the nictitating membrane contraction during a 2 hours period of stimulation shows maximal transmission in ganglion with natural blood supply.

B: Synapse from the stimulated ganglion. Mitochondria (M) are intact, both pre- and postsynaptically. Note the heterogeneity of the synaptic vesicles (sv). ×60,000

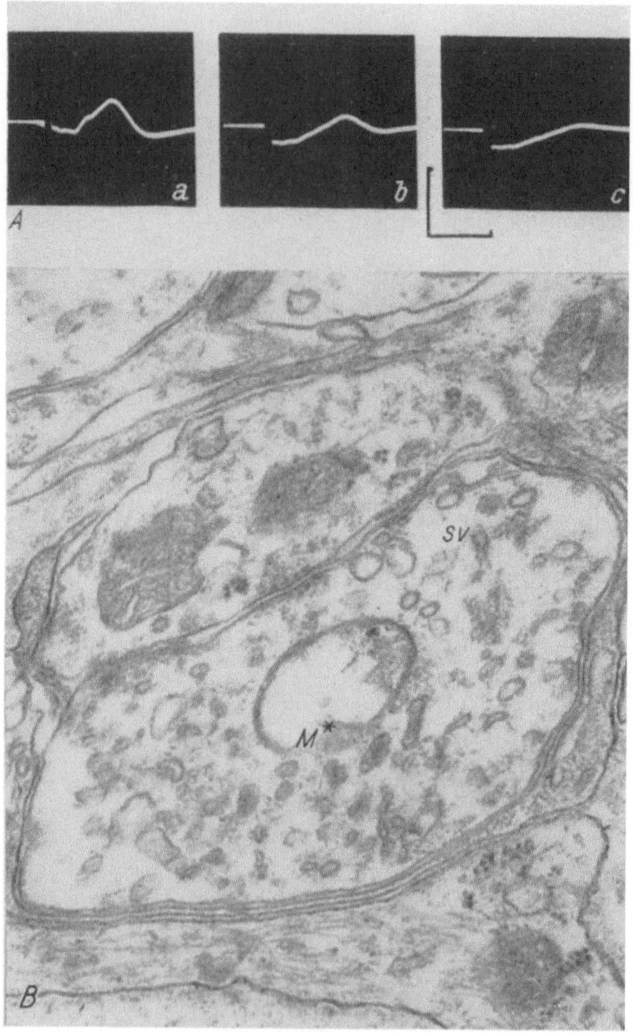

Fig. 2. A: Action potentials led off from the ganglion perfused with choline-free Locke's solution after the first (a), seventh (b) and fifteenth (c) minute of stimulation. The transmission failed completely in the fifteenth minute.

B: After 15 minutes of preganglionic stimulation and perfusion with choline-free Locke's solution, a swollen mitochondrion (M*) and a decreased number of synaptic vesicles (sv) are present in the presynaptic nerve ending. ×60,000

the nictitating membrane were recorded on a kymograph. The cervical sympathetic trunk was stimulated by supramaximal square impulses of the following parameters: frequency, 10—20/sec; duration, 1 msec; amplitude, 6—8 V. In some experiments the action potentials of the ganglion were also recorded.

In some experiments HC-3 was administered during preganglionic stimulation. For ganglia with normal blood supply 10 mg/kg doses of HC-3 were injected intravenously, the cats given HC-3 being kept on artificial respiration. In other cases 10 mg/l HC-3 was dissolved in the perfusion fluid for ganglia. At the end of the experiment the ganglia were excised and subjected to routine electron microscopic examination. This procedure consisted of the following steps: fixation in *Karnovsky*'s (1965) aldehyde fixative (2 h) and then in *Millonig*'s (1961) phosphate buffered osmic acid for 1 h. The samples were dehydrated in alcohol and embedded in Durcupan (Fluka). Thin sections were made on a Porter-Blum ultramicrotome and stained according to *Reynolds*' procedure (1963). For inspection and photography of the sections a Tesla 242D and an Elmiskop 1 Siemens electron microscope were used.

The analyses made were based upon the total count of clear synaptic vesicles in every outlined and planimetrically measured axon terminal in control and different experimental groups. The number of synaptic vesicles per μm^2 was expressed and statistical analysis was performed by using the double "t" probe as significance test.

Results

In the first series of experiments ganglia with normal blood supply were examined. In the resting state the fine structural features of the terminals were typical and the number of synaptic vesicles was $232 \pm 49/\mu m^2$. During stimulation with the above-mentioned parameters no sign of fatigue could be observed. The action potential and the contraction of the nictitating membrane (Fig. 1 A) showed maximum transmission and the fine structure of ganglionic synapses was similar to those in the resting state. Furthermore, there was no significant change in number of synaptic vesicles ($247 \pm 38/\mu m^2$), although more irregular forms could be seen (Fig. 1 B).

In the next series of experiments, the ganglia were perfused with choline-free Locke's solution. In the resting state a smaller vesicle number was observed ($171 \pm 48/\mu m^2$) than that seen with natural blood supply. This difference is statistically significant, but is not accompanied by ultrastructural changes. The excitability of the ganglion also remained normal. Preganglionic stimulation, however, led to a rapid exhaustion and transmission as indicated by the action potentials (Fig. 2 A) ceased completely within 15 minutes. The ganglia

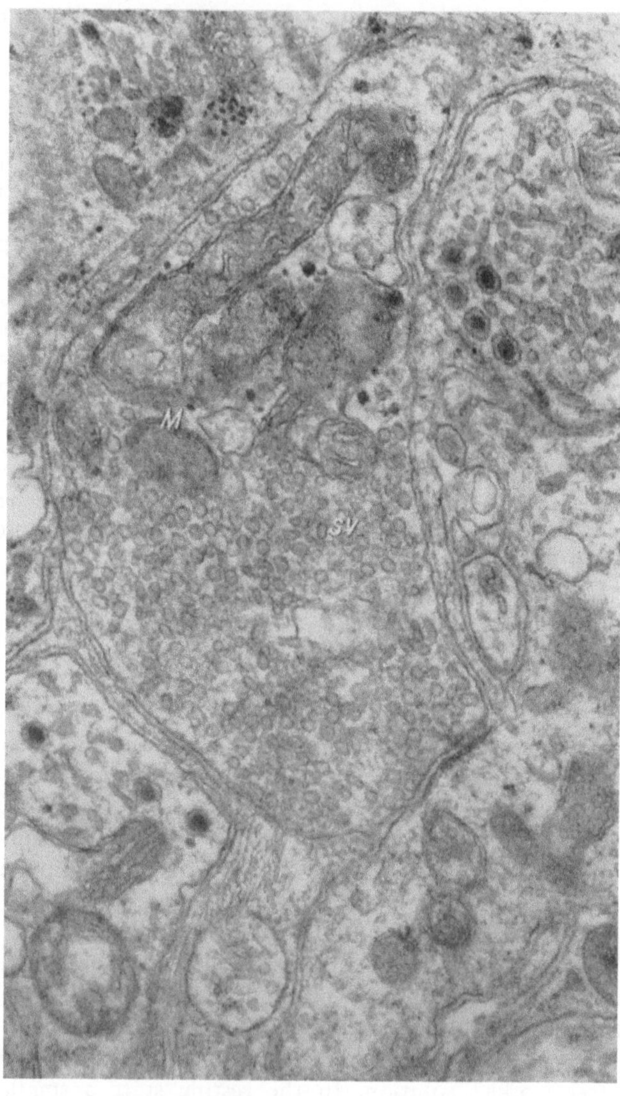

Fig. 3. Synapse of a superior cervical ganglion perfused with Locke's solution, to which choline is added. Electrical stimulation does not lead to exhaustion and there is neither any change in the number of synaptic vesicles (sv) nor in the appearance of presynaptic mitochondria (M). ×54,000

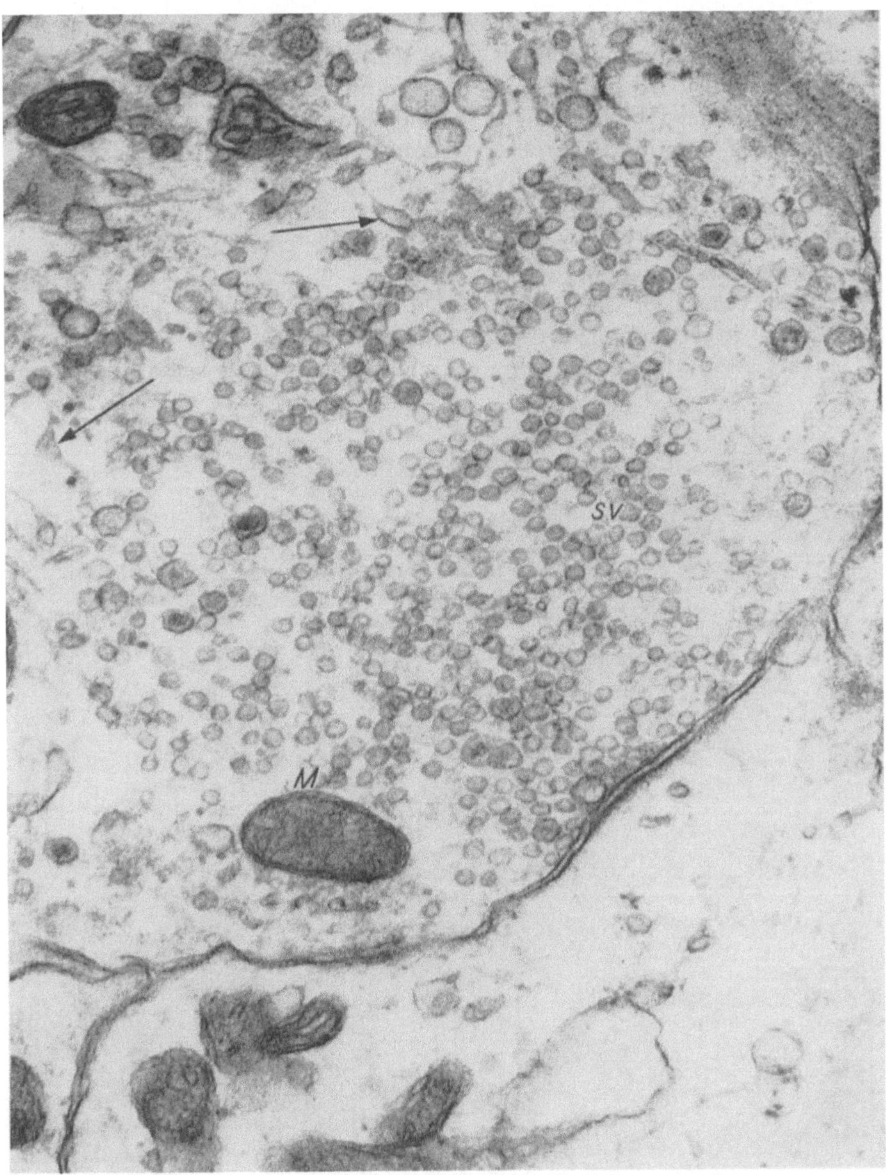

Fig. 4. Synapse of a superior cervical ganglion after i.v. administration of 10 mg/kg HC-3. There is no decrease in number of synaptic vesicles (sv), but many irregular forms can be seen (arrows). The mitochondria (M) are intact. ×60,000

excised after stimulation showed considerable morphological changes in the presynaptic terminals (Fig. 2 B). The number of synaptic vesicles fell to $36 \pm 17/\mu m^2$, which is significantly different from any count obtained before. In the terminals only a few vesicles were crowding around the sites of synaptic contacts. The mitochondria were swollen and structurally damaged, the matrix having partly or completely disappeared and the length and number of cristae having also markedly decreased. The organelles of the postsynaptic elements, however, were unchanged at this time.

Since much of the choline needed for the transmitter synthesis is known to be supplied by the blood stream, it was decided to investigate, whether the choline which could play a part in the

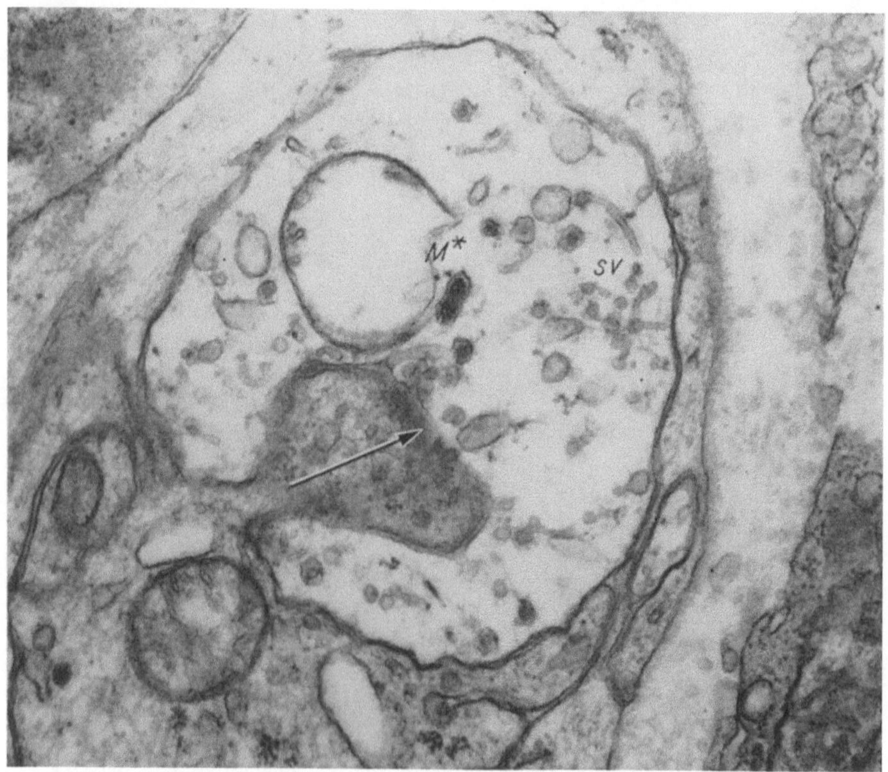

Fig. 5. Synapse from a ganglion, HC-3-treated and stimulated to exhaustion. Structurally damaged mitochondrion (M*) and a few synaptic vesicles (sv) are present in the presynaptic terminal. The arrow points to the membrane thickening indicating the synaptic contact. ×48,000

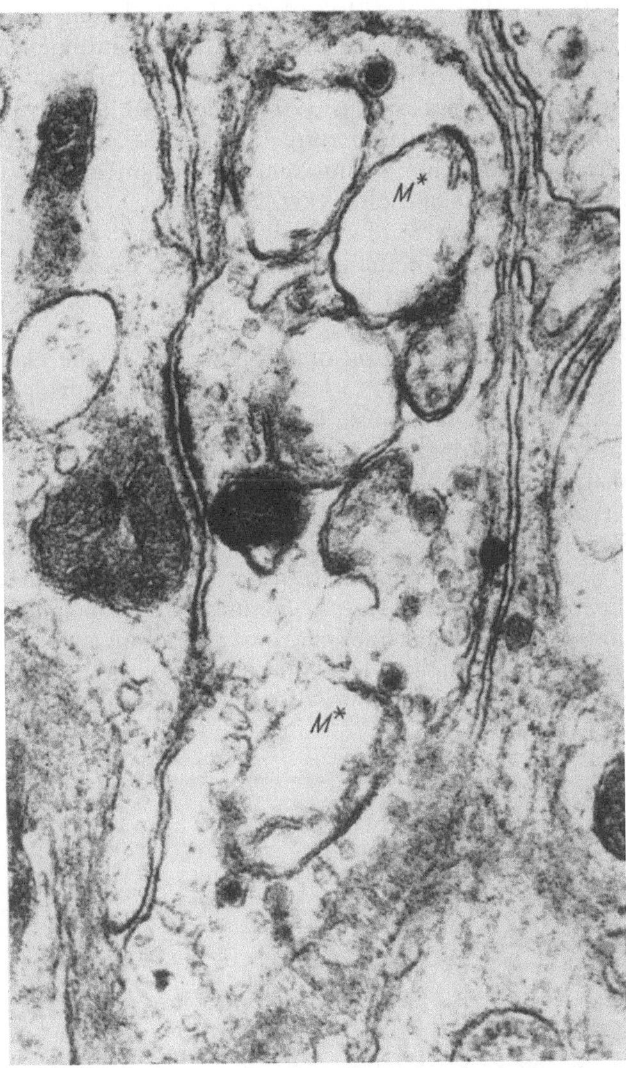

Fig. 6. Synapse from a ganglion perfused with choline containing Locke's solution containing 10 mg/l HC-3. The stimulation causes a decrease in number of synaptic vesicles and damage in the presynaptic mitochondrial structure (M*). ×54,000

maintenance of the normal rate of transmission had any role in preserving the intact fine structural appearance of preganglionic terminals. In ganglia perfused with Locke's solution containing 10 μg/ml choline, transmission, as judged by the nictitating membrane contraction, was unimpaired even 60 minutes after stimulation. The number of synaptic vesicles was $179 \pm 49/\mu m^2$ in the resting and $177 \pm 61/\mu m^2$ in the stimulated state. It must be emphasized that neither perfusion itself nor stimulation resulted in changes of fine structure in the presence of choline (Fig. 3).

For clearing up the role of choline in ganglionic transmission in *in vivo* experiments, one of the hemicholiniums (HC-3) seemed very useful, because this drug has been proved to inhibit the uptake of choline by preganglionic terminals.

Following i.v. administration of 10 mg/kg HC-3, the number of synaptic vesicles decreased to $213 \pm 106/\mu m^2$, which is not significant, but the scattering of data must be mentioned. At the same time obvious alterations appeared in the form and size of synaptic vesicles, many of which had become swollen and irregular in shape, although the mitochondria in the presynaptic terminals remained intact (Fig. 4). When after the administration of HC-3 the preganglionic trunk was stimulated, transmission deteriorated rapidly ceasing completely within 1.5—2 minutes. Electron microscopic examinations showed similar fine structural alterations as in the Locke perfused stimulated ganglia. The mitochondria in the presynaptic terminals were swollen, the number of synaptic vesicles was found to

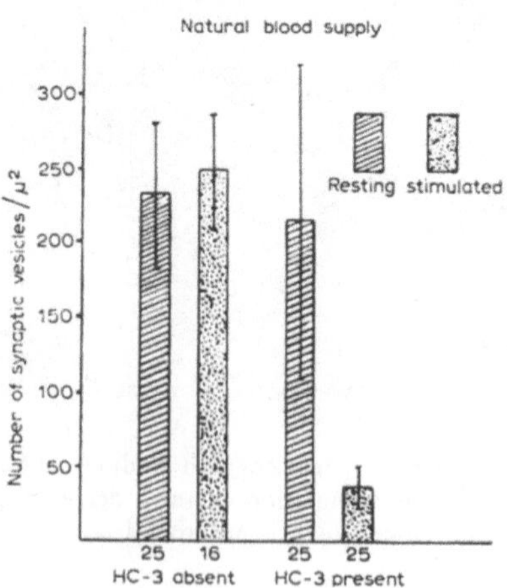

Fig. 7. The effect of stimulation and HC-3 on the number of synaptic vesicles in superior cervical ganglia having a normal blood supply

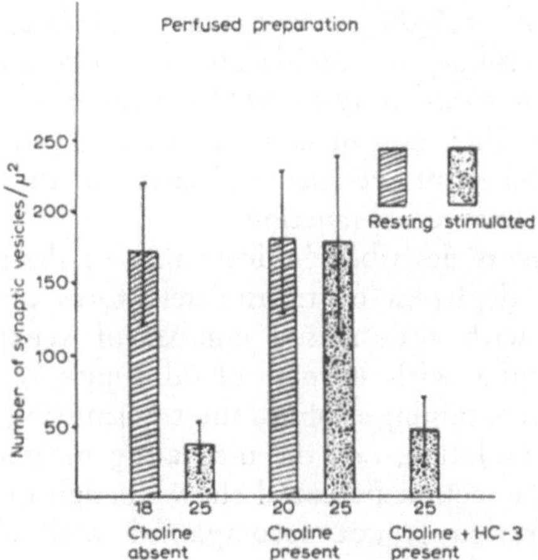

Fig. 8. The effect of stimulation on the number of synaptic vesicles in superior cervical ganglia perfused with Locke's solution

be decreased to $37 \pm 14/\mu m^2$, significantly different from the unstimulated synapses (Fig. 5).

In further experiments choline (10 μg/ml) was combined with HC-3 (10 μg/ml) in the Locke solution. Under these circumstances preganglionic stimulation was effective for only 15 minutes after which the transmission failed completely. The number of synaptic vesicles was $47 \pm 23/\mu m^2$ and the presynaptic terminals all showed the damage characteristic of the HC-3 treatment and that of the Locke-perfused exhausted state (Fig. 6).

The results obtained from the above-described experiments are summarized in figures 7 and 8. The number of axon terminals examined in each case is given.

Conclusion

Although numerous physiological, morphological and biochemical results have already been presented in favour of the vesicle hypothesis, the function of synaptic vesicles in chemically mediated synapses is still far from being fully understood. In this respect, those experiments have to be mentioned in which attempts were made to correlate the fine structural alterations elicited by stimulation with the transmission process. *De Robertis* and *Ferreira* (1957) found changes in the number of synaptic vesicles of splanchnic nerves innervating the adrenal medulla after electrical stimulation. Similarly,

Dyachkowa et al. (1967) observed morphological changes in the ciliary ganglion following stimulation. Furthermore, the results of *Hubbard* and *Kwanbunbumpen* (1968) indicated that in experimental circumstances the increased quantal release can be correlated with a fall in the number of vesicles adjacent to the nerve terminal membrane in the myoneural junction.

Our experiments described indicate that in the superior cervical ganglion of cat depletion of transmitter stores coincides electron-microscopically with a decreased number of synaptic vesicles. In contrast, in ganglia with normal blood supply or perfused with Locke's solution containing choline, the transmission process remains intact during stimulation, *i.e.*, the nictitating membrane contraction and the ganglionic action potential show no sign of exhaustion; the transmitter liberation is not accompanied with significant ultra-structural changes.

The observations in the Locke-perfused and HC-3 treated ganglia can be interpreted by considering, in both cases, the lack of natural choline supply. Namely, in the Locke-perfused ganglia there is no choline in the perfusion fluid, while, on the other hand, the HC-3 strongly inhibits the choline uptake of terminals. In the resting state, as proved by the *in vitro* experiments, the ganglion preserves its transmission capacity for hours without choline (*Nicolescu et al.*, 1966). Our investigations also failed to show fine structural alterations in the presynaptic endings. Stimulation, however, increases the rate of acetylcholine synthesis and liberation (*MacIntosh*, 1963) and leads to the cessation of transmission by depleting the acetylcholine depot. The fine structural features characteristic of the exhausted state are as follows: decreased number of synaptic vesicles and swollen mitochondria in the presynaptic terminals. Since it was possible to preserve both synaptic function and apparently normal fine structure of nerve terminals by addition of choline to the perfusion fluid, it is concluded that destruction of synaptic vesicles and mitochondria occurs only in absence of choline. If the choline supply is provided for the ganglia either by the blood stream or the perfusion fluid, the uptake of choline and release of acetylcholine might not result in any marked change of morphology. If, however, the ganglion is forced to function under circumstances of choline deficiency it may mobilize choline from other, possibly intracellular, sources. As the most striking morphological damages in the exhausted state were seen in the synaptic vesicles and the mitochondria, it seems probable that the choline-containing phospholipids structurally bound to the synaptic vesicles and mitochondria were used for the acetylcholine synthesis.

Such a concept is supported by numerous data derived from different experiments. According to *Friesen et al.* (1967), the preganglionic nerve terminals in the superior cervical ganglion of cat can store appreciable quantities of choline for the synthesis of acetylcholine, not only in free form, but also esterified to phospholipids. This depot is of considerable capacity and the catabolism of these compounds may provide choline for the transmitter synthesis during relatively long periods of time. On the other hand, choline is known to be an essential component of synaptic vesicles and was found to be taken up into synaptosomes and vesicles. According to *Bosmann* and *Hemsworth* (1970), labelled choline appears in the chloroform soluble fraction of synaptic vesicles in 10—100 times the concentration of that in other subsynaptosomal fractions. Therefore, it seems likely that the chloroform soluble fraction is built up at a considerable rate in the presence of choline, while in its absence vesicular choline may be decomposed at a comparable speed.

In interpreting the results obtained from *in vitro* experiments, the absence of choline, according to our views, has also to be taken into consideration. It is conceivable that the fine structural alterations found after stimulation in these experiments were, at least in part, produced by the existing choline deficiency.

It is tempting to assume that situations of relative choline deficiency may also be present in other cases, in particular if the demand for choline is rapidly increased due to the stimulation to provide a greater amount of precursor for the enhanced acetylcholine synthesis.

References

Birks, R. I.: Effects of stimulation on synaptic vesicles in sympathetic ganglia, as shown by fixation in the presence of Mg^{2+}. J. Physiol. *216*, 26—28 P (1971).

Bosmann, H. B., and *B. A. Hemsworth*: Synaptic vesicles. Incorporation of choline by isolated synaptosomes and synaptic vesicles. Biochem. Pharmacol. *19*, 133—141 (1970).

Brown, G. L., and *W. Feldberg*: The acetylcholine metabolism of a sympathetic ganglion. J. Physiol. *88*, 265—283 (1936).

Collier, B., and *C. Lang*: The metabolism of choline by a sympathetic ganglion. Canad. J. Physiol. Pharmacol. *47*, 119—126 (1969).

Collier, B., and *F. C. MacIntosh*: The source of choline for acetylcholine synthesis in a sympathetic ganglion. Canad. J. Physiol. Pharmacol. *47*, 127—135 (1969).

Csillik, B., and *S. Bense*: Function-dependent alterations in the distribution of synaptic vesicles. Acta biol. Acad. Sci. hung. *22*, 131—139 (1971).

Csillik, B., and *F. Joó*: Effect of hemicholinium on the number of synaptic vesicles. Nature (London) *213*, 508—509 (1967).

Del Castillo, J., and *B. Katz*: Biophysical aspects of neuromuscular transmission. Prog. Biophys. *6*, 121—170 (1956).

De Robertis, E., and *A. V. Ferreira*: Submicroscopic changes of the nerve endings in the adrenal medulla after stimulation of the splanchnic nerve. J. biophys. biochem. Cytol. *3*, 611—614 (1957).

Dyachkowa, L. N., *J. Hámori*, and *L. Fedina*: Ultrastructure of synapses in ganglion ciliare of birds after ortho- and antidromic electrical stimulation. Dokl. Acad. Sci. U.S.S.R. *172*, 957—959 (1967). (In Russian.)

Friesen, A. J. D., and *J. C. Khatter*: Effect of stimulation on synaptic vesicles in the superior cervical ganglion of the cat. Experientia *27*, 285—287 (1971).

Friesen, A. J. D., *G. M. Ling*, and *M. Nagai*: Choline and phospholipid choline in a sympathetic ganglion and their relationship to acetylcholine synthesis. Nature (London) *214*, 722—724 (1967).

Hubbard, J. I., and *S. Kwanbunbumpen*: Evidence for the vesicle hypothesis.. J. Physiol. *194*, 407—420 (1968).

Jones, S. F., and *S. Kwanbunbumpen*: The effects of nerve stimulation and hemicholinium on synaptic vesicles at the mammalian neuromuscular junction. J. Physiol. *207*, 31—50 (1970).

Karnovsky, M. J.: Formaldehyde-glutaraldehyde fixative of high osmolarity for use in electron microscopy. J. Cell Biol. *27*, 137 A (1965).

Kibjakov, A. V.: Über humorale Übertragung der Erregung von einem Neuron auf das andere. Arch. ges. Physiol. *232*, 432—443 (1933).

MacIntosh, F. C.: Synthesis and storage of acetylcholine in nervous tissue. Can. J. Biochem. Physiol. *41*, 2555—2571 (1963).

Millonig, G.: Advantages of a phosphate buffer for OsO_4 solutions in fixation. J. appl. Physics. *32*, 1637 (1961).

Nicolescu, P., *M. Dolivo*, *C. Rouiller*, and *C. Foroglou-Kerameus*: The effect of deprivation of glucose on the ultrastructure and function of the superior cervical ganglion of the rat in vitro. J. Cell Biol. *29*, 267—286 (1966).

Párducz, Á., and *O. Fehér*: Fine structural alterations of presynaptic endings in the superior cervical ganglion of the cat after exhausting preganglionic stimulation. Experientia *26*, 629—630 (1970).

Párducz, Á., *O. Fehér*, and *F. Joó*: Effects of stimulation and hemicholinium (HC-3) on the fine structure of nerve endings in the superior cervical ganglion of the cat. Brain Res. *34*, 61—72 (1971).

Paton, W. D. M., and *W. L. M. Perry*: The relationship between depolarisation and block in the cat's superior cervical ganglion. J. Physiol. *119*, 43—57 (1953).

Perri, V., *O. Sacchi*, *E. Raviola*, and *G. Raviola*: Evaluation of the number and distribution of synaptic vesicles at cholinergic nerve-endings after sustained stimulation. Brain Res. *39*, 526—529 (1972).

Reynolds, E. S.: The use of lead citrate at high pH as an electron opaque stain in electron microscopy. J. Cell Biol. *17*, 208—212 (1963).

Whittaker, V. P.: The vesicle hypothesis, in: Excitatory Synaptic Mechanisms (*Andersen, P.*, and *J. K. S. Jansen*, eds.), pp. 67—76. Universitetsforlaget. 1970.

Whittaker, V. P.: Origin and function of synaptic vesicles. Ann. New York Acad. Sci. *183*, 21—32 (1971).

Author's address: Dr. *Á. Párducz*, Electron Microscope Laboratory, Biophysical Institute of the Biological Research Centre of the Hungarian Academy of Sciences, Szeged, Hungary.

Discussion

Storm-Mathisen: I would like to congratulate you on this study! Since you find that hemicholinium interferes with the structure, *e.g.*, of mitochondria and induces a reduction in the number of synaptic vesicles, it is worth noting that hemicholinium has been found to bind to mitochondria and that it is indeed acetylated by choline acetylase. In this connection I would like to ask you to elaborate on the point: Are there any morphological changes in ganglia incubated with hemicholinium for the same period and under the same conditions as the stimulated ones, but without stimulation?

Párducz: The HC-3 treatment itself, as was found by *Rodriguez de Lores Arnaiz et al.* (1970) in central cholinergic synapses, did not result in significant changes of the number of synaptic vesicles in the preganglionic nerve endings. In addition, the mitochondrial structure also remained intact.

Vizi: The problem you raised seems to be crucial in understanding the basic mechanism of neurochemical transmission. The problem is whether the acetylcholine released is bound to vesicles or not. You provided evidence that the impairment of neurochemical transmission in the cervical ganglion in consequence of hemicholinium treatment or choline removal is correlated with a reduction of the number of vesicles. May I ask two questions? First, did you measure the acetylcholine content? Because your evidence for the reduction of ACh is indirect: the ganglionic transmission was reduced, as was measured by an electrophysiological method. In order to prove that the reduction of the number of vesicles corresponds to the reduction of the ACh content, the ACh content has to be measured. Second question: have you tried to measure the number of vesicles when the sodium is removed? You say, no. In this case it would be interesting to see what happens to the vesicles because under this condition the synthesis of ACh is blocked completely. Even if you replace sodium by lithium the ACh synthesis is blocked and its content is extremely low (*Vizi, Illés, Rónai* and *Knoll*, Int. J. Neuropharmacology *11*, 1972). Possibly the removal of choline is not the best way to study this problem because glucose is still a good precursor for ACh synthesis.

Párducz: As to the changes in the ACh content of the ganglia occuring during stimulation and in experimental circumstances I may quote the papers of *Birks* and *MacIntosh* (1961) and of *Friesen* and *Khatter* (1971). The results obtained by these authors can well be correlated with our findings. We think it indeed important to determine not only the ACh content of the ganglia, but also the amount of the released ACh. At present we are going to extend our investigations to this field.

As to your second question: we are well aware of those investigations in which the effect of sodium deprivation was studied, but we are intending to depress ACh synthesis by other ways.

We cannot entirely agree with your last remark concerning the role of choline because, as I have mentioned, it is well known from previous investigations that in superior cervical ganglia extracellular choline supply is needed for the maintenance of the optimal transmitter synthesis and release during prolonged activity. This means that, besides glucose, choline is the other precursor of ACh synthesis and this is why we think the removal of choline from the perfusion fluid to be suitable for studying this problem.

Nozdrachev: Dr. Párducz, I would like some information about the conditions of electric stimulation in your investigation of the superior cervical ganglion: characters of the stimulus, frequency, length of stimulation. This is a very important side of your experimentation, since the functional possibilities of the ganglion have a range of characteristics which are different from those of structures of the central nervous system.

Judging by the responses that you obtained by the oscillographic technique, fatigue of the ganglionary apparatus appeared relatively early, though, even in conditions of perfusion it is possible to follow the response during many hours. In our laboratory, in similar conditions of stimulation, the response was registered continuously during 18—20 hours. It is true that, in our case, the material used was the caudal mesenteric ganglion.

Párducz: The parameters of the stimulation were as follows: frequency: 10—20/sec.; duration: 1 msec; amplitude: 6—8 V. We should like to emphasize that in case of normal blood supply and perfusion with choline-containing Locke's solution no sign of fatigue was observed even after two hours of stimulation. The exhaustion occurred only in experiments in which the ganglia were perfused with choline-free Locke's solution or treated with HC-3.

Journal of Neural Transmission, Suppl. XI, 315—323 (1974)

Concluding Remarks and Summing Up

J. Szentágothai

First Department of Anatomy, Semmelweis University Medical School, Budapest, Hungary

With 1 Figure

It is, first of all, my pleasant duty to express my thanks on behalf of the Biological Section of the Hungarian Academy of Sciences to all foreign participants, especially the officers of the Neurovegetative Society and the invited speakers, who have taken the trouble to come to Tihany and who have shared with us the benefits of their studies and experience; our special thanks are due to Professor Csillik and his staff, who have been able, against serious odds—particulary with respect to narrow time limits—to organize this Conference successfully; last but not least I have to thank our local hosts, Inst. Dir. Professor J. Salánki and his staff for their hospitality, and all participants and discussants.

I hope that the participants do not expect me to give a complete recital of all lectures and the highlights of the discussions. I would, instead consider it as my task to attempt to indicate the fundamental aspects and/or concepts of his topic that have been dealt with—or should have been dealt with—and, particularly, to give an outlook in order to predict at least some of the directions or goals towards which the study of the Neurovegetative System (NVS) might most probably move on in the next few years.

Let us start, though, with a brief consideration of what parts of the nervous system might (or should) be considered as belonging to the vegetative system.

1. Conventional neuroanatomy considers as vegetative—*sensu stricto*—the preganglionic neurones, the peripheral ganglia, and their distal connexions with the innervated tissues.

2. It is readily agreed by most students of the NVS that the immediate (and perhaps also some of the secondary and tertiary)

"supranuclear" neuron systems of the preganglionic nuclei should be included. But here we run immediately into major difficulties, since even the primary supranuclear connexions of all preganglionic neurons are virtually unknown—not to speak of the respective afferent systems (if any)—so that nothing really can be said presently of the neuronal organization of the supranuclear pathways. With respect to ascending mechanisms it is not even clear, whether a separation of NVS and somatic ascending systems would be *a priori* possible and/or meaningful.

3. There is already less agreement in the question, whether (and, if so, how far) the hypothalamic mechanisms controlling (and partly exercizing) pituitary functions—and we must not forget the pineal body the relations of which (admittedly the other way round) to the sympathetic system are much more intimate than those of the pituitary—ought to be included into the frame of the NVS.

However, if question (3) were to be answered in the affirmative —and particularly if the mechanisms and neuron systems involved in the control and production of the specific hypothalamic neurohumors (releasing factors, inhibiting factors, ADH-oxytocin system) were included in the NVS—we would have immediately to enlarge the sphere of this system to a much larger and more general (macro-) part of the CNS: the limbic system.

The continuation of this reasoning would lead us ultimately to the inescapable conclusion that the separation of neurovegetative mechanisms from other spheres of the nervous system is entirely artificial and less possible the more one ascends to higher and more complex structural and functional levels. Nevertheless—and in this probably most of us will readily agree—it is necessary to uphold the concept of the NVS, with all its difficulties in definition and separation, for practical purposes especially in consideration of the vast clinical implications of the sphere.

The complexity of the NVS, which encompasses various structures, from primitive neuron nets found at the lowest phylogenetical levels of neural organization up to very highly differentiated brainstem, hypothalamic, limbic and to neocortical structures asks certainly for broad general and imaginative outlooks such as that offered by the hypothesis of Dr. Sakharow.

Being a neuroanatomist, let me now go along again in the same ascending sequence through the main problems touched upon.

A. Periphery

A few of those being here may probably share this acute sense of not only great progress, but in fact complete breakthrough, that some of us—belonging now to the more elderly generation—may certainly feel, who experienced—mainly during the "Thirties" but to some extent also in the "Forties"—the complete preponderance of the strange concepts concentrated around the "Terminalreticulum" of Stöhr and other similar but more realistic ideas proposed by other authors, whom I, therefore, would not consider fair to mention in this negative context. It is not, unfortunately, to the credit of European histology and clinical sciences that these ideas received so much acclamation in their time. However, this experience might make us more critical towards "fashions" in research, in comparison to which ladies fashions are quite unnecessarily commented upon so unfavourably although they are—in contrast to those in scientific pursuits—harmless indeed.

Although in my own mind there has never been any shadow of doubt as to the real microscopic nature of the vegetative terminal formations, that could be very well—albeit indirectly—deduced from degeneration results (*Schimert*, 1936, 1937, 1938; *Szentágothai*, 1957), the final breakthrough came through the use of the electron microscope and the new catecholamine-fluorescence techniques (*Falck et al.*, 1962). It is by now fairly clear that the terminal nerve formations of the vegetative system do not differ essentially from any other "free" terminal arborization, both in the centre and in the periphery. With "free" terminal arborization a structure is meant in which the axons arborize without any specifically localized terminal apparatus, and in which numerous sites of transmission occur along the terminal branches, indicated generally (in efferent terminals) by fusiform swellings containing accumulated synaptic vesicles. The more proximal part of the terminal arborizations is embedded more completely in Schwann cells as shown in figure 1 A, although some of the axons, especially their varicose swellings, emerge at the surface of the Schwann cell and may be active in releasing transmitter substances. In more distal portions (or since the Schwann cells form meshworks in the finer meshes of the terminal plexus) the swellings of the axons appear to be loosely attached only and to protrude from a central Schwann cell process (Fig. 1 B). Eventually, the finest branches may completely lose all contact with the Schwann cell processes. However, in most cases a thin tongue-like process can be found following the otherwise free axon (Fig. 1 C). The presumably active —with respect to transmission function—swellings of the axon rarely if ever are in immediate contact with the innervated tissue elements. No really convincing evidence of specific synaptic contacts has so far been found. Since the distance between axon swelling and innervated tissue elements is between 200—2000 Å, it is probably most reasonable to assume that the transmitters are released into the common intercellular space. There is non fair evidence available for the fact that adrenergic and cholinergic terminal axons are embedded side by side into the same Schwann cell process and they can be differentiated both on the basis of their electronmicroscopic structure,

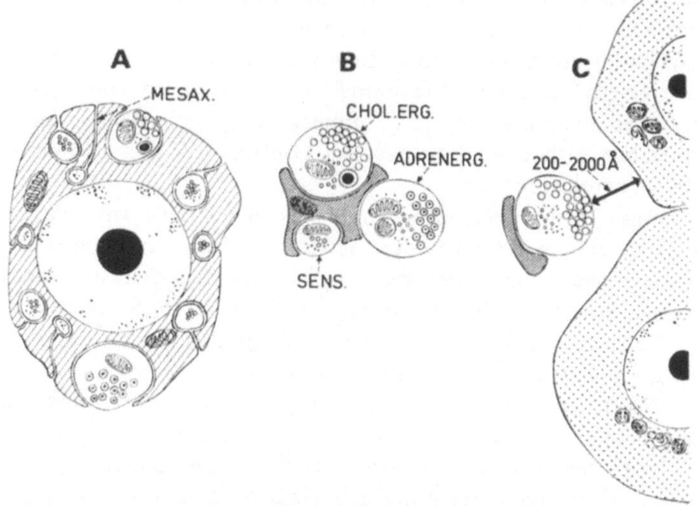

Fig. 1. Diagram explaining the EM structure of various portions of vegetative terminal formations the "ground plexus" of the light microscope era. A. Numerous axons embedded in nuclear portion of Schwann cell (hatched), some of the synaptically active thickenings (containing synaptic vesicles) emerge at the surface of the Schwann cell. B. Distal (finer) portion of plexus with one cholinergic and one adrenergic axon superficially attached to Schwann cell process (hatched), the thin non-synaptic axon might be interpreted as being occasionally of a sensory nature. C. Single, almost free cholinergic axon still maintaining contact with tongue-shaped Schwann process. Two adjacent smooth muscle cells are shown to indicate the distance of the terminal axon from the innervated tissue

selective uptake of their respective mediator, and on the basis of other histochemical criteria (*e.g.*, catecholamine fluorescence). I might also add that, at the electron microscope level, degeneration gives now some support to my old assumption that freely arborizing sensory fibres may also be present in the terminal plexus (*Ungváry* and *Léránth*, 1970). The problem of the peripheral vegetative innervation has been thoroughly covered by several contributions of our Conference. Among these I should like to mention the beautiful histological observations of Dr. Stach which have brought back into focus the interstitial cells, an important issue that has been so elegantly studied some ten years ago by Dr. Taxi, but that somehow has faded out recently from our visual field, due to the remarkable recent development due to the catecholamine fluorescence techniques. Now it appears that the interstitial cells may have a come-back, and that there is more complexity in the structure of the peripheral vegetative end-formations than what might appear from the simplified picture that I gave above. Dr. Csillik's histochemical analysis has made good use of the possibilities offered by peripheral vegetative terminal formations, I will, however, return to this in Section C dealing with the more fundamental aspects of synaptic structures and transmission mechanisms. The most elegant analysis of Dr. Häggendal has given us a vivid impression of the power of the modern techniques available. The differentiation of two different pools of norepinephrine, one easily mobilizable and another much more sluggish in its reactions, affords insights of fundamental importance into the mechanisms of neurotransmitter metabolism. The report by Dr. Vizi on the possibility of presynaptic action by norepinephrine on the release of acetylcholine is also of crucial importance.

B. Ganglia

If anybody would have had any doubt concerning the continued importance of vegetative ganglia as useful models for the study of synaptic transmission in general, he would certainly have to change his mind after the lectures given on this Conference. The lectures given by Drs. Párducz, Joó and Kiss indicate the possibilities offered by various ganglia for investigations into neurobiological mechanisms of general significance. Very unfortunately, we also have to recognize that the crude simplification of considering most vegetative ganglia as a single synaptic relay does not hold true in many cases. The most interesting findings of Dr. Taxi show that—as recognized recently in many central nervous structures—accumulations of synaptic vesicles, often with adjacent membrane specializations, occur both in dendrites and in cell bodies. Synaptic articulation, hence, is not so simple and straightforward as hitherto assumed.

The studies of Dr. Stach on the intramural ganglia of the intestine have given us an important picture of the still entirely unexplained diversity of neurones in these plexuses. I for myself have always favoured a view, somewhat modified from the classical concepts of *Dogiel* (1896), assuming that the Dogiel type I cell would be a higher differentiated efferent neuron distributing excitation received from the extrinsic vegetative nerves (vagus, sacral parasympathetic nerves, and possibly the sympathetic). Conversely, the Dogiel type II would be a lowly differentiated "all purpose" cell the processes of which would invade both the innervated tissue, other nerve cell groups, and which would even send processes back to the extrinsic (prevertebral) ganglia. I assumed that these cells form true neuron networks in which direction of transmission could vary according to the circumstances, for example, from one peripheric site to intramural ganglion, to another peripheric site (muscle tissue), or the reverse, as generally assumed of the nerve nets of lower invertebrates. The same processes or cells, thus, would be afferent in one situation and efferent in another. As seen from the report of Dr. Ungváry, nerve cell processes from the local (intramural) neuronal plexus may reach the prevertebral ganglia and may establish synapses there. We always assumed that these processes belonged to the Dogiel type II, and that at least some of these cells might act as real afferent elements. Now after hearing the results of Dr. Stach I am not so certain whether my earlier interpretation is correct. We may prepare ourselves to face the possibility of a much higher diversity of nerve cells in these plexuses, for example: serotoninergic and peptidic neurones in addition to the hitherto known types of cholinergic and adrenergic neurones.

Very unfortunately Dr. Milochin and collaborators from Leningrad could not come. They could have given us their most recent account of an entirely different afferent mechanism, based on the presence of sensory terminals in vegetative ganglia (*Milochin*, 1967). The number of sensory fibres entering the sympathetic chain is quite considerable, and their diameters are widely variable (*Kiss* and *Zádory*, 1941). There are certainly more of them than could be accounted for by the sparse sensory terminals found in most internal organs. The explanation of this discrepancy is explained on the one hand by probably quite numerous freely arborizing sensory axons that remain unrecognized in the peripherial vegetative endformations, but can be identified by degeneration under the light microscope (*Schimert*, 1936, 1937) and much better under the electron microscope (*Ungváry* and *Léránth*, 1972). Another explanation, particularly for the large calibre afferents joining the sympathetic, is the idea of Milochin, according to which these would terminate partly in the vegetative ganglia themselves. This possibility was envisaged already much earlier (around 1934) by Podhradszky (personal communication made to me at that time but never published).

These comments might suffice to show what important issues are waiting for further elucidation with the powerful methods that are available today.

C. Synaptic Structure and Transmission Mechanisms

Although much of the new information that has been summarized at our conference is not immediately related to the NVS *sensu stricto*, everybody will obviously agree in my evaluation that the lectures dealing with synaptic structure and transmission mechanisms in general were the true highlights of our conference. The excellent reviews given by Drs. Akert, Csillik, Gerebtzoff, Storm-Mathisen and Malinsky gave us a vivid impression where research in ultrastructure, histochemistry and biochemistry of the synapse is standing today. I do not even try to summarize these excellent reviews, and wish only to indicate that the attempts by Dr. Storm-Mathisen to localize GAD more exactly and that by Dr. Gerebtzoff to pinpoint sites of activity of cyclic AMP seem to offer great possibilities for the future. I am certain that application of all these powerful techniques specifically to nuclei, ganglia or other elements of the NVS, will probably yield a host of new information not only on the vegetative system itself but for neurobiology in gėneral.

D. Preganglionic Neurones and their Immediate Supranuclear Connexions

The preganglionic neurones of the NVS have been neglected up until the last years both by physiologists and particularly by anatomists. After the paper given by Dr. Réthelyi we may now become aware of the fact that the sympathetic neurones in the intermedio-lateral nucleus of the spinal cord have some very remarkable features in dendritic and synaptic architecture that make them well worth of more detailed ultrastructural and histochemical studies. Although the cells of the spinal vegetative centres—both ortho- and parasympathetic—can now be well localized and unequivocally identified, this is not so simple in the case of the lower brain stem parasympathetic nuclei. It would be too long to enter here into a discussion of the abundant but rather patchy literature dealing with this subject. I myself have devoted years to this study (*Szentágothai*, 1942, 1952) but the method of orthograde degeneration I used is not too good in the case of such longitudinally oriented colums of rather scattered cells. However, what was defeating me—and I am afraid also other authors (although most of them did not realize this)—was that many of the vagal preganglionic fibres do not show the usual signs of secondary degeneration with the silver stains then available. It would be well worthwhile to take up this pursuit again in systematic studies with improved methods.

The situation is even worse regarding the immediate supranuclear connexions of the preganglionic neurones. One might think that, after the exact electron microscopal identification of the sympathetic preganglionic cells, a degeneration analysis of their supranuclear pathways might easily be accomplished. Unfortunately Dr. Réthelyi's preliminary efforts showed only relatively short range connexions. This does not mean, of course, that no long descending pathways terminate directly on these cells. A more parsimonious explanation would be to assume that we miss the degeneration only because we have not yet found its appropriate time course. Remarkably, norepinephrine fluorescence diminishes very considerably after lateral funiculus lesions at the cervical level (*Dahlström* and *Fuxe*, 1965), especially after bilateral lesions within five—six days. This would be at variance with the assumption that there are no long descending monosynaptic pathways to the intermedio-lateral nucleus, or else, one would have to assume that the disappearance of fluorescence is a transneuronal effect. However this may be, even the very preliminary results available show that the organization of supranuclear vegetative pathways as well as the synaptic organization of the nuclei giving rise to them are far from being simple.

E. Hypothalamic Mechanisms

The timing of our conference was very unfortunate with respect to the simultaneously held International Congress of Endocrinology in Washington, so that our originally listed speakers on hypothalamic neuroscretory mechanisms were not available. We are, therefore, very grateful to Dr. Endrőczi who included some of these aspects into his excellent review on the role of the limbic (septal, hippocampal, amygdala) mechanisms in the higher level control of endocrine mechanisms as related particularly to the organization of behaviour. The wealth of important new information about these mechanisms indicate clearly how fruitful the application of the most recent powerful ultrastructural, histo- and biochemical methods to the neurosecretory systems might be. — Dr. Pfeiffer's paper on the brain stem catecholamine mechanisms gave an impressive sample of the possibilities of modern neuropharmacology, both in the study of basic mechanisms and in immediate practical applicability.

F. Clinical Aspects

Dr. Lassmann's paper has given an excellent example of how useful modern structural and histochemical methods have become in clinical diagnostics. Important insight into the clinical relevance of neurovegetative studies, up to a level of complexity reflected in the EEG, have been given by the group of Professor Shimoda. A certain split between basic research and applied clinical science has always been—and still is—conspicous in the neurovegetative sphere. The reasons for this are quite obvious: there is a deplorable discrepancy between the demand for the explanation of clinically very relevant phenomena, as compared to the meagre material in hard facts. It is the noble task of the Neurovegetative Society to trigger and coordinate efforts in the study of basic neurovegetative mechanisms in order to bring them up to the level at which neurobiological research today is being generally conducted.

References

Dahlström, A., and *K. Fuxe*: Evidence for the existence of monoamine neurons in the central nervous system. II. Experimentally induced changes in the intraneuronal amine levels in bulbospinal neuron systems. Acta physiol. Scand. *64*, Suppl. 247, 7—85 (1965).

Dogiel, A. S.: Zwei Arten sympathischer Nervenzellen. Anat. Anz. *11*, 679—687 (1896).

Falck, B., *N.-A. Hillarp*, *G. Thieme*, and *A. Torp*: Fluorescence of catecholamines and related compounds condensed with formaldehyde. J. Histochem. Cytochem. *10*, 348—354 (1962).

Kiss, F., and *E. Zádory*: Experimentell-morphologische Analyse der Rami communicantes. Anat. Anz. *91*, 209—225 (1941).

Milochin, A. A.: Die sensible Innervation der vegetativen Neuronen. Nauka, Leningrad. 1967. (Russian with German summary.)

Schimert, J.: Untersuchungen über den Ursprung und die Endausbreitung der Nerven der Iris. Z. Zellforsch. *1936*, 247—258.

Schimert, J.: Die Nervenversorgung des Myokards. Z. Zellforsch. *27*, 246—266 (1937).

Schimert, J.: Die „syncytielle Natur" des vegetativen Nervensystems. Z. mikr.-anat. Forsch. *44*, 85—118 (1938).

Szentágothai, J.: Die zentrale Leitungsbahn des Lichtreflexes der Pupillen. Arch. Psychiatr. Nervenkr. *115*, 136—156 (1942).

Szentágothai, J.: The general visceral efferent column of the brain stem. Acta morph. Acad. Sci. hung. 2, 313—328 (1952).

Szentágothai, J.: Einige Bemerkungen zur Struktur der peripheren Endausbreitung vegetativer Nerven. Acta Neuroveg. *15*, 417—445 (1957).

Ungváry, Gy., and *Cs. Léránth*: Termination in the prevertebral abdominal sympathetic ganglia of axons arising from the local (terminal) vegetative plexus of visceral organs. Peripheral reflex arc. Z. Zellforsch. *110*, 185—191 (1970).

Ungváry, Gy., and *Cs. Léránth*: Innervation of the hepatic vein sphincters. Acta Anat. *83*, 619—632 (1972).

Author's address: Prof. Dr. *J. Szentágothai*, First Department of Anatomy, Semmelweis University Medical School, Budapest, Hungary.

Subject Index

→ effect upon
← effected by